# Apple Files

## David Miller

**RESTON PUBLISHING COMPANY**

*A Prentice-Hall Company*
Reston, Virginia

**Library of Congress Cataloging in Publication Data**

Miller, David
    Apple files.

    Includes index.
    1. Apple II (Computer)–Programming. 2. File organization (Computer science) I. Title.
QA76.8.A662M54      001.64'25      82-5432
ISBN 0-8359-0192-0              AACR2
ISBN 0-8359-0191-2 (pbk.)

Editing and interior design
by Ginger Sasser DeLacey.
Cover design by Joyce Thompson.

Apple® is a registered trademark of Apple Computer, Inc.
VisiCalc® is a registered trademark of VisiCorp.

© 1982 by
Reston Publishing Company, Inc.
*A Prentice-Hall Company*
Reston, Virginia

All rights reserved. No part of this book may
be reproduced in any way, or by any means,
without permission in writing from the publisher.

10  9  8  7  6  5  4  3  2

Printed in the United States of America

*As our lives are dedicated to the service of the Lord,
so also is this work.*

*"Whither thou goest..."
To Emmy—the best wife and proofreader ever.*

*"An Apple a day..."
To the Apple II—the only computer with personality.*

# Apple IIe

There are several differences between the Apple II or II Plus and the Apple IIe that are of importance to IIe owners reading this book. The most obvious is the inclusion of lower-case display capability on the IIe. Older versions of the II did not have that built-in capability. For that reason, when Apple created its DOS and BASIC, lower-case DOS and BASIC commands were not allowed. To maintain compatibility, the IIe functions in the same way. Therefore, users of the IIe should place the computer in the CAPS LOCK position while programming. (An alternative for the experienced programmer is the "Restricted-case mode.") All programs in this book were created to recognize only upper-case. Unless modified, these programs should be run with the CAPS LOCK key engaged.

Older versions of the II came with a 40-character-per-line display. The IIe can easily be modified to display 80 characters per line. If such a modification is done, be certain to read the accompanying 80-Column Text Card manual carefully, and make note of any program suggestions included. The programs in this book are set up for a 40-character-per-line display but can easily be changed to take advantage of the 80-character-per-line capability.

If you have a IIe with the 80-Column Text Card, you must issue a PR#3 command to turn on the 80-column display. If you have turned on the 80-column display, the programs in this book must be changed wherever a PR# (n) command has been used. Most of these incidents occur when the output is to go to a printer. The following code steps should replace the Apple II and II Plus standard "PR#1–PR#0" code to activate and deactivate the printer for Apple IIe owners using the 80-column display. First, deactivate the 80-column card with PRINT CHR$(21). Second, activate the printer with PRINT CHR$(4);"PR#1". Third, when you are finished printing, deactivate the printer and reactivate the 80-column card with PRINT CHR$(4);"PR#3". Please remember, this change needs to be done only by Apple IIe owners using the 80-column card at the time they want to send output to another peripheral device such as a printer.

Knowledge of and familiarity with the keyboard are essential for error-free operation of just about every computer. The IIe is no exception. The IIe keyboard has more keys, a somewhat different placement of certain keys, and a different spelling of one key. Other versions of the II contain a key labeled CTRL. On the IIe, the same key is now spelled out as CONTROL.

When you are told in this book to type PR#6 or to turn the computer off and then back on, IIe users should, instead, press and hold the CONTROL and OPEN-APPLE keys, then press and release the RESET key. This sequence is known as a "COLD BOOT" and has the same effect as turning an older II off and then back on.

Each character key on the IIe keyboard has an auto repeat capability. This can be very helpful at certain times but at other times may cause some people problems. For instance, if a person presses the RETURN key for too long, more than one INPUT variable may be filled with what is called a "nul string" or simply no visible characters. In other words, the computer may think you have entered something when all you did was hold down the RETURN key too long. Such an action can cause errors in certain programs.

The Apple IIe has a slightly different slot arrangement. At the rear of the computer are seven slots numbered 1 thru 7, instead of the eight slots numbered 0 thru 7 on older Apple IIs. But the IIe has an 8th slot located on the left side toward the middle of the main logic board. In effect, this slot is a duplicate of slot 3 and is intended to contain Apple's 80-Column Text (or Text and memory) Card.

The standard cursor looks somewhat different on the IIe than on the other IIs. And depending upon the particular mode you are in, the cursor may be entirely differrnet than any cursor on older Apple II computers. The IIe manual can best explain the purpose for the different shaped cursors.

As of this writing (February 1983), these are the known differences between the II and IIe versions that have any bearing on the material in this book.

# Contents

|  |  |  |
|---|---|---|
|  | *Preface* . . . . . . . . . . . . . . . . . . . . . | ix |
|  | *Introduction* . . . . . . . . . . . . . . . . . . . | xi |
| Chapter 1 | Apple's Four File Types . . . . . . . . . . . . | 1 |
| Chapter 2 | Applesoft and Integer Files . . . . . . . . . . | 6 |
| Chapter 3 | Text File Introduction . . . . . . . . . . . . . | 15 |
| Chapter 4 | Creating Sequential Files . . . . . . . . . . . | 27 |
| Chapter 5 | Appending Sequential Files . . . . . . . . . . | 40 |
| Chapter 6 | Displaying Sequential Files . . . . . . . . . . | 56 |
| Chapter 7 | Correcting Sequential Files . . . . . . . . . . | 82 |
| Chapter 8 | EXEC and CHAIN . . . . . . . . . . . . . . . | 96 |
| Chapter 9 | Additional Sequential File Techniques . . . . . | 114 |
| Chapter 10 | DIF Files . . . . . . . . . . . . . . . . . . . | 144 |
| Chapter 11 | Random File Introduction . . . . . . . . . . . | 167 |
| Chapter 12 | Home Inventory System . . . . . . . . . . . . | 190 |
| Chapter 13 | Planning a File System . . . . . . . . . . . . | 238 |
| Chapter 14 | Binary Files . . . . . . . . . . . . . . . . . . | 258 |
|  | *Appendices* . . . . . . . . . . . . . . . . . . . | 269 |
| Appendix A | RWTS (Read/Write Track/Sector) . . . . . . . | 270 |
| Appendix B | Tape Files . . . . . . . . . . . . . . . . . . . | 274 |
| Appendix C | Mailing List System Programs . . . . . . . . . | 278 |
| Appendix D | Math System Programs . . . . . . . . . . . . | 308 |
| Appendix E | Recipe and Drill & Practice Programs . . . . . | 328 |

| | | |
|---|---|---|
| Appendix F | DIF Programs | 335 |
| Appendix G | Medical Records System Programs | 345 |
| Appendix H | Home Inventory and Back Order System Programs | 356 |
| Appendix I | Stockmarket Programs | 393 |
| Appendix J | Miscellaneous Programs | 409 |
| Appendix K | Miscellaneous Information | 410 |
| | *Index* | 411 |

# Preface

The purpose of this book is to take some of the misery and mystery out of using Apple Computer's file structure. The book is aimed at people who know some BASIC and would like to learn to use the computer to assist them at home or at work by using the file capabilities of the Apple. *Apple Files* is designed as a step-by-step tutorial. The book explains some things that, without adequate manuals, take many painful hours of trial and error to learn. Progress has been made in creating better file-handling techniques on the Apple, and an explanation of some of these techniques is included.

Upon completion of the book, you should fully understand what files are and how to use them. You will be able to create your own sequential or random access files. Examples of both of these file types are included throughout the book. Program examples include file creation programs for: the stock market, mailing lists, inventories, grades, recipes, and medical records.

There are some very good data base programs available commercially. If your needs require an elaborate data base structure, you should probably use one of those programs or pay a programmer to create one for you. Reading this book will not make you capable of creating complete data base programs, but with practice, you will be able to effectively create and use any type of file you want.

I really enjoy programming and creating programs for my own use. I like the freedom programming gives me, because I can easily change or add to what the program does. I hope this book conveys some of that enjoyment and freedom.

# Introduction

No book is magic in that, by possessing the book, you possess the knowledge of that book. Yet, I have tried to make it relatively easy for *anyone* to learn to use the Apple computer.

No single book will suffice for everyone, and this book makes no claim to being the exception. Nevertheless, I have attempted to make it useful for the beginner as well as the more experienced Apple user. In answer to the often-asked question, "What can I do with it besides play games?" the program examples cover the areas of home, education, business, hobby, or investment.

Computer vocabulary has been introduced very gradually. Readers somewhat knowledgeable about the vocabulary may find the process repetitious at first, but I have found this to be the best method for acquiring a working knowledge of the multitude of "jargon."

The "system" approach has been used so the reader would not be overwhelmed with a large number of different application programs. The programs presented are intended to be useful as well as instructive. The programs build upon themselves so that something that may appear awkward to an experienced programmer is used to help explain a concept needed in later chapters.

Information for the more experienced Apple user includes a thorough discussion of the "B" parameter with random access and sequential files. Other items are: EXEC files, tape files, Binary files, DIF files with application programs, and RWTS. The section on Tape files briefly explains how this book can be useful to someone without a disk drive.

You cannot just absorb this information. You must read the book and plan to re-read and/or study the text and programs of parts that are at first unclear. Invest time in learning how to get the most out of the Apple. Experienced Apple users may find that they can either skip parts or proceed quickly through certain sections. I would encourage everyone to finish the book.

Finally, a disk containing all the programs presented in the book will be available. You can make the disk yourself by typing in all the programs, but if you just want to see the programs in operation, then you may want to purchase the disk. I sincerely hope you enjoy the book and find it instructive.

The programs in this book are available on disk
with additional documentation and suggestions.
To order your copy, please send $15.00
($17.50 outside North America) to:
AEN, 9525 Lucerne Street, Ventura, CA 93004.
California residents, add 6% sales tax.

# I
# Apple's Four File Types

There are as many definitions of the word file as there are kinds of files. You can quickly become confused if your understanding of the term differs from an author's intended use, and dictionary definitions are of little use in the computer world of today. Before becoming involved with the computer, my understanding of a file was limited to information that was kept in a folder in a file cabinet. I think we often learn best by trying to fit that which is new into something we already understand. Therefore, following this idea, I will try to explain Apple® file structure in terms of a file cabinet.

In a four-drawer file cabinet, one drawer might be for accounts payable, while another could be for accounts receivable, a third for personnel information, and the fourth for inventory information. These are used only as examples to show that each drawer might contain different file types. The file cabinet just as easily can contain game instructions in one drawer, receipts in another, name and address information in a third, and medical records in the fourth. The idea is of four drawers containing four different types of information. *The Apple's file cabinet (the disk and diskette) is mainly a four-drawer cabinet.* (Actually Apple has eight possible file types, but only four are usually used.) The first drawer or type of file contains Applesoft BASIC program files and is marked with an "A". The second drawer contains Integer BASIC program files and is marked with an "I". The third holds binary information and is marked with a "B", and the fourth (and the one we will look at in detail in this book) holds text information and is mark with a "T". Therefore, each diskette you use is like that four-drawer file cabinet. It is set up to accept information

---

Apple® is a registered trademark of Apple Computer, Inc.

in any of these drawers, but it does not have to have information in all of them—just like the real filing cabinet doesn't have to have something in all four drawers. If you never have any binary, or B, information, you are not required to have B files on your diskette. Some diskettes may not show some of the four file types.

How do you know what files are on your diskettes? We will begin the tutorial part of this book by going through all the steps necessary in order to find out just what files are on your diskette. (If you are already acquainted with the procedure used to start up the computer and disk drive—sometimes called "booting the system"—you can skip the rest of this paragraph.) Take your SYSTEM MASTER Diskette (or almost any other diskette), and insert it into the disk drive. (If you have more than one drive, make sure to put the diskette into drive one.) Then do one of the following: (1) Turn on the computer, or (2) If the computer is already on, type PR# 6 (or whatever slot number your disk interface card is in) and press the key marked RETURN, or (3) Turn the computer off and back on again. The disk drive should make some noise, and the IN USE light on the disk drive should come on. If you have not already shut the disk drive door, shut it carefully at this time. Soon the disk drive will stop, the IN USE light will go out, and there should be a blinking box somewhere on the screen. (If you have not been able to get to this point with the computer, try another diskette, preferably the diskette marked "SYSTEM MASTER". If switching diskettes does not work, you will need to check the manuals for your particular system.)

Now, type CATALOG and press the key marked RETURN. The disk drive will come back on and you will see a list with three things on each line of text:

1. a single letter,
2. a three digit number, and
3. a name made up of letters and/or numbers.

The single letter tells what type of file it is: usually either an A, B, I, or T. The three digit number gives the size of the file, which we will discuss in more detail later. The name, consisting of letters and/or possibly numbers, is the actual name of the specific file. In our file cabinet example, this is the same information that might appear on each folder. You might label each folder with (1) a single letter indicating which drawer it goes in, (2) how much information it contains, and (3) a name for the information within the folder. If you were able to see all the names of the folders in the four-drawer file cabinet, all of the A type files would be together, as would the B, then I, and finally T type files. In reality, most file cabinets do not have a list of all the files stored within them. It would be a time-consuming job to update that list every time you added, changed or threw away a file, but Apple's file manager system does just that and does it automatically. That list of everything in the file cabinet (on the

CHAPTER 1   APPLE'S FOUR FILE TYPES   3

diskette is what you get when you type CATALOG. *Typing CATALOG shows the list previously created by the file management system.* Since it doesn't much matter to the disk where the information is stored, an A file may be followed by a T file, followed by another A file or B file or I file or another T file. In other words, the list of the files on the disk is not separated by file type. Instead, usually the list (CATALOG) shows the order in which the files are created.

How are these files used? By now you know the four main file types used by Apple, but you don't know much about them. What are they? How are they different? How are they used? Returning to the four-drawer file cabinet, the person in charge of that cabinet might put some rules or locks on the drawers. In other words, he or she might say that the A type file could be used only in a certain way or only by certain people. The same could be true of the other file types or drawers. This is exactly what Apple's file management system has done. Each file type is used differently and can be used only in certain specific ways. A *and* I *files are instructions (also called programs) to the computer to do something.* Examples of such instructions would be:

20 HOME
40 DIM X$ (30)
50 GOTO 100
100 PRINT "HELLO"

*Most* T *files are not computer instructions but contain information of value to people* such as names, addresses, zip codes, payroll deductions, pay rates, and book titles. Often T files are just lists of such information. Such lists, of course, would not make sense as instructions to the computer which is the reason T files cannot be RUN like A or I files (programs) can be. *Binary files can be either computer instructions or human information (lists), or sometimes a combination of both.*

It should be clear that Apple files are used to store information just as you or I use a filing cabinet, that there are different types of files and that they are used for different purposes. In the next chapter, we will look more closely at the four file types and how they are used. In subsequent chapters, we will look inside the drawer of one of those file types, the T file, and examine how information is kept, how that information is used, and how those files are created. The latter, creating files, is the main emphasis of this book and will occupy the remaining chapters. If you want to know how to create the A or I files (programs), you will need to learn "programming." Effectively using the T files requires some knowledge of how to program in either Applesoft or Integer BASIC. The programs discussed in this book will, for the most part, be Applesoft BASIC programs, and the discussion will be such that anyone willing to try the examples should learn to program as well as learn to create and use

T files. In other words, although the main emphasis of this book is on the T files, you will learn a certain amount of programming---A or I BASIC program files, in order to be able to create, display and change T files. And I repeat, *anyone willing to try all the examples, and read carefully through the discussion of the examples, can and will learn to program and thus make effective use of the T files.* Individuals, no matter what their age, background or experience, can learn the information presented in this book. Programming and file manipulation are a matter of learning how to give instructions to the computer in a manner the computer can understand, or, more simply, programming is learning how to talk to the computer and tell it what you want it to do.

## QUESTIONS

1. How many main file types does Apple have?
2. What letter is used in the CATALOG of a diskette for Integer files?
3. What is a "B" file?
4. Which file type will this book concentrate on?
5. What word would you need to type in order to see a list of the files on a diskette?
6. Which files contain instructions to the computer?
7. Which files contain information of value to people?
8. Which file can contain information of use to both computers and people?

## ANSWERS

1. 4
2. I
3. BINARY
4. T or TEXT
5. CATALOG
6. A and I
7. T and B (Programmers aren't human!)
8. BINARY or B

# 2
# Applesoft and Integer Files

In this chapter, we are going to take a closer look at the first two main file types. This would be the same as opening our file cabinet drawers for a quick look at what is kept inside.

We begin with what are probably the two most common types of files, the "A" or Applesoft BASIC files, and the "I" or Integer BASIC files. Some of you might already be confused because you have always referred to Applesoft BASIC or Integer BASIC as "programs" rather than files. In reality, they are both. Suppose one drawer in the file cabinet is used for games. Each folder contains the rules or instructions for playing a different game. Most of the time, you would simply refer to the folders as games, not files, yet they are really both games and files. When you have taken one folder out of the file cabinet and are using the instructions to play the game, it is not a file; however, when you are finished with the game and want to put it back in its place, it becomes a file---one of many game files. The same is true of Applesoft BASIC program files. The A drawer contains only Applesoft BASIC programs or computer instructions (rules). When the computer is using the instructions in one of those Applesoft "folders," the instructions are a program, but when the computer is not using the instructions, the instructions are stored as files. The important thing to understand is that the A drawer, or Applesoft files, contain only computer instructions (programs). Some of those files contain larger or longer sets of computer instructions than others, but an A file can only be a set of instructions for the computer or, therefore, a computer program. The second limitation of the A files is that they can only be a certain kind of computer program---an Applesoft BASIC computer program---not an Integer BASIC computer program or a Fortran computer program or a Cobol computer program, etc. (Fortran

## CHAPTER 2   APPLESOFT AND INTEGER FILES

and Cobol are two other computer languages, just as BASIC is a computer language.) They will always be what is sometimes called floating point programs. The main difference between Integer BASIC programs and Applesoft BASIC programs is in the way they work with numbers. Applesoft BASIC programs are able to work directly with numbers that contain decimal points such as 23.45 or $59.85, while Integer programs are not able to work directly with decimals. The number 23.45 in Integer would be treated as 23. The amount $59.85 would simply be $59. There are other differences, but this explanation is sufficient for our needs now.

We can see that two of our four drawers contain only computer instructions or programs. The A drawer is for Applesoft BASIC program files, and the I drawer is for Integer BASIC program files. Let's look at the rules for using either of these files, the A or I files. In our mythical office, we have three main secretaries that can use either the A or I files. Secretary number one can only go and get the file (LOAD). Secretary number two can only put the file away (SAVE). Secretary number three can go and get the file and immediately begin execution or operation of the program (RUN). *These three secretaries or commands do not have access to the other two drawers or file types.*

Secretary number three (RUN) does two jobs by loading a program into memory from a diskette and then beginning the operation or execution of that program. In other words, the RUN command goes and gets a copy of the file from the disk, puts it in the computer's memory and tells the computer to begin operation according to the instructions in the file (now program).

Secretary number two (SAVE) can only put the file (program) currently in the computer's memory in the file cabinet (disk).

Secretary number one (LOAD) is only able to get the file (program) from the file cabinet (disk) and put it on the boss's desk (in the computer's memory). The LOAD command goes to the diskette and gets a specific file. In order to know which file to get, the LOAD command must be given the specific name of the file: LOAD MATH DRILL or LOAD CHECKER GAME. If the file name is not spelled exactly (including spaces) the way it is spelled on the CATALOG (the list of all the files in the file cabinet), then the LOAD command or secretary won't be able to find the file and will come back and tell you "FILE NOT FOUND". On the other hand, if the LOAD command does have the exact name it will go to the disk and get a copy of the file. Notice the use of the word "copy." The LOAD command does not actually go and remove the file from the diskette like a secretary would remove a file from the file cabinet. The LOAD command takes only a copy so that the original always remains on the disk. The copy of the file is loaded into the computer's memory, similar to a secretary getting a file and putting it on the boss's desk. You, the boss, must then decide what you want to do with the file. If you

8  CHAPTER 2  APPLESOFT AND INTEGER FILES

want to open it and look at it, you use some form of the LIST command. (Type LIST or LIST 100-200 or LIST 400,500 etc.) If you want to see the program in operation, type RUN.

Secretaries, or Disk Operating System commands, must be given a specific file name. Usually, after using the LOAD command, you will want to look at the instructions (LIST) and perhaps change, add, or remove some instructions. When you have finished, you may want to keep what you have done by giving the file to secretary number two (SAVE) and telling the secretary the exact name you want this file kept under. If you have made changes but still want to keep the original currently on the disk, then the secretary must be informed of a new file name. If the secretary uses the same file name as the file currently on the disk, the secretary will throw out the file on disk (erase it) and replace it with the one you have changed. This may or may not be what you want, so be careful what name you use (the SAVE command).

Let's actually try some of these commands. Use your SYSTEM MASTER diskette to get the computer working; i.e., to boot the system. (Review the procedure in Chapter 1, if necessary.) When you get the blinking box (called cursor) on the screen, type the following carefully, and remember to press the key marked RETURN after each entry:

```
NEW
20 PRINT "HELLO"
40 PRINT "MY NAME IS APPLE II"
```

Check your typing to be sure you have typed everything exactly the way it is shown above. The word NEW erases anything that is already in the computer's memory. It does not do anything to the diskette or any information stored on a diskette. The numbers 20 and 40 are line numbers in an Applesoft BASIC program. Line numbers can be any number from 0 to 65535. Usually, the numbers chosen are not consecutive in order to allow other lines to be added, if necessary. The word PRINT instructs the computer to display on the screen whatever follows it and is between the quotation marks. Now type the word RUN and press the key marked RETURN. You should see just the words between the quotation marks.

```
HELLO
MY NAME IS APPLE II
```

You have just written and executed an Applesoft BASIC program. The program is still in the computer's memory, but if you were to turn off the computer now, you would lose that program. It would be lost because it had not been saved anywhere permanently; i.e., on tape or disk. We will take care of that with our next step.

This next step is very important. *Remove the SYSTEM MASTER diskette and insert a blank or new diskette.* If you do not remove the

CHAPTER 2  APPLESOFT AND INTEGER FILES

diskette you used to boot the system, you will destroy the information on it when you follow the next series of instructions. If you do not have a new diskette, you should: (1) Wait until you do have a new diskette before doing the next steps, or (2) Use a diskette that you know has some room on it and skip the first step below, or (3) Use an old diskette that contains information you no longer need. Once you have chosen and inserted your diskette, type the following:

    INIT HELLO

Remember, this first step destroys any existing information on an old diskette. This step is used to "format" (INITialize) a new diskette or re-format an old diskette so that the diskette can store files. The INIT command is usually used only once on each diskette. A second use erases whatever is currently on the diskette. The INIT command will only work with a program in memory. The entire process can take about 30 seconds for DOS 3.3 users to one-and-a-half minutes for DOS 3.2 users. The disk drive makes a noise and the IN USE light comes on. The computer is transferring numerical information onto the diskette to enable the computer to later find locations on that diskette. It is also saving our program as the first file on this diskette.

When the IN USE light goes out and the blinking box (cursor) reappears on the screen, type:

    CATALOG

(Remember to press the key marked RETURN.)
You should see:

    A 002 HELLO

Then type:

    LIST

You should see:

    20 PRINT "HELLO"
    40 PRINT "MY NAME IS APPLE II"

LIST is similar to CATALOG in that CATALOG shows what is on a diskette and LIST shows what is in the computer's memory.

Now type:

    NEW
    LIST

The program is now gone, and there is nothing in the computer's memory. Type:

    LOAD HELLO

LIST

and the program is back.

Immediately after the LOAD HELLO, the disk drive comes on for a brief time. The computer is instructed to go to the diskette, bring in a copy of the file called HELLO and store that copy in its memory. When you type LIST, you are telling the computer to show you what it has in its memory. Therefore, the program now actually exists in two places: (1) in the computer's memory and (2) as the first file on the diskette.

Type carefully and add a third line like this:

60 PRINT "I AM A SMART COMPUTER"

Then:

SAVE HELLO 2
CATALOG

Now the list shows:

A 002 HELLO
A 002 HELLO 2

There are two files on the diskette. Both are Applesoft BASIC program files. After you type SAVE HELLO 2, the disk drive comes on briefly while the computer transfers a copy of the contents of its memory to the diskette. CATALOG shows the new list of files on the diskette.

Finally type:

NEW
LIST

The program is gone. Type:

RUN HELLO 2

and the screen shows:

HELLO
MY NAME IS APPLE II
I AM A SMART COMPUTER

Now type:

LIST

and the full program is back.
Type:

RUN

(this time without a file name, since the instructions are already in memory) and you should get the same message.

## CHAPTER 2 APPLESOFT AND INTEGER FILES

First, you erase the program in the computer's memory (NEW) and then ask to see if there is anything left in the computer's memory (LIST) just to verify what you did. Next, RUN HELLO 2 tells the computer to access the diskette, load the file called HELLO 2 into its memory and begin operation according to the file's instructions. LIST shows that the program is back in memory. To prove it, RUN without a file name tells the computer to again operate according to the program's instructions.

Let's review from the viewpoint of the secretaries and file cabinet. Remember, so far we have three main secretaries: number one (LOAD), number two (SAVE), and number three (RUN). In order for these secretaries to do anything, they must be given a file name:

```
LOAD MATH DRILL
RUN CHECKER GAME
SAVE ANYTHING
```

Disk Operating System commands (DOS commands) must be given a specific file name.

There are five other disk operating commands that can be used with files A, I, T, and B. Commands four and five (LOCK and UNLOCK) are used to protect and unprotect files. If you have a file you never want changed or erased, you can lock that specific file: LOCK CHECKERS; LOCK MATH DRILL. Files that are locked will have an asterisk by their file types in the catalog:

```
*A 007 MATH DRILL
*I 045 CHECKERS
*T 023 NAME FILE
```

If you change your mind and do want to make changes, then command number five (UNLOCK) will remove the asterisk and make it possible for you to change or rename a file or erase it.

Command number six (RENAME) can change the name of any file. Command number seven (DELETE) can remove or erase any file. Command number eight (VERIFY) can check on the condition of the file in the file cabinet (on the disk) to see that it is really there.

I have used the concept of secretaries for two reasons. First, I believe it gives the impression that the Disk Operating System is there to help you. In the examples, the secretaries are really Disk Operating System or DOS commands. DOS does certain things for you that a number of personal secretaries might do. The only limitation is that you must be exact and specific with the secretaries (DOS commands). DOS commands must be used with a specific file name.

Second, the concept of secretaries may help to clarify the use of duplicate terms. Some of those DOS commands use the same words, but in a different manner, as Applesoft BASIC and/or Apple II computer commands; i.e., RUN, SAVE, and LOAD. When these words are used without a file name they are no longer seen as secretaries or DOS commands.

Typing SAVE, without a file name, is an Apple II computer command telling the computer to save whatever is in its program memory to TAPE (assuming a tape recorder is properly connected to the computer). Yet typing SAVE, with a file name, is a Disk Operating System command telling the computer to save whatever is in its program memory to the DISK. Typing LOAD, without a file name, is an Apple II computer command telling the computer to attempt to bring a program into its memory from TAPE. But typing LOAD, with a file name, is a Disk Operating System command telling the computer to bring a program into its memory from a DISK. Typing RUN, without a file name, instructs the computer to execute the instructions currently in memory. RUN, with a file name, instructs the computer to access the DISK, load the specified program into memory and begin operation according to that program's instructions.

We've covered a lot of new information in this chapter. If something is not clear, you should go back over it and use the Apple and disk drive to better understand these concepts.

CHAPTER 2 APPLESOFT AND INTEGER FILES

## QUESTIONS

1. TRUE or FALSE: Applesoft BASIC and Integer BASIC programs are stored on disk as files.
2. The main difference between Applesoft and Integer programs is in the way they work with _____.
3. What does DOS stand for?
4. How many main DOS commands are used with Applesoft BASIC program files?
5. Which DOS command gets the program from disk and immediately begins execution or operation of the program?
6. Which DOS command stores programs on the disk as files?
7. TRUE or FALSE: The LOAD command actually removes the program from the disk and loads it into the computer's memory.
8. What happens when you save back to disk a program you have changed, and you save it under the same name?
9. Name the other five DOS commands that can be used with all four file types.
10. TRUE or FALSE: LIST shows what is on the diskette.
11. LOAD and SAVE, without a file name, attempt to access which—(A) DISK or (B) TAPE?
12. TRUE or FALSE: Applesoft and Integer are programs, never files.
13. Explain what NEW does.

## ANSWERS

1. TRUE
2. NUMBERS or DECIMALS
3. Disk Operating System
4. 3
5. RUN with a file name
6. SAVE with a file name
7. FALSE, it takes a copy.
8. The previous version is erased and replaced with the new version.
9. LOCK, UNLOCK, RENAME, DELETE, VERIFY.
10. FALSE
11. B or TAPE
12. FALSE
13. Erases whatever is in the computer's memory.

# 3
# Text File Introduction

If you go back to the file cabinet example used in the last two chapters, this chapter is a quick look inside the T drawer and a superficial look inside the two different kinds of file folders in this drawer. We will examine the characteristics that are common to both kinds of TEXT files and look at how you can get at those files.

Of the four main types of files, we have seen that two types are instructions (called programs) for the computer: Applesoft or Integer BASIC program files. One of the remaining two types usually contains information for people rather than machines. By this, I do not mean that the computer cannot make use of the information, but that the information usually is not in the form of direct instructions for the computer. An example of an instruction for the computer is:

    20 PRINT "HELLO, HOW ARE YOU?"

An example of information that is not in the form of a computer instruction would be:

    Title: APPLE FILES
    Author: David Miller
    Publisher: Reston Publishing
    Address: Reston, Va.

This last example is the kind of information usually kept in a TEXT file. Before we get into the process of actually storing and retrieving text files, we need to understand the main difference between the two kinds of TEXT files.

Text files have two ways of storing and retrieving information. (Remember that the information really stays on the diskette and we are just getting a copy of the information!) These two ways of storing and retrieving information are sequential access and random access. "Sequential access text files" basically means that the information stored in the file is kept in sequential order. "Random access text files" usually means that each part of the file is divided equally and can be reached directly and at random instead of going through all previous records. The process of looking at each record in order (sequence) to decide if it is the record you want is a characteristic of sequential files and can require more time than the direct method of random access files.

The basic difference between sequential text files and random text files is somewhat like the difference between a cassette tape and a phonograph record. If I want to find a specific song on a cassette tape, even using the best available tape deck, I must begin at the current location of the tape and proceed either forward or backward, passing over all intervening songs until I have found the song I want. The process proceeds in sequence, one song after another. For example, if I want to only play the fourth song on the tape, I would have to advance the tape through the first, second and third songs until I get to the fourth one. On the other hand, if the songs are on a phonograph record, all I would have to do to play the fourth song would be to place the phono cartridge containing the needle at the start of the fourth division instead of at the start of the first song. I can do that because I am able to clearly see the divisions between songs and because those individual songs are directly accessible. I do not have to go through the grooves of the first three songs in order to get to the fourth. And moving the needle by hand takes only seconds. So imagine that the T drawer contains two basic divisions: the first division contains files that operate in a way similar to cassette tapes, while the second division contains files that operate like phonograph records in the way described.

But these two kinds of TEXT files do have things in common, just like tapes have things in common with phono records. The most obvious common characteristic is that they both usually contain information that is not in the form of instructions for the computer. In other words, they contain information like lists of things, addresses, receipts, and inventories. Second, since they are in the same drawer, they require the same "key" to get to them: control 'D'. Third, both files make use of the same DOS commands, but with different parameters.

Because these TEXT files are not computer instructions, they cannot be used in the same manner as Applesoft or Integer program files. In other words, you cannot RUN a TEXT file, SAVE or LOAD it. Those three commands, when combined with a file name, are the computer's means of access to Applesoft or Integer disk files. The obvious question, then, is that if you cannot use RUN, SAVE, or LOAD with TEXT files, how does the computer get the information on the disk in a TEXT file or back off the disk from a TEXT file? The answer begins with a "key."

## CHAPTER 3  TEXT FILE INTRODUCTION

We will begin by finding out about this key, the control 'D'. Imagine, for a moment, that our file cabinet has only one drawer that can lock. In order to get into that drawer, we would need the key, and the key would have to be used in the right manner. This is similar to what Apple has done with TEXT files. As we have said, since these files are not programs you cannot get at them by using the DOS commands RUN, LOAD, and SAVE. In fact, it turns out that there is little that can be done to access TEXT files directly. Usually, TEXT files must be accessed from within a computer program. In order to let the computer know that we want information, either to go to the disk or to come from the disk (rather than to or from the tape or to or from the keyboard), we must use the infamous control 'D' as our key.

On the last page of the original Disk II manual, we were told that in order to use DOS commands within a program, we must first print a string that contained a "CTRL D" followed by a command. What this means is that, in order to use any DOS command from within a program, several things need to be done. First, they chose the control 'D' as the method to tell the computer that the lines in the program that followed were to access the disk rather than the tape, the keyboard, or the screen.

Outside a program, in what is called the immediate mode (where what you type is acted upon immediately after the RETURN key is pressed), the control 'D' is relatively easy to use. Just like any other control key, you must press down on the key marked "CTRL" and while holding that key down, press on one of the other letter keys. Therefore, outside of a program, the control 'D' key is used by holding down the "CTRL" key and then, at the same time, pressing the 'D' key. If you want to see how this works and actually hear the results, hold down the CTRL key and then press the 'G' key at the same time. You will not see anything appear on the screen because control keys are usually non-printing characters, but the control 'G' key also produces a bell-like sound.

There are several methods of making use of this control 'D' from within a program. Programs operate in what is called the deferred mode, because the computer does not act according to the instructions in the program immediately. Rather it waits until it is told to RUN the program. It defers action until it is specifically told to act. This is the reason for the line numbers in programs. Line numbers tell the computer the exact sequence by which the computer is to follow the program instructions. The CTRL 'D' procedure described above can be used within a program, but nothing is printed on the screen using this method. If you go back to a program later, you cannot be sure if you remembered to include the control 'D' or not because you cannot see it. An example of this method of using the control 'D' would be:

```
20 PRINT "OPEN ADDRESS FILE"
```

The control 'D' should be between the quotation marks, but no one can tell if the control 'D' key is actually included because control keys are non-printing characters.

Another method, but still not very good, is to set a string variable equal to a control 'D', and then use the string variable with the DOS command. A string variable is identified in BASIC by a letter or letters followed by the dollar sign—$. Variables are names assigned to locations in the computer's memory. String variables can contain just about any value: i.e., numbers, letters, punctuation, and so forth. They are referred to as "variables" because their value may vary within a program; i.e., it is not constant. An example of a string variable might be NAME$ where the first value of NAME$ is "ANDY", the second value is "MARY", the third value is "PAUL", and the fourth value is "JANE". Using a string variable (D$), the example would look like this:

```
20 D$ = " "
40 PRINT D$; "OPEN ADDRESS FILE"
```

Now, at least, you can return to a program and see if you included something before the DOS command, but there is still no way of seeing exactly what was in line 20. You might have accidentally hit a control 'S' instead of a control 'D'. (Another reason for using a string variable is to substitute a visible value for an invisible value, or a value that is easier to type or shorter than some other value, especially if the longer value has to be typed repeatedly.)

Finally, another method for using control 'D' within a program suggests that line 20 should contain the value of control 'D' set by the code for all characters. The code used in the Apple and most micros is called ASCII. It sets a decimal and hexadecimal value for all characters used. If you look on page 138 of the APPLESOFT II manual and go down the page under the column WHAT TO TYPE, until you find "CTRL D," you will find three columns to the left. We are concerned only with the column labeled DEC. The value for a control 'D' in the DEC (decimal) column is "4". Now we have a value but need to know how to use that value. Applesoft BASIC provides the method we need. The reserved characters CHR$ are used with parentheses and a DECimal value to display non-printing characters on the screen. (Integer BASIC does not have the CHR$ command.) Now our example would look like this:

```
20 D$ = CHR$ (4)
40 PRINT D$; "OPEN ADDRESS FILE"
```

Now we have a method of telling exactly what "D$" stands for. This method is less than ideal since you must remember the ASCII value of control 'D' every time you begin a program. It is not difficult to remember that a control 'D' is a CHR$(4), but it is somewhat bothersome. A very

CHAPTER 3  TEXT FILE INTRODUCTION  19

good habit to get into when programming is to label anything that is not completely self-evident. BASIC allows a programmer to label something by using the REM (remark) command. Following this rule, line 20 now would read:

20 D$ = CHR$ (4) : REM CONTROL D

(For Integer users the line should be:

20 D$ = " " : REM CONTROL D BETWEEN QUOTES)

We now have the key and know how to use it, but we need to know with what DOS commands to use it.

In order to gain access to TEXT files, you must use certain DOS commands in specific ways, depending on the kind of TEXT file you are accessing. Both sequential and random access text files primarily use four DOS commands: OPEN, CLOSE, READ, and WRITE. Future chapters will examine in detail how each of these are to be used for either of the two kinds of TEXT files. For now, you need only to understand the essential task of each command. Again, the example of the filing cabinet is useful. In much the same way that a secretary must open a file folder, so also must all text files be opened before the information they contain can be put to use. And as the secretary should properly close the file folder before replacing it in the filing cabinet, all text files should be closed before ending the program or turning off the computer. If a secretary does not close the file folder, some information might drop out and get lost. The same is true if text files are not properly closed. This is usually only the case after new information has been written to the file and the file not closed. Loss of information should not occur after a text file has only been read and not closed. As implied by their descriptive command names, READ and WRITE are the processes by which information is either read from or written to the file. If you only want to see information already in a text file, the DOS READ command is the command you would use. If you want to add information to the file or create a new file, the DOS WRITE command is the one to use.

At this point, let's try out some of this information on the computer. Take the diskette that you initialized in the last chapter, place it in the disk drive, then either turn on the computer, or, if the computer is already on, type PR# and the number of the slot your disk interface card is in—usually slot 6. (Refer to Chapters 1 and 2 if you are not sure what to do.) When the cursor (blinking box) appears, type carefully and remember to press the RETURN key after each entry. Type:

CATALOG

And you should see:

A 002 HELLO
A 002 HELLO 2

(Remember, CATALOG shows the names of the files on the diskette.) Now type:

NEW

(This clears the computer's memory.)

20 D$ = CHR$ (4) : REM CONTROL D
40 PRINT D$; "OPEN ADDRESS FILE"
60 PRINT D$; "WRITE ADDRESS FILE"
80 PRINT "APPLE II IS A BRIGHT COMPUTER"
100 PRINT D$; "CLOSE ADDRESS FILE"

Check your typing carefully, then type:

SAVE APPLE

(The program is now saved on the diskette under the name APPLE.) Type:

CATALOG

and you now see:

A 002 HELLO
A 002 HELLO 2
A 002 APPLE

Next type:

LIST

to see that the program is in the computer's memory. Then type:

RUN

The disk drive comes on but little happens on the screen. Once again type:

CATALOG

This time you get:

A 002 HELLO
A 002 HELLO 2
A 002 APPLE
T 002 ADDRESS FILE

We have created a text file! Even though you did not actually see the text file being written to the diskette, that is exactly what happened immediately after you typed RUN and pressed the RETURN key. The reason you did not see anything on the screen is because the control 'D' key told the computer to print our information to the disk rather than to the screen.

We will look at this program to see what each line does and what the correct syntax for each should be. Line 20 sets the string variable (D $) equal to the ASCII value of a control 'D'. It acts the same as any other string variable when used within a program. The colon (:) is the statement divider allowing more than one BASIC statement on this line. REM is a BASIC reserved word meaning remark, indicating that what follows is only a comment by the programmer and will not be executed by the computer. The words "CONTROL D" are the actual comment. Line 40 begins with a PRINT statement. Apple chose the BASIC reserved word (or command) PRINT to be used with the control 'D' key. D $ is set equal to a control 'D' in line 20 so that in line 40, the first thing that is done is to PRINT a control 'D'. This tells the computer that we want to access the disk in some way. In other words, we have used the key to gain access to the file cabinet's T drawer. The rest of line 40 tells the computer what we are going to do with the disk. We are going to open a file called "ADDRESS FILE". If the file already exists, the computer gets ready to use that file. If no such file exists, the computer first creates such a file and gives it the name of "ADDRESS FILE". Notice the quotation marks around the words OPEN ADDRESS FILE. These quotation marks must be included in the program statements. That is how the computer understands where the specific disk instruction begins and ends. One other observation should be made. It was first thought necessary to include the semi-colon between the string variable and the specific instruction, but as can be discovered through experimenting, the semi-colon is not necessary. I include it because it helps set off the key to the disk (the control 'D') and the specific DOS instruction and makes the program easier to read. Line 60 again uses the control 'D' key. (Yes, it must be used on every line that contains a DOS command.) Next comes the optional semi-colon followed by the required quotation mark. Then, we tell the computer that we are going to be adding information to the file with the WRITE command. Since more than one file could possibly be opened at a time, we must also include the file name with the WRITE command. The line is concluded with the closing quotation mark. Line 80 tells the computer what information to put in the text file. Anything between the quotation marks will be written to the text file on the disk. (The two exceptions are the comma and the colon which will be explained in a later chapter.) The PRINT statement in this line does not print the string on the screen when the program is RUN, as would normally be expected, since this PRINT statement follows the two previous lines that contain information telling the computer to print to the disk rather than to the screen. Line 100 contains the key to the disk, control 'D', followed by the semi-colon, the opening quotation mark, the DOS command CLOSE, the file name ADDRESS FILE, and the closing quotation mark. This time the file name is optional since a PRINT D$;"CLOSE" will close *all* text files that are open. If there is only one file open, then that file will be closed with this simplified CLOSE statement. For clarity's sake, it is a better programming habit to include the specific file name with the CLOSE command.

We have now put information onto a diskette. The next task is to be able to read back from the diskette what we wrote. There are a number of different ways that we can read back the information, but for now, we will use the single program approach and simply add more lines to the program that wrote the information to the disk. Type:

LIST

(The program should still be in the computer's memory.)
Add the following lines carefully:

120 HOME
140 PRINT D$; "OPEN ADDRESS FILE"
160 PRINT D$; "READ ADDRESS FILE"
180 INPUT NAME$
200 PRINT D$; "CLOSE ADDRESS FILE"
220 VTAB 10
240 PRINT NAME$

Check your typing. Then type:

SAVE APPLE

(This replaces the previous and shorter version of our program.)

Finally type:

RUN

No file name is necessary since the program, besides being on the diskette, is also still in the computer's memory. This time the disk drive will come on for a brief time, then the screen will go blank, and the words "APPLE IS A BRIGHT COMPUTER" will be printed ten lines from the top of the screen.

Let's examine each line of the additional program lines. Line 120 clears the screen of anything left on the screen and places the blinking cursor in the upper left-hand corner of the screen. Line 140 is exactly the same as line 40. We need to re-open the file in order to start at the beginning of the file so that we can read from the file what we have just written. Line 160 is similar to line 60 except that this time we want to tell the computer we will be reading from the file rather than writing to it. Line 180 is the opposite of line 80. INPUT brings a copy of the information in the file called ADDRESS FILE into the computer and stores it in a string memory location we have labeled NAME$. INPUT brings information into the computer, usually from the keyboard, but following a control 'D' and the DOS command READ, INPUT brings information into the computer from the specified disk file. If there had been more than one piece of information, we would not have gotten it. We simply asked the computer to get one string variable and store that string in memory location NAME$. Just like line 80, line 180 accesses the disk rather than the keyboard because the two previous lines have given the computer instructions to that effect. I might add that following line 80 we could have included other PRINT statements that would have written other information to the disk text file ADDRESS FILE. Those statements could have taken the same format of line 80 or a different format altogether. The same would then be true following line 180. We could have had more INPUT statements if there was more information actually in the file. If there was not and we asked for more with another INPUT, the computer would have responded with a beep and "OUT OF DATA ERROR" on the screen. The program would have stopped at that line number. Line 200 is exactly the same as line 100. Line 220 simply readies the screen for line 240 by spacing down from the top of the screen 10 lines (Vertical tab). Line 240 prints the value contained in memory location NAME$ on the screen 10 lines from the top. The computer prints this information on the screen because we have not told it that we wanted the information to go anywhere else. The screen is the default for PRINT statements. Default in this case means that a certain value, the screen, has been predetermined and unless the value is specifically changed by a control 'D' and a DOS command, the predetermined value is taken as the desired value. Since we have closed our TEXT file and not included the control 'D' key in the two previous statements, the computer understands that we want the information printed on the screen.

This program is merely to give a brief explanation of the four basic DOS commands used with either sequential or random access text files. It is not intended to be a meaningful or useful program in any other sense. Such programs will begin in the next chapter. Please review this chapter and the program example with explanation until you are confident you fully understand what each of the DOS text file commands, OPEN, CLOSE, READ, and WRITE, does and how each relates to the use of the control 'D' within a BASIC statement.

Finally, let's "clean up" our diskette so that we can put some serious programs on it in the next chapter. By following the instructions given below, you will gain practice in the use of two other DOS commands. If you do not wish to erase these programs, you can skip what follows and start the next chapter with a fresh diskette, providing you remember to first initialize the new diskette. (Refer to Chapter 1 if necessary.)

Type:

CATALOG

The list shows:

A 002 HELLO
A 002 HELLO 2
A 003 APPLE
T 002 ADDRESS FILE

Now carefully type just the following:

DELETE ADDRESS FILE

Then again type:

CATALOG

and

ADDRESS FILE

should be gone!
Next type:

DELETE HELLO
RENAME HELLO 2, HELLO

We have erased one more file and changed the name of another one. As the last step, type:

LOAD HELLO
DELETE HELLO
SAVE HELLO

This step puts HELLO back as the first file on the diskette. If you have followed these instructions carefully, the CATALOG should show just two files:

A 002 HELLO
A 003 APPLE

## QUESTIONS

1. Name the type of file that usually contains lists of information rather than computer instructions.
2. Give the two kinds of text files.
3. Which kind of text file is similar to a cassette tape?
4. What is the "key" to accessing text files?
5. TRUE or FALSE: You can RUN a text file just as you RUN an Applesoft program.
6. Give the number of modes in which the computer can operate and name them.
7. Name the four DOS commands usually used with text files.
8. What does REM stand for?
9. Explain what the BASIC reserved word HOME does.
10. Explain what the BASIC reserved word INPUT does without a control 'D' and a DOS command in the preceding line.
11. What symbol is used to designate string variables?
12. What are variables?

## ANSWERS

1. TEXT FILES
2. SEQUENTIAL ACCESS, RANDOM ACCESS
3. SEQUENTIAL ACCESS
4. CONTROL D
5. FALSE
6. 2; IMMEDIATE and DEFERRED
7. OPEN, CLOSE, READ, WRITE
8. REMARK
9. Clears the screen, places the cursor in the upper left corner of the screen
10. Brings information into the computer from the KEYBOARD
11. $
12. Names of locations in the computer's memory where values can be stored

# 4
# Creating Sequential Files

We begin to get into the heart of our study with the first part of our examination of sequential text files. First, let's review briefly. We have seen that there are four main types of files: Applesoft program files, Integer program files, Binary files and Text files. You should understand by now that Applesoft program files and Integer program files are files that contain specific instructions for the computer. Binary files may be files that contain specific computer instructions, but they may also contain lists of data. Text files usually contain only lists or data, not computer instructions. We have also seen that there are two kinds of text files, sequential text files and random access text files. The difference between these two kinds of text files lies in the way the information within them is accessed—sequential requires accessing one record after the other, and random allows access to any record directly and immediately. In the last chapter, you were introduced to the control 'D' character which acts as a key allowing access to the disk from within either an Applesoft or Integer program. A new set of DOS (Disk Operating System) commands common to both sequential and random text files was also introduced. Now we are ready to put some of this knowledge to work and come up with some useful programs.

We will begin by taking a closer look at the program given in the last chapter and modifying it to make it more useful.

```
20 D$ = CHR$ (4) : REM CONTROL D
40 PRINT D$; "OPEN ADDRESS FILE"
60 PRINT D$; "WRITE ADDRESS FILE"
80 PRINT "APPLE II IS A BRIGHT COMPUTER"
100 PRINT D$; "CLOSE ADDRESS FILE"
120 HOME
```

```
140 PRINT D$;"OPEN ADDRESS FILE"
160 PRINT D$;"READ ADDRESS FILE"
180 INPUT NAME$
200 PRINT D$;"CLOSE ADDRESS FILE"
220 VTAB 10
240 PRINT NAME$
```

This is the same program from the last chapter. Take your diskette that you have been using and place it in the disk drive. Start the computer (boot the system) in one of the ways mentioned in the previous chapters. When the cursor appears, type the following to see the program instructions:

```
LOAD APPLE
LIST
```

(If you have erased the program, use the listing above and type it in again.) Type:

```
RUN
```

Again, you should see the screen clear, and ten lines from the top the following will appear:

```
APPLE II IS A BRIGHT COMPUTER
```

Type:

```
CATALOG
```

```
ADDRESS FILE is back!
```

Then change line 80 to:

```
80 PRINT "APPLE IS OK"
```

Now type:

```
LIST
```

to make sure the change was made. Then type:

```
RUN (Yes, again)
```

You should see:

```
APPLE IS OK
```

But that is not everything the ADDRESS FILE now contains. To see if ADDRESS FILE contains anything else, add the following lines to the program:

```
190 INPUT EXTRA$
260 PRINT EXTRA$
```

## CHAPTER 4 CREATING SEQUENTIAL FILES

Then type RUN once more! You should now see:

APPLE IS OK
A BRIGHT COMPUTER

Why did we get the added phrase "A BRIGHT COMPUTER"? We changed line 80 so that it only said "APPLE IS OK". Yet, by just adding an extra INPUT statement (line 190) and an extra display statement (line 260), we see the last part of our original sentence (APPLE II IS A BRIGHT COMPUTER).

In our first example, line 80 wrote 29 characters to the disk. By changing line 80 to a shorter string, we wrote less than 29 characters to the same file—ADDRESS FILE. But the difference between 29 and the length of the shorter string left that amount of extraneous characters still on the disk in the same file—ADDRESS FILE.

In this simple example, extra data in the file is not likely to cause any trouble, but in useful programs, extraneous information may prove disastrous. An illustration of this would be the confusion that would result if the numbers in an address file were overwritten but the street name was not. You might end up with the correct numbers going to the wrong street. There is an easy way to make certain that this overwriting does not occur. Add:

50 PRINT D$; "DELETE ADDRESS FILE"
55 PRINT D$; "OPEN ADDRESS FILE"

Now when you type RUN, you should get an OUT OF DATA error in line 190 because there is no more extraneous information in the file. If you get the error message as you should, remove line 190 and 260. If you did not get the error message, you should go back over every step until you do get the error message. In this case, the error message indicates that you have been following along correctly and are ready to go on.

After removing lines 190 and 260, RUN the program one last time. (You can remove unwanted program lines by just typing the line number and pressing the RETURN key.) This time you should only get the "APPLE IS OK" phrase.

By first opening the file and deleting it, then opening it again and proceeding as before, we get rid of the extra characters by eliminating the previous file. There is an obvious danger in doing this since we have erased what we did the first time. In this working example, it makes no difference, but with a permanent file, we might want to add to the file rather than erase it. DOS provides a word for just such additions to sequential files: APPEND. We will soon use APPEND, but let's go back and make a useful address file program.

I am not going to get involved in the discussion over flow-charting or structured programming, but as in building anything, you should have a plan. In this mailing list system, we will need several programs. The first program should create the file. That first program should probably have at least three different parts: an input routine, a correction routine, and a file creation routine. With this minimum plan, let's begin. We start with what might be called the housekeeping statements.

```
10 REM **--MAILING LIST CREATOR--**
20 D$ = CHR$ (4) : REM CONTROL D
40 DIM NAME$ (20)
60 K = 1 : REM LINE COUNTER
```

Line 10 simply gives a name to the program. Line 20 is our line from the first program setting D string equal to a control 'D'. Line 40 uses the DIM (dimension) statement to reserve space in the computer's memory for 20 lines of 255 characters per line. This number of lines should be enough for a starter program, but you can change the 20 to a larger or smaller number. Line 60 uses a counter "K" to keep track of the lines used.

```
70 REM **--INPUT ROUTINE--**
80 HOME: VTAB 5
100 PRINT "TYPE NAME AND ADDRESS AS IF ADDRESSING AN
     ENVELOPE. ";
120 PRINT "DO NOT USE A COMMA OR COLON."
140 PRINT: PRINT "TYPE 'END' WHEN FINISHED."
```

Line 70 labels the routine and line 80 uses the BASIC reserved word HOME to clear the screen and place the cursor in the upper left-hand corner. VTAB 5 provides for spacing 5 lines down from the top of the screen. Lines 100-140 provide instructions to the person entering information: in this case, names, addresses, and phone numbers. All of these lines are optional, but I have included them because they help the user to enter the information correctly. And, of course, the instructions themselves can be re-worded to your own preference. The warning concerning the use of a comma or colon is included because they act as de-limiters when used in an INPUT statement. (De-limiters define the limit or length of the string.)

```
160 PRINT: PRINT "TYPE IN LINE "; K
180 INPUT NAME$ (K)
200 IF NAME$ (K) = "END" THEN 300
220 K = K + 1
240 GOTO 160: REM GO BACK FOR ANOTHER LINE
```

These lines are the heart of the input routine. Line 160 instructs the user which line is being typed in. The first PRINT, prints a blank line. Next, the phrase "TYPE IN LINE " is displayed. Notice the space between the E and the closing quotation mark. There is a reason for that space.

CHAPTER 4   CREATING SEQUENTIAL FILES                          31

The semi-colon instructs the computer to print the value of the variable K immediately next to the end of the quotation. If there were no space, the value of K would be printed next to the E in LINE like this: TYPE IN LINE1. Line 180 accepts what the individual types in and stores it in the reserved memory, depending upon the value of K. Remember that we told the computer to reserve space for 20 possible lines of NAME$. (Reserving multiple space for a variable creates an ARRAY for that variable.) Line 200 checks what was typed in to see if it equals the word "END". If it does equal "END", the computer is instructed to jump ahead to line 300 immediately. If it does not equal "END", this line is ignored and the computer goes to the next instruction in line 220. Line 220 is a method of increasing the line count. The first time through, K will equal 1, so the formula really is: K = 1 + 1 or 2. Once we have increased our line count, we want to go back and get another line, which is exactly what line 240 does.

```
300 NAME$(K) = "*": REM SEPARATOR FOR PHONE NUMBER
320 K = K + 1
340 PRINT "PHONE: ";: PRINT "TYPE 'RETURN' IF NONE. "
360 INPUT NAME$(K)
380 K = K + 1
400 NAME$(K) = "!": REM SEPARATOR BETWEEN SETS OF
    INFORMATION
```

These may be the most confusing lines to understand. In order to easily separate the name and address from the phone number, I have included a separator, "*", on a line by itself. The reason for separating the phone number from the rest of the information is that now we can use the first part of our information to produce mailing labels. I have also included a separator, "!", to easily differentiate between the name, address, and phone number of one person and the name, address, and phone number of the next person. Therefore, line 300 sets the Kth line of NAME$ equal to "*". At this point, if the first line contains the name; the second line the address; and the third line the city, state and zip code; then the fourth line will contain the word "END," and K will be equal to 4. By making the fourth line equal to "*", we have actually accomplished two tasks: eliminating the word "END" and establishing a one-character separator before the phone number. We could have required the user to type the "*" when he/she finished entering the name and address, but I prefer to have the user type something natural within the context. Line 320 increases the line count by one for the phone number. Line 340 gives instructions about typing in the phone number. Line 360 accepts whatever format the individual uses to type in the phone number and stores the information in the string reserved memory. Line 380 again increases the count by one, this time for the separator between sets of information. Line 400 makes the Kth line of NAME$ equal to "!". If the fourth line of NAME$ is "*" and the fifth line is the phone number, then K would be 6 and the sixth line would equal "!". This concludes the input routine.

```
410 REM **--CORRECTION ROUTINE--**
420 HOME: VTAB 5
440 PRINT "DO NOT CHANGE THE LINE WITH THE '*'"
460 PRINT "THIS SYMBOL IS USED AS A SEPARATOR."
480 PRINT
```

These lines explain what the routine is, set the format for the correction routine, and give instructions to the user about the separator "*". Line 410 labels the routine, and line 420 clears the screen and places the cursor in the upper left-hand corner five lines from the top. Lines 440 and 460 print the instructions for the user. Line 480 prints a blank line after the instructions.

```
500 FOR I = 1 TO K-1
520 PRINT I; " "; NAME$(I)
540 NEXT I
```

Lines 500-540 make a loop used to get the information stored in the string-reserved memory and print that information on the screen. Line 500 is the first line of a FOR-NEXT loop. It uses a counter (I) that starts with the value of 1 and counts to the value of K minus 1. In our example above, the sixth line was the last line and was therefore set equal to "!". Since that line should not be changed, there is no reason to display the line. Therefore, the counter only goes to K-1. Line 520 prints the current value of the counter, a blank space, and then the information contained in NAME$(I) stored in the computer's memory. Line 540 increases the counter by one until the counter equals the value of K-1.

```
560 PRINT
580 INPUT "CHANGE ANY LINE? TYPE 'Y' OR 'N' "; YES$
600 IF YES$ = "Y" THEN 640
620 GOTO 740: REM GO TO FILE CREATION ROUTINE
640 INPUT "CHANGE WHICH LINE "; LINE
660 IF LINE > K-1 THEN PRINT "NUMBER TOO LARGE": GOTO 640
680 PRINT "OLD LINE = "; NAME$(LINE)
700 INPUT "CORRECT LINE = "; NAME$(LINE)
720 GOTO 420
```

These lines are fairly standard correction routine lines. Line 560 prints a blank line before the question in 580. Line 580 asks the necessary question and provides instructions for answering it. The user's response is stored in the string variable YES$. Line 600 checks the answer, stored in YES$, to see if it equals "Y". If it does, the computer is instructed to jump to line

640 and proceed. If it does not equal "Y", the computer goes to the next instruction in line 620. Line 620 tells the computer to go to line 740 which will be the start of the file creation routine. It does this because, when the computer gets to this line, the user has typed something other than a "Y" in answer to the question of line 580. Line 640 requests the number of the line that needs changing and stores that value in the variable "LINE". (Notice that there is no dollar sign following LINE. This indicates that this variable is a numeric variable rather than a string variable. Numeric variables can only contain numbers.) Line 660 checks to see if the user has typed a number larger than the total number of lines displayed. If that is the case, a message is printed and the computer returns to line 640 to ask again for the number of the information-line to be changed. Line 680 prints the originally typed line, and 700 waits for the user to type in the correct information. Finally, 720 returns to line 420 to begin the correction process over again. The correction process will be repeated until the user answers the question in 580 with something other than a "Y". There are a number of other lines or checks that could have been included, but for our present needs, these lines are sufficient.

```
730 REM **--FILE CREATION ROUTINE--**
740 PRINT D$;"OPEN ADDRESS FILE"
760 PRINT D$;"DELETE ADDRESS FILE"
780 PRINT D$;"OPEN ADDRESS FILE"
800 PRINT D$;"WRITE ADDRESS FILE"
820 PRINT K:REM NUMBER OF LINES
840 PRINT "     ":REM 5 SPACES FOR INCREASING COUNTER
860 FOR I=1 TO K
880 PRINT NAME$(I)
900 NEXT I
920 PRINT D$;"CLOSE ADDRESS FILE"
940 END
```

We are finally down to the actual file handling routine. As you can see, the routine is quite short. The key to filing system programs is often in proper planning. If you have tried to anticipate and provide for all possible requirements, present and future, your text files can become very powerful and useful. If you are not careful in your planning, however, you may find that some of the information you thought you had in the file has been overwritten, lost, or practically unavailable. This is the reason for including the two single character separators and the reason for lines 820 and 840 in this routine. Lines 740 through 800 are the same sequence we used in our original work program. Line 740 opens the file; 760 deletes the file in case there was a previous file by the same name. Line 780 opens the file again, and 800 tells the computer that we are going to put information in the file. Now, we come to the two unfamiliar lines. Line 820 prints the current value of K which, in our example, should be a "6". This is done to keep track of the total number of lines used so that we know how many lines to read back into the computer with other programs. The six is a single-digit number and, therefore, takes up only one space on the disk. (Actually, it takes up two since a 'RETURN' character is automatically added to the end of most PRINT statements.) In this program, it doesn't really matter how many digits represent the value of K, since we are only going to write this value once. But in the program that will follow, which is used to add more names to this file, the number of digits in the value of K will be very important. There is no problem as long as the value of K remains at one digit. But when it goes to two, three, and possibly four digits, we will have a problem. If we were to begin writing the first person's name immediately after the value of K, when K reached two digits, the first letter of the name would be overwritten. Therefore, we print a line of five blank spaces in line 840 to provide for overwriting. If you wanted to add more spaces, you certainly could, but five blank spaces leave enough room for a six-digit number, and that many lines would take a very long time to read back into the computer even if they would fit in memory. Again, there are other ways of keeping track, but in a sequential file, this is one of the easiest and clearest. Lines 860 through 900 are essentially the same loop as lines 500 to 540, but this time, the information is printed to the disk instead of just the screen. This time we do want to print the separator "!", so the counter goes from 1 to the value of K. Finally, we close the file in line 920 and end the program.

If you have been following along and typing in the program, you should save this program on disk by giving it a name such as the name in line 10. Remember that to save a program, you type the word "SAVE" and the program file name. Like this:

SAVE MAILING LIST CREATOR

Now type the word "CATALOG" and see if the file name is listed. It should be listed something like this:

A 006 MAILING LIST CREATOR

At this point, you can run the program and enter your own name, address, and phone number if you wish. When you type the word "RUN," the screen will clear and five lines down the message from lines 100-160 appears. Type in a name and hit the "RETURN" key. You will then be told to type in line number 2. If you want to type in a title for the name in the first line, you can. If no title is needed, then simply type in the street address and press "RETURN". This process should continue until you type in the word "END". When you do type in the word "END", you will be asked for the phone number and told that if there is no phone number, you should just hit "RETURN". After the phone number has been typed, you are shown a

## CHAPTER 4　CREATING SEQUENTIAL FILES

list of the lines you have typed and asked if you wish to change any of those lines. If you do want to change a line, you must answer the question with a "Y". If you do not, either type an "N" and the "RETURN" key or just the "RETURN" key. If you need to make changes and have typed a "Y", you will then be asked which line number you want to change. Respond with a number on the screen. You will then be shown the originally typed line and asked for the correct information for this line. After typing in the new line and pressing "RETURN", you will be shown the list of lines again with the new line in place of the old line this time. You can make as many changes as you wish. When you are satisfied and do not wish to make any more changes, a response of "N" to the question about changes will instruct the computer to write the information out to the disk. Now, if you type "CATALOG", you should see:

```
A 006 MAILING LIST CREATOR
T 002 ADDRESS FILE
```

But several questions now present themselves. How do I add more names to this file? And how do I actually see what is in the file? As you may have realized by now, there are a number of possible answers. One answer would be to add more lines to this program so that the program reads back what it just wrote to disk. Another answer is to write a separate program and possibly a program menu that would be able to switch easily between programs that write information and programs that read the information. In the next chapter, we will explore a number of these possibilities and see a little of what can be done with the information once it is safely and correctly on disk. But to conclude this chapter, I am going to show a quick way to see what is in sequential text files. Type the following:

```
NEW
```

(But make sure you have saved your program first!)

```
MON I, O, C
```

(This is a special computer instruction that allows us to actually see displayed on the screen what is either written to or read from the disk.)

```
EXEC ADDRESS FILE
```

Do not be upset by the computer beeping at you or the "SYNTAX ERROR" messages that appear. The EXEC command was not intended for this purpose but does work on sequential files (although not on random access files). The EXEC command will be discussed in later chapters.

## QUESTIONS

1. TRUE or FALSE: Information in a sequential access file can be overwritten by additional information.
2. What DOS command is used to avoid overwriting to a sequential text file?
3. Name the three main parts (or routines) in the MAILING LIST CREATOR program.
4. What does reserving multiple space for a variable create?
5. What symbol did we use to separate sets of information?
6. What does DIM stand for?
7. "FOR I = 1 TO K" is the first line in what kind of loop?
8. The program user's response is tested by what kind of BASIC statement?
9. What computer instruction allows us to see what goes between the computer and the disk?
10. What DOS command can read a sequential text file in the immediate mode?

## ANSWERS

1. TRUE
2. APPEND
3. Input, Correction, and File routines
4. An ARRAY
5. !
6. Dimension
7. A FOR-NEXT loop
8. An IF...THEN statement. Check lines 200, 600, and 660
9. MON I,O,C. The reverse is NOMON I,O,C which turns off the display of what is going between the computer and the disk.
10. EXEC

## MAILING LIST CREATOR

```
10  REM **--MAILING LIST CREATOR--**
11  :
12  :
20  D$ = CHR$(4): REM CONTROL D
40  DIM NAME$(20)
60  K = 1: REM LINE COUNTER
65  :
66  :
70  REM **--INPUT ROUTINE--**
80  HOME : VTAB 5
100 PRINT "TYPE NAME AND ADDRESS AS IF ADDRESSING  AN
    ENVELOPE. ";
120 PRINT "DO NOT USE A COMMA OR COLON. "
140 PRINT : PRINT "TYPE 'END' WHEN FINISHED"
160 PRINT : PRINT "TYPE IN LINE ";K
180 INPUT NAME$(K)
200 IF NAME$(K) = "END" THEN 300
220 K = K + 1
240 GOTO 160: REM GO BACK FOR ANOTHER LINE
300 NAME$(K) = "*": REM SEPARATOR FOR PHONE NUMBER
320 K = K + 1
340 PRINT "PHONE: ";: PRINT "TYPE 'RETURN' IF NONE. "
360 INPUT NAME$(K)
380 K = K + 1
400 NAME$(K) = "!": REM SEPARATOR BETWEEN SETS OF
    INFORMATION
405 :
406 :
410 REM **--CORRECTION ROUTINE--**
420 HOME : VTAB 5
440 PRINT "DO NOT CHANGE THE LINE WITH THE '*'"
460 PRINT "THIS SYMBOL IS USED AS A SEPARATOR. "
480 PRINT
500 FOR I = 1 TO K - 1
520 PRINT I;" ";NAME$(I)
540 NEXT I
560 PRINT
580 INPUT "CHANGE ANY LINE? TYPE 'Y' OR 'N' ";YES$
600 IF YES$ = "Y" THEN 640
620 GOTO 740: REM GO TO FILE CREATION ROUTINE
640 INPUT "CHANGE WHICH LINE ";LINE
```

# MAILING LIST CREATOR

```
660 IF LINE > K - 1 THEN PRINT "NUMBER TOO LARGE"
    : GOTO 640
680 PRINT "OLD LINE = "; NAME$(LINE)
700 INPUT "CORRECT LINE = "; NAME$(LINE)
720 GOTO 420
725 :
726 :
730 REM **--FILE CREATION ROUTINE--**
740 PRINT D$; "OPEN ADDRESS FILE"
760 PRINT D$; "DELETE ADDRESS FILE"
780 PRINT D$; "OPEN ADDRESS FILE"
800 PRINT D$; "WRITE ADDRESS FILE"
820 PRINT K: REM NUMBER OF LINES
840 PRINT "     ": REM 5 SPACES FOR INCREASING COUNTER
860 FOR I = 1 TO K
880 PRINT NAME$(I)
900 NEXT I
920 PRINT D$; "CLOSE ADDRESS FILE"
940 END
```

# 5
# Appending Sequential Files

Now the fun begins. We have created a file, but as you will soon see, the creation is one of the easiest parts of file manipulation. There are two things we would like to do immediately with this file: add to the file, and read what is in the file. Both tasks are easy to do, but because the job of reading is simpler to explain and more rewarding, we will go over a short program to read the file first.

```
10 REM **--MAILING LIST READER--**
20 D$ = CHR$ (4) : REM CONTROL D
30 REM INPUT ROUTINE
40 PRINT D$; "OPEN ADDRESS FILE"
60 PRINT D$; "READ ADDRESS FILE"
80 INPUT K
100 DIM NAME$ (K)
120 FOR I = 1 TO K
140 INPUT NAME$ (I)
160 NEXT I
180 PRINT D$; "CLOSE ADDRESS FILE"
```

By now, lines 10 and 20 should be clear. They name the program and set the D string equal to a control 'D'. Line 40 uses the D string (control 'D') to tell the computer that we want to use the disk and that we want to open a file called "ADDRESS FILE". Line 60 advises the computer that we want to read from the "ADDRESS FILE". We tell the computer exactly what we want to read in line 80. Here we are asking for the number of lines we wrote to the disk in our file creation program. If you are not clear on this, check back to the explanation of lines 820 and 840 in the previous chapter. We are simply reading back the number written

CHAPTER 5   APPENDING SEQUENTIAL FILES                           41

in those lines. Line 100 is our line that reserves space in the computer for the information we will be bringing in off the disk. Since we are not sure of the exact number of lines and the number will change every time we add information, we should use the variable "K" which will always equal the number of lines that have been written. Now we can bring in a copy of the information contained in the file. Lines 120, 140 and 160 bring in that data. Line 120 establishes the boundaries for the loop we want. We want the count to go from the first line to the last line represented by the variable K. Because we have previously used the control 'D' key, the computer understands that the INPUT statement in line 140 refers to the disk and not the keyboard. This line actually goes to the disk and obtains a copy of the information contained in the line specified by the variable I. The operation is on the same principle as was the logic of lines 860 to 900 in our file creation program. But this time, we are bringing information into the computer from the disk instead of transferring information from the computer to the disk. Now the information physically exists in two locations. One location is in the computer's memory, and the other location is still out on the disk. By bringing the information into the computer, we have not erased that information on the disk. Merely READing a file does not disturb the contents of that file. Finally, line 180 closes the file. If you run this program, you are not likely to see anything happen except perhaps a little flickering below the word RUN. We need a routine that will display the information. We will become deeply involved in different ways of displaying our information a little bit later, but for now the following routine will get the job done.

```
200 REM **--DISPLAY ROUTINE--**
220 HOME: VTAB 5
240 FOR I = 1 TO K
260 PRINT NAME$(I)
280 NEXT I
```

Our input routine can be used in a number of different programs to bring in all the information from the file, but our display routine will not be really functional in very many situations. We will alter this routine later to make it more useable (see Chapter 6). Save the program now as MAILING LIST READER, then type CATALOG to see the list of files now on the disk. Type:

```
SAVE MAILING LIST READER
CATALOG
```

and you should see:

```
A 002 HELLO
A 006 MAILING LIST CREATOR
```

CHAPTER 5  APPENDING SEQUENTIAL FILES

    T 002 ADDRESS FILE
    A 003 MAILING LIST READER

To this point, you have created a file, ADDRESS FILE, and written the first group of lines containing information to that file. Now, you have read that information back and displayed it. Next, we need to be able to add more information to the file. If you ran the file creation program again and used a different set of lines, what would happen? Would the new information be added to the file? Would the old information be replaced? If you do not know for certain what would happen, look at line 760 of the file creation program (MAILING LIST CREATOR). Every time this program is RUN, the computer is instructed to erase or delete "ADDRESS FILE" before writing any new information. What happens to the first set of lines already on the disk if you try to use this program to add a second set of lines? That first set is erased, so we need a third program to add more lines of information to our ADDRESS FILE. (For those of you itching to put all these programs into one large program, have patience. I will eventually explain how these programs can be tied together without actually existing as one large program.) This third program is really just a modification of the file creation program. But the modification needs to be done, or, as we have seen, the results will be worthless. The modification is relatively simple if you follow closely the instructions given below.

Down to line 720 of the MAILING LIST CREATOR program (see the complete listing at the end of Chapter 4), the new program can be the same with two minor changes. Load the MAILING LIST CREATOR program first and then list it to see a complete listing of the instructions in this program.

    LOAD MAILING LIST CREATOR
    LIST

The program will not all fit on the screen at one time. The first instructions disappear from view off the top of the screen. In Applesoft, you can list to a certain instruction-line number, or from a certain instruction-line number to the end of the program, or from one line number to another, like this:

    LIST - 200
    LIST 350-
    LIST 100-800

Line 10 should be changed to read:

    10 REM MAILING LIST ADDER;

and line 620 should become:

    620 GOTO 2000: REM GO TO FILE CREATION ROUTINE.

CHAPTER 5   APPENDING SEQUENTIAL FILES                             43

Except for those two changes, the MAILING LIST CREATOR program works fine down to line 720 for our new MAILING LIST ADDER program, so be certain to make those changes before continuing. Next, you should delete lines 730 through 940. Type:

    DEL 730,940

The logic for the placement of these next two routines will become clear a little later. For now, add these lines:

    1000 REM **--REPEAT ROUTINE--**
    1020 HOME : VTAB 5
    1040 PRINT "DO YOU WANT TO ADD MORE INFO? "
    1060 INPUT "TYPE 'Y' OR 'N' ";YES$
    1080 IF YES$ = "Y" THEN RUN
    1100 END

Line 1000 names the routine. Line 1020, as we have seen, clears the screen and spaces down five lines. Line 1040 prints the question about additional information, and line 1060 uses the INPUT statement to wait for a response from the user. The computer knows to wait for the response from the keyboard because we have not used the control 'D' key to the disk before this input statement. Line 1080 checks the response. If the response equals "Y", then the computer is instructed to "RUN" the program again. If the response is anything else, the computer goes to the next line for its instruction. Line 1100 tells the computer to end the program and go no further, but neither of these choices writes anything to the disk. That is the reason we need another routine, our file addition routine. This routine will do three things: first, check to see how many information-lines are currently in the ADDRESS FILE; second, add the number of information-lines in the file to the number of new information-lines for a revised information-line total; and finally, actually append the new information-lines to the ADDRESS FILE.

    2000 REM **--FILE ADDITION ROUTINE--**
    2020 PRINT D$;"OPEN ADDRESS FILE"
    2040 PRINT D$;"READ ADDRESS FILE"
    2060 INPUT REC
    2080 PRINT D$;"CLOSE ADDRESS FILE"

This part of the routine should look somewhat familiar. It is almost exactly the same as the first few lines in our program to read the file (MAILING LIST READER). Line 1000 again names the routine. Line 2020 uses D string (control 'D') as the key to the disk and opens the file. Line 2040 tells the computer we want to read the file called ADDRESS FILE. Line 2060 inputs the same number as line 60 in the MAILING LIST READER program but this time stores its value in the variable REC (for record). We cannot use the variable K because we are already using it in the first part of our program and want to continue using it. Once again, this number represents the number of lines of information already in the file. Line 2080 then closes the file.

```
2100 REC = REC + K
2120 PRINT D$;"OPEN ADDRESS FILE"
2140 PRINT D$;"WRITE ADDRESS FILE"
2160 PRINT REC
2180 PRINT D$;"CLOSE ADDRESS FILE"
```

These lines simply add up the number of lines already on the disk with the number of new lines we have just typed into the computer. This new total is then written back out to the disk over the previous total. Line 2100 provides the method for totaling the previous line count with the additional number of new lines. The logic for this line is the same as for our now standard "K=K+1" lines. If you are not clear about this logic, it is best to just accept that this is one way the computer totals things. The other lines should be clear. Line 2120 opens the file; 2140 prepares to write to the file; 2160 actually writes the new total of the line count to the disk; and 2180 closes the file. It is necessary to open the file again in order to write the new line count total in its correct location, at the beginning of the file.

```
2200 PRINT D$;"APPEND ADDRESS FILE"
2220 PRINT D$;"WRITE ADDRESS FILE"
2240 FOR I = 1 TO K
2260 PRINT NAME$(I)
2280 NEXT I
2300 PRINT D$;"CLOSE ADDRESS FILE"
2320 GOTO 1000: REM GO TO REPEAT ROUTINE
```

This is the actual part of the routine that adds our new information-lines to the existing ADDRESS FILE. Line 2200 uses a new DOS command APPEND, to do just what it says. It appends or adds to the file rather than overwriting any of the information already in the file. This line tells the computer to prepare to add to the file. Line 2220 seems like a superfluous line since you cannot add to the file except by writing to it, but the line is necessary. From line 2240 on, the routine is the same as the routine in the MAILING LIST CREATOR program (lines 860 to 920). Line 2240 sets up the loop. Line 2260 prints the information in line "I" to the disk after the information already on the disk. Line 2280 goes back for another line of information. Line 2300 closes the file. Line 2320 directs the computer to jump back up to the instruction in line 1000 which is the start of the routine asking if the user wants to add more information. Thus, we are finished with the program to add more information to our ADDRESS FILE.

## CHAPTER 5 APPENDING SEQUENTIAL FILES

You should now save this new program to the disk as MAILING LIST ADDER. You can check your typing by going over the complete listing of the program given at the end of this chapter. Please remember that it is very important to follow along by typing the necessary lines on your Apple.

We now have three complete programs: MAILING LIST CREATOR, MAILING LIST READER, and MAILING LIST ADDER. The combination of these programs will create a file, add information to the file, and read information back from that file. The three adequately demonstrate the procedures used to accomplish these tasks, but the programs are not really very useful or practical as they now exist. For instance, every time you run the MAILING LIST READER program, you will read the entire file and display the entire file. This happens even if you only want just one name and address. And after just a few names and addresses are added to the file, the list begins to disappear off the top of the screen during display. It is quite obvious that more modification needs to be done in order to make these programs useful. If you are already a good programmer in BASIC, you probably have some ideas about "features" you would like to see in one or more of these programs. If you have little experience in programming, you will soon become much more experienced. I am going to add a few "features" to these programs and fully explain each additional step. If you would like to include these features and become more experienced at programming, especially with file information data, follow closely with the different programming lines and explanations given. If you don't need these features or want to create your own, you might want to skip ahead to the chapters on advanced sequential text file manipulation. If you have had enough of sequential text files, you might want to jump immediately to the chapters on random access text files. We will use some of these same routines in the chapters on random access, but I will not go into the same kind of detail as in these chapters. In other words, the routines will be repeated, but the full explanations will not be repeated. Let's begin adding "features" to our three programs and making them more useful.

We will begin with the MAILING LIST ADDER program. If you have used this program to enter a number of names and addresses, you will have noticed that the disk operates every time you have accepted a set of information-lines as correct. This disk operation may not bother you if you are somewhat slow in typing or are in no hurry to enter a large number of names and addresses, but there is no reason that the disk needs to operate after every name. Why not write the information out to disk only after we have finished entering all of our information lines? Such a change is clearly a preference "feature" that proponents and opponents often argue about. In this situation, I prefer to enter all of my information before writing any of it to the disk. But even before typing in a second set of information lines, I may want to print a mailing label of the information I have just entered. This is obviously a preference "feature," and it will do you no good if you do not have a printer. If you do not have a printer, the routine will still be of interest because we will be formatting our display in a new way. So the first two additional routines will be to the MAILING LIST ADDER program. The first will consist of adding lines of computer instructions to allow the user to print out in a mailing label the information the user just entered. The second includes the computer instructions necessary so that the information will be written to the disk only after all information for the current session has been entered and corrected. The additional computer instructions necessary to include both of these "features" are fairly small in number.

First, we will begin with the PRINT LABEL ROUTINE. Add the following lines to the program:

```
800 REM **--PRINT LABEL ROUTINE--**
810 PRINT "DO YOU WANT TO PRINT A LABEL NOW?"
820 INPUT "TYPE 'Y' OR 'N' ";YES$
840 IF YES$ = "Y" THEN 880
860 GOTO 1000: REM IF NOT 'Y' THEN REPEAT ROUTINE
880 PR#1: REM PRINTER SLOT
900 FOR I = 1 TO K
920 IF NAME$(I) = "*" THEN I = I + 1: GOTO 980
940 IF NAME$(I) = "!" THEN 980
960 PRINT NAME$(I)
980 NEXT I
990 PR#0: GOTO 800: REM RETURN TO SCREEN ONLY
```

Change line 620 to:

```
620 GOTO 800: REM PRINT LABEL ROUTINE
```

With these few additional instructions, we can now print a mailing label of the information just entered. Line 800 gives the title of the routine, and line 810 prints the question while line 820 waits for the response from the keyboard. Line 840 checks the response. If the response is positive, which indicates that a printed label is desired, the computer is instructed to jump over the next line and go on to the rest of the routine. Line 860 is only reached if the response is something other than "Y." In other words, the computer will treat everything but a "Y" as a negative response, indicating that the user does not want a printed label. If the user wants to press the RETURN key instead of either the "Y" or "N" keys, the computer will view that as a negative response. Negative responses end the routine and transfer the computer to the instruction at line 1000 for the REPEAT ROUTINE. Line 880 tells the computer to send its information to a printer interface card located in slot 1 of the APPLE II computer. If there is no card there, or if the printer is turned off, the computer will "hang;" that is, it will stop operating and appear to have a major problem. (Many people have been certain that their computer was broken when all that was wrong was that the program they were running was looking for an interface card and couldn't find the card or found that the peripheral device (usually a printer) was not turned on.) If you do have a printer but have the interface card in some other slot, you need to change line 880 so that the number after the "#" symbol matches the number of the slot in which your printer interface card is located. As you face the APPLE, the numbers of the slots go from left to right beginning with number 0 and ending with number 7. The instruction in line 900 is the now familiar beginning of the loop. Lines 920 and 940 are different from anything we have had so far. Line 920 checks the contents of each NAME$ string for the "*" symbol. If, and only if, it locates that symbol, it instructs the computer to add 1 to the value of the variable I and then to proceed to the instruction at line 980. The reason for this is simple but hard to explain. When the computer comes to an asterisk, we do not want that asterisk printed nor do we want the phone number printed in a mailing label. So we skip printing the asterisk and the phone number by adding one to the counter (I) and jumping to the end of the loop, line number 980. Line number 980 increases the counter by one more, so we have skipped two lines in the file: the lines that contained the separator symbol "*" and the phone number. I think this will become clear, if not already so, when you type in and try the routine. Line 940 does much the same thing. It tells the computer to jump over the print statement and go to the instruction that increases the counter. Thus, we have the effect of "skipping over" the "!" separator symbol and not printing it either. Line 960 does the printing. It prints the contents of every string in NAME$ that is not either a "*", "!" or phone number, unless the phone number has been typed in before the "*" symbol. Line 980, as I have said, increases the counter. When the loop is finished, when all lines of information have either been printed or skipped and the I counter has reached the value of K, then the computer can go on to the next instruction "outside" the loop. Line 990 is the first instruction outside the loop and because we are finished printing, we need to "turn off the output" to the printer. That means we are telling the computer to stop sending information to the printer interface card indicated in line 880. "PR#0" is the method used to do this. That is all there is to the PRINT LABEL ROUTINE. The new repeat routine is even shorter and easier.

Add to line 40:

",LINE$(100)"

Line 40 now reads:

40 DIM NAME$(20),LINE$(100)

Add the following:

```
1023 FOR I = 1 TO K
1026 LINE$(TK+I) = NAME$(I)
1033 NEXT I
1036 TK = TK + K
```

Change line 1080 to read:

```
1080 IF YES$ = "Y" THEN GOTO 60
```

Delete line 1100 and 2320, and change the lines to:

```
2100 REC = REC + TK
2240 FOR I = 1 TO TK
2260 PRINT LINE$(I)
```

It looks like such a small change, but when viewed by the computer, the changed and added instructions make quite a difference in the way the program works. The program now becomes more practical. First, we have added another string variable, and therefore, we must have the computer reserve memory for the contents of this new variable. That is what the added code does in line 40. We have DIMensioned the string variable LINE$ so that we can now add 100 lines of information before we need to write the information to the disk. You can make this number smaller or larger. The number 100 is a completely arbitrary choice. Next, we have added four lines of code: 1023, 1026, 1033, and 1036. The choice of these line numbers is also arbitrary. They could just as easily have been any combination between 1021 and 1039. Line 1023 is the start of a loop. Line 1026 is the real reason for the loop and the instruction that allows us to continue entering information without the necessity of writing each set of information to the disk separately. Again, the logic is fairly easy. We are going to keep the contents of the string variable NAME$ in the string variable LINE$ also. Then, since we have the information stored in two locations in the computer, we can use the NAME$ string variable over again. In other words, we have moved the information from one memory location to another memory location. We have moved it from NAME$(1) to LINE$(1), and from NAME$(2) to LINE$(2), etc. The instruction at 1036 helps us keep track of all the lines that are typed in. TK (which may stand for Total K) is a cumulative total of all the lines of information typed in during one session. For the first set of information lines, the value of TK is a zero since we have not previously given TK any value. After the first set of information lines, TK becomes the value of K or the number of lines of information in the first set of information. And in the loop, we have now moved the contents of the second NAME$(1) to LINE$(TK + 1), NAME$(2) to LINE$(TK + 2), etc. This process can continue until we have accumulated 100 lines of information (or more if you have dimensioned LINE$ to more than 100). If you are adding more lines of information, line 1080 must direct the computer to begin again with the input routine. Finally, line 2260 is changed to write the information contained in the string variable LINE$ instead of the contents of NAME$. This last is a very important change and will mess up the file badly if it is not made. In the next chapter, we will add some "features" to our display program and combine all our programs so that they can operate together. When you have made the necessary changes, be sure to save this new version. There are several ways of saving a new version of the same program. But for now, type the following:

```
SAVE MAILING LIST ADDER2
RENAME MAILING LIST ADDER, MAILING LIST ADDER1
```

Then type CATALOG and you should see:

```
A 002 HELLO
A 003 APPLE
A 006 MAILING LIST CREATOR
T 002 ADDRESS FILE
A 003 MAILING LIST READER
A 007 MAILING LIST ADDER1
A 008 MAILING LIST ADDER2
```

Do not worry if your numbers are not the same as the numbers I have given here. These numbers are the number of DOS 3.3 sectors (sections) of the diskette used by each file. Each diskette has a maximum of 560 sectors, although the number of sectors available for any type of file use is less. DOS takes up some of the sectors.

## QUESTIONS

1. TRUE or FALSE: Running the File Creation Program a second time with new information does no harm to the first information stored in the ADDRESS FILE.
2. Give the name of the BASIC command that erases program lines from the computer's memory.
3. To what does line 2260 in the MAILING LIST ADDER1 program print the information?
4. What is the BASIC word used to tell the computer to jump to a certain line number?
5. In a DOS 3.3 CATALOG listing, what does the three-digit number represent?

## **ANSWERS**

1. FALSE
2. DEL
3. DISK
4. GOTO
5. SECTORS

## MAILING LIST READER1

```
10  REM ***--MAILING LIST READER--***
11 :
12 :
20  D$ = CHR$(4):REM CONTROL D
25 :
26 :
30  REM **--INPUT ROUTINE--**
40  PRINT D$;"OPEN ADDRESS FILE"
60  PRINT D$;"READ ADDRESS FILE"
80  INPUT K
100  DIM NAME$(K)
120  FOR I = 1 TO K
140  INPUT NAME$(I)
160  NEXT I
180  PRINT D$;"CLOSE ADDRESS FILE"
190 :
191 :
200  REM **--DISPLAY ROUTINE--**
220  HOME : VTAB 5
240  FOR I = 1 TO K
260  PRINT NAME$(I)
280  NEXT I
```

## MAILING LIST ADDER1

```
10  REM **--MAILING LIST ADDER1--**
11 :
12 :
20  D$ = CHR$(4):REM CONTROL D
40  DIM NAME$(20)
60  K = 1:REM LINE COUNTER
65 :
66 :
70  REM **--INPUT ROUTINE--**
80  HOME : VTAB 5
100  PRINT "TYPE NAME AND ADDRESS AS IF ADDRESSING AN ENVELOPE.";
120  PRINT "DO NOT USE A COMMA OR COLON."
140  PRINT : PRINT "TYPE 'END' WHEN FINISHED"
160  PRINT : PRINT "TYPE IN LINE ";K
```

```
180  INPUT NAME$(K)
200  IF NAME$(K) = "END" THEN 300
220  K = K + 1
240  GOTO 160: REM GO BACK FOR ANOTHER LINE
300  NAME$(K) = "*": REM SEPARATOR FOR PHONE NUMBER
320  K = K + 1
340  PRINT "PHONE: ";: PRINT "TYPE 'RETURN' IF NONE."
360  INPUT NAME$(K)
380  K = K + 1
400  NAME$(K) = "!": REM SEPARATOR BETWEEN SETS OF
     INFORMATION
405  :
406  :
410  REM **--CORRECTION ROUTINE--**
420  HOME : VTAB 5
440  PRINT "DO NOT CHANGE THE LINE WITH THE '*'"
460  PRINT "THIS SYMBOL IS USED AS A SEPARATOR."
480  PRINT
500  FOR I = 1 TO K - 1
520  PRINT I;" ";NAME$(I)
540  NEXT I
560  PRINT
580  INPUT "CHANGE ANY LINE? TYPE 'Y' OR 'N' ";YES$
600  IF YES$ = "Y" THEN 640
620  GOTO 2000: REM GO TO FILE CREATION ROUTINE
640  INPUT "CHANGE WHICH LINE ";LINE
660  IF LINE > K - 1 THEN PRINT "NUMBER TOO LARGE" :
     GOTO 640
680  PRINT "OLD LINE = ";NAME$(LINE)
700  INPUT "CORRECT LINE = ";NAME$(LINE)
720  GOTO 410
740  :
760  :
1000 REM **--REPEAT ROUTINE--**
1020 HOME : VTAB 5
1040 PRINT "DO YOU WANT TO ADD MORE INFO?"
1060 INPUT "TYPE 'Y' OR 'N' ";YES$
1080 IF YES$ = "Y" THEN RUN
1100 PRINT D$;"RUN MENU"
1120 :
1140 :
2000 REM **--FILE ADDITION ROUTINE--**
2020 PRINT D$;"OPEN ADDRESS FILE"
2040 PRINT D$;"READ ADDRESS FILE"
```

```
2060  INPUT REC
2080  PRINT D$; "CLOSE ADDRESS FILE"
2100  REC = REC + K
2120  PRINT D$; "OPEN ADDRESS FILE"
2140  PRINT D$; "WRITE ADDRESS FILE"
2160  PRINT REC
2180  PRINT D$; "CLOSE ADDRESS FILE"
2200  PRINT D$; "APPEND ADDRESS FILE"
2220  PRINT D$; "WRITE ADDRESS FILE"
2240  FOR I = 1 TO K
2260  PRINT NAME$(I)
2280  NEXT I
2300  PRINT D$; "CLOSE ADDRESS FILE"
2320  GOTO 1000: REM REPEAT ROUTINE
```

## MAILING LIST ADDER2

```
10   REM ***--MAILING LIST ADDER2--***
11  :
12  :
20   D$ = CHR$ (4) : REM CONTROL D
40   DIM NAME$(20), LINE$(100)
60   K = 1: REM LINE COUNTER
65  :
66  :
70   REM **--INPUT ROUTINE--**
80   HOME : VTAB 5
100  PRINT "TYPE NAME AND ADDRESS AS IF ADDRESSING AN ENVELOPE.";
120  PRINT "DO NOT USE A COMMA OR COLON."
140  PRINT : PRINT "TYPE 'END' WHEN FINISHED"
160  PRINT : PRINT "TYPE IN LINE ";K
180  INPUT NAME$(K)
200  IF NAME$(K) = "END" THEN 300
220  K = K + 1
240  GOTO 160: REM GO BACK FOR ANOTHER LINE
300  NAME$(K) = "*": REM SEPARATOR FOR PHONE NUMBER
320  K = K + 1
340  PRINT "PHONE: ";: PRINT "TYPE 'RETURN' IF NONE."
360  INPUT NAME$(K)
380  K = K + 1
400  NAME$(K) = "!" : REM SEPARATOR BETWEEN SETS OF INFORMATION
```

```
405 :
406 :
410 REM **--CORRECTION ROUTINE--**
420 HOME : VTAB 5
440 PRINT "DO NOT CHANGE THE LINE WITH THE '*'"
460 PRINT "THIS SYMBOL IS USED AS A SEPARATOR."
480 PRINT
500 FOR I = 1 TO K - 1
520 PRINT I; " "; NAME$(I)
540 NEXT I
560 PRINT
580 INPUT "CHANGE ANY LINE? TYPE 'Y' OR 'N' "; YES$
600 IF YES$ = "Y" THEN 640
620 GOTO 800: REM PRINT LABEL ROUTINE
640 INPUT "CHANGE WHICH LINE "; LINE
660 IF LINE > K - 1 THEN PRINT "NUMBER TOO LARGE": GOTO 640
680 PRINT "OLD LINE = "; NAME$(LINE)
700 INPUT "CORRECT LINE = "; NAME$(LINE)
720 GOTO 410
740 :
760 :
800 REM **--PRINT LABEL ROUTINE--**
810 PRINT "DO YOU WANT TO PRINT A LABEL NOW"
820 INPUT "TYPE 'Y' OR 'N' "; YES$
840 IF YES$ = "Y" THEN 880
860 GOTO 1000: REM REPEAT ROUTINE
880 PR#1
900 FOR I = 1 TO K
920 IF NAME$(I) = "*" THEN I = I + 1: GOTO 980
940 IF NAME$(I) = "!" THEN 980
960 PRINT NAME$(I)
980 NEXT I
990 PR#0: GOTO 800
995 :
996 :
1000 REM **--REPEAT ROUTINE--**
1020 HOME : VTAB 5
1023 FOR I = 1 TO K
1026 LINE$(TK + I) = NAME$(I)
1033 NEXT I
1036 TK = TK + K
1040 PRINT "DO YOU WANT TO ADD MORE INFO? "
1060 INPUT "TYPE 'Y' OR 'N' "; YES$
1080 IF YES$ = "Y" THEN GOTO 60
```

```
1100 :
1111 :
2000 REM **--FILE ADDITION ROUTINE--**
2020 PRINT D$;"OPEN ADDRESS FILE"
2040 PRINT D$;"READ ADDRESS FILE"
2060  INPUT REC
2080 PRINT D$;"CLOSE ADDRESS FILE"
2100 REC = REC + TK
2120 PRINT D$;"OPEN ADDRESS FILE"
2140 PRINT D$;"WRITE ADDRESS FILE"
2160 PRINT REC
2180 PRINT D$;"CLOSE ADDRESS FILE"
2200 PRINT D$;"APPEND ADDRESS FILE"
2220 PRINT D$;"WRITE ADDRESS FILE"
2240 FOR I = 1 TO TK
2260 PRINT LINE$(I)
2280 NEXT I
2300 PRINT D$;"CLOSE ADDRESS FILE"
3000 PRINT D$;"RUN MENU"
```

# 6
# Displaying Sequential Files

By now, if you started with a new diskette, you should have four program files (or five if you have kept both versions of the ADDER program), the text file ADDRESS FILE, and the HELLO program on your diskette. When you type CATALOG, your catalog should look similar to the following except the order and numbers might not be exactly the same:

```
A 002 HELLO
A 003 APPLE
A 007 MAILING LIST CREATOR
T 002 ADDRESS FILE
A 003 MAILING LIST READER
A 007 MAILING LIST ADDER1
A 008 MAILING LIST ADDER2
```

When you want to use the MAILING LIST READER program, you must type: RUN MAILING LIST READER. When you are ready to add to the file, you need to type: RUN MAILING LIST ADDER2 (or "1" depending upon your preference.) For occasional use, that amount of typing is not a problem, but if you are going to use the programs quite often, the necessity of typing RUN and the file name can become bothersome. Besides, the computer can help eliminate the need to type that, so why not let it do so? All that is needed is another program. You will still have to type RUN and the name of this new program. The difference is that when properly set up, you may need to do the typing only once, and then you will be able to switch back and forth between programs with little typing other than a number. Let's see how this can work. Make sure any program currently in memory is saved on the disk, and then type the following:

# CHAPTER 6  DISPLAYING SEQUENTIAL FILES

```
NEW
10 REM **--MAILING LIST PROGRAM MENU--**
11 :
12 :
20 D$ = CHR$(4): REM CONTROL D
25 :
26 :
30 REM **--MENU ROUTINE--**
40 HOME: VTAB 5
60 HTAB 17: PRINT "PROGRAM MENU"
80 PRINT:PRINT
100 HTAB 8:PRINT "1. FILE CREATION PROGRAM"
120 PRINT
140 HTAB 8:PRINT "2. FILE ADDITION PROGRAM"
160 PRINT
180 HTAB 8:PRINT "3. FILE DISPLAY PROGRAM"
200 PRINT
300 HTAB 8:PRINT "4. CATALOG"
320 PRINT
340 HTAB 8:PRINT "5. END"
360 PRINT: PRINT
380 HTAB 8: INPUT "WHICH PROGRAM NUMBER? ";NUMBER
400 IF NUMBER < 1 OR NUMBER > 5 THEN 380
420 IF NUMBER = 1 THEN 1000
440 IF NUMBER = 2 THEN 2000
460 IF NUMBER = 3 THEN PRINT D$;"RUN MAILING LIST
    READER"
480 IF NUMBER = 4 THEN PRINT D$;"CATALOG": INPUT "HIT
    RETURN TO GOTO MENU"; L$: GOTO 40
500 IF NUMBER = 5 THEN END
600 :
700 :
1000 REM **--FILE CREATOR PROGRAM--**
1020 PRINT:PRINT "IF THE ADDRESS FILE ALREADY
     EXISTS"
1040 PRINT:PRINT "DO NOT RUN THIS PROGRAM!!"
1060 PRINT:PRINT "DO YOU WANT THE FILE CREATION
     PROGRAM?"
1070 PRINT
1080 INPUT "TYPE 'YES' IF YOU DO: ";YES$
1100 IF YES$= "YES" THEN PRINT D$;"RUN MAILING LIST
     CREATOR"
1120 GOTO 40
1500 :
```

58   CHAPTER 6   DISPLAYING SEQUENTIAL FILES

```
1600 :
2000 REM **--FILE ADDITION PROGRAM--**
2020 PRINT:PRINT "YOU WANT TO ADD TO THE EXISTING"
2040 PRINT:PRINT "ADDRESS FILE. IS THIS CORRECT?"
2060 PRINT:INPUT "TYPE 'YES' IF IT IS. ";YES$
2080 IF YES$= "YES" THEN PRINT D$;"RUN MAILING LIST
     ADDER2"
2100 GOTO 40
```

When you have all the program lines typed in, save it to disk as MENU by typing:

SAVE MENU

Now all that is needed to run any of our programs is: RUN MENU. Then choose a number and let the computer do the rest. The program should be fairly easy to understand by now. Line 10 is the name of the program (a name we have considerably shortened for the CATALOG). Lines 20 and 40 are our standard housekeeping lines. Line 60 uses a new BASIC statement—HTAB. This tells the computer to horizontally tab over 17 spaces and then print the words "PROGRAM MENU". We tab over 17 spaces to place the words we want printed in the approximate center of the screen. Line 80 gives us two blank horizontal lines after our title. Lines 100 through 340 set up the actual menu of choices with a blank line between each choice. Notice that I have left room between 200 and 300, between choice 3 and choice 4. The reason for this is that you may want to add other programs to this PROGRAM MENU. If you do add more choices, make certain that all lines relating to specific numbers are changed. Line 380 requests which program to run. The number typed is stored in the numeric variable "NUMBER". This value is first checked (line 400) to see that it is really within the range of actual possibilities. If the number is either less than 1 or greater than 5, the computer is instructed to go back to line 380 and request another number. Therefore, if the computer reaches the instruction at line 420, we know that we have a number somewhere between (or including) 1 and 5. Instructions 420 to 500 are checking which number is contained in the numeric variable NUMBER. If, for instance, the value is 3, indicating that the user wants to read the ADDRESS FILE, we simply instruct the computer to go to the disk (remember we used the control 'D' key) and RUN the appropriate program. Control is then transferred to the program MAILING LIST READER and the computer receives and follows the instructions contained in that program. Such transfer of control erases the PROGRAM MENU computer instructions from the computer's memory, replacing those instructions with the instructions in the selected program. That is why it is important to SAVE this MENU to disk before RUNning it. It is also the reason why,

CHAPTER 6   DISPLAYING SEQUENTIAL FILES           59

when you are finished adding information, for example, and the information has been written out to the disk, that the program simply ends. You are not returned to the PROGRAM MENU. This is not convenient and certainly does not allow us to easily switch between programs. To do the switching, it is necessary to make a one line modification to each of our previous programs.

```
MAILING LIST CREATOR:
940 PRINT D$; "RUN MENU"
MAILING LIST ADDER1:
1100 PRINT D$; "RUN MENU"
MAILING LIST ADDER2:
3000 PRINT D; "RUN MENU"
MAILING LIST READER:
500 PRINT D$; "RUN MENU"
```

If you load each of the programs, make those changes, and then save each of the revised programs, you will have a SYSTEM of programs that work together and are controlled by one MASTER program.

```
LOAD MAILING LIST CREATOR
940 PRINT D$; "RUN MENU"
SAVE MAILING LIST CREATOR
```

This same sequence is necessary for the other programs. There is no need to go through all the antics that are sometimes involved in order to build one large program. In addition, it is much easier to make changes to individual programs than to change something in a large program that might have an unnoticed effect. This borders somewhat on programmer preference, but I have found this method to be easy.

We now have a system that will create a file, add to that file, and in a primitive way, read the file. Two main tasks are left: improving the display features of the MAILING LIST READER program, and coming up with a program that will change and delete information in the file. One other "feature" that we will add within our display program is the re-formatting of our data and possible creation of a new file for this re-formatted data. We will begin with the program to increase our display options.

All our present program does is display every line in the ADDRESS FILE, including the two separator symbols. You don't really want to see those symbols, so eliminating them should be one of the first tasks in creating a new display program. What else would be nice or useful to have in this display program? The computer could display a list of just the names of the individuals in the file. How about a list of the names and addresses without the phone numbers? Can we get a display of a single name, address, and phone number? How about a single name and address, without a phone number? What about an alphabetical list? Can we have a range of names and addresses displayed rather than just the entire list or a single individual? The answer to all these questions is "yes;" we can do these things and others also. With all these possibilities, the obvious solution would be to have a menu for these choices.

If you LOAD MAILING LIST READER and LIST it, the lines should go from 10 to 500 if you have made the MENU change (280 if you have not made the change). We can keep lines 10 through 180 pretty much the same. But line 100 must have some additional variables dimensioned. So line 100 should read:

```
100 DIM NAME$ (K) , ND$ (K) , L (K) , R (K) , AD$ (K)
```

and add the following lines:

```
200 REM **--MENU ROUTINE--**
210 HOME: VTAB 5
220 HTAB 17: PRINT "DISPLAY MENU"
230 PRINT: PRINT
240 HTAB 3: PRINT "1. DISPLAY INFO--ORIG. ORDER"
250 PRINT
260 HTAB 3: PRINT "2. DISPLAY NAMES ONLY"
270 PRINT
280 HTAB 3: PRINT "3. DISPLAY INFO--NO PHONE"
290 PRINT
300 HTAB 3: PRINT "4. DISPLAY SPECIFIC NAME"
310 PRINT
320 HTAB 3: PRINT "5. DISPLAY SPECIFIC NAME--NO
    PHONE"
330 PRINT
340 HTAB 3: PRINT "6. DISPLAY INFO--RANGE"
350 PRINT
360 HTAB 3: PRINT "7. DISPLAY INFO--ALPHABETICAL"
365 PRINT
370 HTAB 3: PRINT "8. RETURN TO PROGRAM MENU"
375 PRINT
380 HTAB 3: INPUT "WHICH NUMBER? "; NUMBER
385 IF NUMBER < 1 OR NUMBER > 8 THEN PRINT "INCORRECT
    NUMBER": GOTO 380
```

If you have been following along with our programs, these lines of code should now be easy to understand. We are doing the same sequence of programming we did when we created the PROGRAM MENU program. We format the menu display, request a number, and check to see that the number is within the actual possibilities. The next series of program lines are familiar also.

# ORIGINAL ORDER ROUTINE

```
410 IF NUMBER = 1 THEN 1000
420 IF NUMBER = 2 THEN 2000
430 IF NUMBER = 3 THEN 3000
440 IF NUMBER = 4 THEN 4000
450 IF NUMBER = 5 THEN 5000
460 IF NUMBER = 6 THEN 6000
470 IF NUMBER = 7 THEN 7000
480 IF NUMBER = 8 THEN PRINT D$; "RUN MENU"
```

Now you have the basic structure for the rest of the program. All that is necessary is to fill in the code for each routine. The routines get progressively more difficult to follow, and it is not the intent of this book to teach the concepts behind routines such as sorting and searching, but it is within its scope to present examples of such routines so that readers can make use of these routines in their own file manipulation programs.

## ORIGINAL ORDER ROUTINE

```
1000 REM **--ORIGINAL ORDER ROUTINE--**
1020 GOSUB 10000: REM PRINTER ROUTINE
1040 HOME: VTAB 5
1060 FOR I = 1 TO K
1080 IF NAME$(I) = "*" THEN 1140
1100 IF NAME$(I) = "!" THEN PRINT: GOTO 1140
1120 PRINT NAME$(I)
1140 NEXT I
1160 GOTO 20000: REM RETURN TO MENU ROUTINE
```

If you look closely, this routine is very similar to the original MAILING LIST READER Display Routine (lines 200 to 280). All this routine does is display all the information lines in the file in the order they were entered. With lines 1080 and 1100, we have eliminated the display of the separator symbols "*" and "!". There are two additional lines that are new. Lines 1020 and 1160 direct the computer to separate routines used by each of the main routines. Line 1020, as indicated by the REM statement, will be the code that asks if the user wants the information printed on a printer. This instruction uses a GOSUB statement which directs the computer to go to the instructions that begin at line 10000 and follow those instructions until the computer encounters a RETURN statement. At that point, the computer returns to the instruction following the GOSUB instruction. Line 1160 is the instruction that directs the computer to the routine that returns the user to the DISPLAY MENU when the user is ready.

## NAME ONLY ROUTINE

```
2000 REM **--NAME ONLY ROUTINE--**
2020 GOSUB 10000: REM PRINTER ROUTINE
2040 HOME: VTAB 5
2060 FOR I = 1 TO (K - 1)
2080 IF NAME$(I) = NAME$(2) THEN PRINT I;
     " "; NAME$(I)
2100 IF NAME$(I) = "!" THEN PRINT I;" ";NAME$
     (I + 1)
2120 NEXT I
2140 GOTO 20000: REM RETURN TO MENU ROUTINE
```

This routine should not be very difficult to understand. We want to print only those lines that follow the "!" separator. We print those lines because those are the lines that should contain the name of the individual. We need the instruction at 2080 because there is no separator for the first name. We use K - 1 because we do not want to get to the last "!" separator since there is no name to follow it yet.

## NO PHONE ROUTINE

```
3000 REM **--NO PHONE ROUTINE--**
3020 GOSUB 10000: REM PRINTER ROUTINE
3040 HOME: VTAB 5
3060 FOR I = 1 TO K
3080 IF NAME$(I) = "*" THEN I = I + 1: GOTO 3140
3100 IF NAME$(I) = "!" THEN PRINT: GOTO 3140
3120 PRINT NAME$(I)
3140 NEXT I
3160 GOTO 20000: REM RETURN TO MENU ROUTINE
```

This routine should look completely familiar. It is practically the same routine we used to print a label in MAILING LIST ADDER2 (lines 900 to 980 in that program). The effect is the same here also. We can print a mailing label for every person in our file with this routine. Because just about every type of printer handles things differently, you will probably need to add some code to this routine to get the labels spaced properly. One method of spacing would be to find out the number of lines on the label and between labels and then adjust the routine to always space just exactly that number of lines regardless of the number of lines to be printed. That method would always start the printer at the top of the label and not center the material on the label, but it is probably the easiest method to develop.

## SEARCH ROUTINE

```
4000 REM **--SEARCH ROUTINE--**
4020 HOME: VTAB 5
4040 PRINT "TYPE 'END' WHEN FINISHED."
4050 INPUT "NAME TO FIND? ";FIND$
4060 IF FIND$ = "END" THEN 4400
4070 GOSUB 10000: REM PRINTER ROUTINE
4080 PRINT
4100 FOR I = 1 TO K
4120 IF NAME$(I) = FIND$ THEN 4160
4140 GOTO 4340
4160 IF NAME$(I) = "*" THEN 4340
4180 IF NAME$(I) = "!" THEN PRINT: GOTO 4340
4200 PRINT NAME$(I)
4220 PRINT NAME$(I + 1)
4240 PRINT NAME$(I + 2)
4260 IF NAME$(I + 3) < > "*" THEN PRINT NAME$(I + 3)
4280 IF NAME$(I + 4) = "*" THEN 4320
4300 PRINT NAME$(I + 4): GOTO 4340
4320 PRINT NAME$(I + 5)
4340 NEXT I
4360 PRINT
4380 GOTO 4040: REM REPEAT ROUTINE
4400 GOTO 20000: REM RETURN TO MENU ROUTINE
```

The routines begin to get more difficult now. Up to this point, we have not really made any assumptions about the number of lines of information in each set. But with this routine, we make the assumption that there are a maximum of 6 lines of information in any set. If you want a greater maximum, then additional code will have to be added to print out those other lines. The additional code would follow the pattern of 4260 to 4320. We begin in the same way with 4000 and 4020 (our routine name and clear screen lines). Line 4040 gives instructions to the user to type the word "END" when the user is finished looking for a specific name. Line 4050 requests the name from the user and stores that name in the string variable "FIND$". Line 4060 checks the contents of "FIND $" to see if it contains the word "END". If it does, the computer is directed to go to line 4400 which further directs the computer to go to the "RETURN TO MENU ROUTINE". One might logically ask why 4060 does not instruct the computer to go directly to the "RETURN TO MENU ROUTINE". The reason lies in the necessity of structuring the various routines in the same way so that any programmer can locate the exit point of the routine easily. There are a number of GOTO statements in this routine, but all of them direct the computer (and any programmer) to various lines within this routine. In following the logic of this and all other routines, you never need to look outside the routine, except for the print routine and exit routine which are common to all other routines. The idea is to keep the flow of logic in one place as much as possible. You enter at the top of the routine and exit at the base of the routine. This is the case for all the routines.

Lines 4100 to 4340 are the heart of this routine. They are also the boundaries of the loop used to find and print the information associated with a specific name. Line 4120 checks the contents of NAME$(I) to see if it equals the contents of FIND$. If it does, the computer is instructed to jump over the next instruction. If it does not, the next instruction is executed. Line 4140 is reached only if the contents of NAME$(I) and FIND$ do not match, and 4160 is reached only if they do match. Lines 4160 and 4180 check for the separators and skip them when they are found. At this point in the routine, we have found the name we are looking for and now want to print out the information associated with this name. We assume that the first three lines will not contain a separator and, therefore, will automatically print those lines. Lines 4200, 4220, and 4240 accomplish this task. Lines 4260 through 4320 are lines of code that require some thought. If the fourth information-line does not contain the separator "*", then we want to print this line also (4260), but if it does contain the separator, we do not want the fourth information-line printed. Rather we know that the fifth information-line contains something to be printed (the line following the "*" will have the phone number if there is a phone number). Line 4300 prints this fifth information-line. Line 4280 first checks the fifth information-line to see if it contains the asterisk separator. If it does contain the separator, then we need to jump over 4300 (the instruction that prints that fifth information-line) and instead, print the sixth information-line (4320). Go back through the explanation if you are not certain you understand. We use this same routine, combined with the previous one, for our next routine.

## SEARCH ROUTINE--NO PHONE

```
5000 REM **- -SEARCH ROUTINE- -NO PHONE- -**
5020 HOME: VTAB 5
5040 PRINT "TYPE 'END' WHEN FINISHED"
5050 INPUT "NAME TO FIND? ";FIND$
5060 IF FIND$ = "END" THEN 5400
5070 GOSUB 10000:REM PRINTER ROUTINE
5080 PRINT
```

## SEARCH ROUTINE–NO PHONE

```
5100 FOR I = 1 TO K
5120 IF NAME$(I) = FIND$ THEN 5160
5140 GOTO 5340
5160 IF NAME$(I) = "*" THEN I = I + 1: GOTO 5340
5180 IF NAME$(I) = "!" THEN PRINT: GOTO 5340
5200 PRINT NAME$(I)
5220 PRINT NAME$(I + 1)
5240 PRINT NAME$(I + 2)
5260 IF NAME$(I + 3) < > "*" THEN PRINT NAME$(I + 3)
5270 IF NAME$(I + 3) = "*" THEN I = I + 1: GOTO 5340
5280 IF NAME$(I + 4) = "*" THEN I = I + 1: GOTO 5340
5300 PRINT NAME$(I + 4): GOTO 5340
5320 PRINT NAME$(I + 5)
5340 NEXT I
5360 PRINT
5380 GOTO 5040: REM REPEAT ROUTINE
5400 GOTO 20000: REM RETURN TO MENU ROUTINE
```

I included this routine for a number of reasons. First, it is a very useful routine because with a printer, one can print out a specific mailing label. Second, it shows how two routines can be combined into a third routine. This latter point is the most important reason. Very few programs will do everything anyone could ever want of them, but if a person understands these separate routines, then combining two or more to form others should be possible. There are quite a number of combinations that are possible and might be useful to some people. As you can see, this routine is exactly the same as the previous one down to the instruction at 5160. The only difference is that when we find the "*" separator, we add one to I, thus skipping the phone number. Lines 5180 through 5260 are the same instructions as 4180 through 4260. The instructions at 5270 and 5280 are the only different instructions. Both of those instructions are simply checking to see which information line contains the separator symbol and then advancing the counter by one. The end of the routine is the same as the end of the previous routine.

With the routines at 4000 and 5000, you have the ability to search for a specific name and display that name, either with the phone number or without the phone number. But both of these routines require that you know and type in the exact spelling of the name, including spaces. That presents a reason for our next routine, the range routine. With this routine, you will only need to know the starting and ending information-line numbers to be able to display the information you want. You can obtain those numbers from the DISPLAY NAMES ONLY routine. I will present the range routine only, but you might want to combine this routine with the DISPLAY NAMES ONLY routine and possibly some others also.

## RANGE ROUTINE

```
6000 REM **--RANGE ROUTINE--**
6020 HOME: VTAB 5
6040 INPUT "TYPE BEGINNING LINE NUMBER ";BL
6060 PRINT
6080 IF BL < 2 THEN PRINT "NUMBER TOO SMALL":
     GOTO 6040
6100 INPUT "TYPE ENDING LINE NUMBER ";EL
6120 PRINT
6140 IF EL > K THEN PRINT "NUMBER TOO LARGE":
     GOTO 6100
6160 GOSUB 10000: REM PRINTER ROUTINE
6180 FOR I = BL TO EL
6200 IF NAME$(I) = "*" THEN I = I + 1: GOTO 6260
6220 IF NAME$(I) = "!" THEN PRINT: GOTO 6260
6240 PRINT NAME$(I)
6260 NEXT I
6280 GOTO 20000: REM RETURN TO MENU ROUTINE
```

Line 6040 asks for the beginning information-line number. Remember that you can check the numbers first by using the DISPLAY NAMES ONLY routine or by actually including that routine at the beginning of this one. Line 6080 checks the number typed to see if it is less than 2, the number of the first information-line. If it is too small, a message is printed and the user is again asked for a beginning number. Line 6100 requests the ending information-line number and goes through the same checking process, this time for a number larger than the maximum number of information-lines. Then comes our loop (6180 to 6260). I have included the code for printing the information without the phone number (6200), thus providing a routine that can print out a selected range of mailing labels.

I have tried to show how you can take various routines and combine them in just about any way you might want. With the addition of each new routine, the number of possible combinations of routines increases so much that no single programmer could include all possibilities within one program, but, with a minimum of understanding, everyone can create combinations of routines necessary for their needs.

## ALPHABETICAL ORDER ROUTINE

We come now to the most complex of our routines. I will not even attempt to explain the logic involved in all parts of this alphabetizing routine since complete books have been written on various sorting techniques.

## ALPHABETICAL ORDER ROUTINE

The sort method I chose to include is sometimes called the QUICKSORT technique. There are a number of other public domain sorting routines that I could have used, such as the bubble sort or the Shell-Metzner sort, but I decided on the Quicksort because it is very efficient and somewhat less publicized. I modified the sort to enable it to work with string variables. Otherwise, the sort subroutine is a standard routine that can be used in a number of different ways to order lists composed of numbers or letters. For example, if you want to display the information in the ADDRESS FILE in zip code order, you first need to access the zip codes and then use the Quicksort subroutine to arrange the zip codes and their associated information-lines in either ascending or descending order. The creation of such a routine would require that you completely understand another feature of this routine: the flexibility possible with string variables and the manner of utilizing that flexibility. This is another subject that complete books have been written on. Again, I will not try to fully explain the logic or programming power behind the BASIC statements of LEFT$, MID$, or RIGHT$. I will strongly encourage you to learn as much as possible about these BASIC statements and how they can be used to take string variables apart and put them back together in just about any way you want.

This alphabetizing routine will be presented in two sections. The first section makes use of this string variable flexibility to: access the last section of characters in that first information-line; reverse the order of that information-line placing the last section of characters first; and then combine all other information-lines associated with this first line into one long string variable, AD$(I). The second section alphabetizes the list now stored in the string variable ND$(J).

```
7000 REM **--ALPHABETICAL ORDER ROUTINE--**
7040 HOME: VTAB 5
7060 PRINT "WORKING--PLEASE DON'T TOUCH!!"
7065 :
7066 :
7070 REM GET FIRST INFO-LINE
7080 FOR I = 2 TO K - 1
7100 IF NAME$(I) = NAME$(2) THEN 7160
7120 IF NAME$(I) = "!" THEN I = I + 1: GOTO 7160
7140 GOTO 7340
7145 :
7146 :
7150 REM REVERSE ORDER
7160 LN = LEN (NAME$(I))
7180 FOR J1 = 1 TO LN: IF MID$(NAME$(I),J1,1) = " "
     THEN J2=J1
7200 NEXT J1
```

```
7210 IF J2 = 0 OR J2 > LN THEN AD$(I) = NAME$(I) :
     GOTO 7240
7220 AD$(I) = MID$ (NAME$ (I), J2+1, LN-J2) + " "
     + LEFT$ (NAME$ (I), J2)
7240 AD$(I) = AD$(I) + "**" + NAME$(I+1) + "**" + NAME$
     (I+2)
7260 IF NAME$ (I+3) <> "*" THEN AD$(I) = AD$(I) + "**"
     + NAME$(I+3)
7280 IF NAME$(I+4) = "*" THEN 7320
7300 AD$(I) = AD$(I) + "**" + NAME$(I+4) :
     GOTO 7340
7320 AD$(I) = AD$(I) = "**" + NAME$(I+5)
7340 NEXT I
7345 :
7346 :
7350 REM RENUMBER FOR SORT
7360 J=1
7380 FOR I = 1 TO K
7400 IF LEN (AD$(I)) > 0 THEN ND$(J) = AD$(I) :
     J = J + 1
7420 NEXT I
7440 N = J - 1
```

As I said, the routines get more complex. If you do not understand the LEFT$, MID$ and RIGHT$ statements, the best thing to do is to get a clear definition of them from a book devoted to teaching BASIC and then practice their uses. Essentially, they perform the functions for which they are named. The LEFT$ statement will retrieve a specified number of characters beginning at the left side of a string variable. LEFT$(A$,4) gets the first four characters in the string variable A$. The RIGHT$ statement retrieves a specified number of characters from the right-most character of a string variable. RIGHT$(B$,3) gets the last three characters in the string variable B$. The MID$ statement retrieves a specified number of characters from a specified position within a string variable. MID$(C$,2,6) gets the next six characters beginning at the second character in the string variable C$. Therefore, the instructions in 7080 to 7140 identify the first information-line in each set of data. Lines 7160 through 7340 reverse the order of the first information-line and then combine all the other information-lines associated with it. Finally, 7360 to 7440 are the instructions that renumber the sets of information in such a way that the sort subroutine can function.

```
7460 REM ***--QUICKSORT--***
7480 S1=1
7500 PRINT "WORKING--PLEASE DON'T TOUCH!!"
```

## ALPHABETICAL ORDER ROUTINE

```
7520 L(1) = 1
7540 R(1) = N
7560 L1 = L(S1)
7580 R1 = R(S1)
7600 S1 = S1-1
7620 L2 = L1
7640 R2 = R1
7660 X$ = ND$(INT((L1+R1)/2))
7680 C = C + 1
7700 IF ND$(L2) = X$ OR ND$(L2) > X$ THEN 7760
7720 L2 = L2 +1
7740 GOTO 7680
7760 C = C1
7780 IF X$ = ND$(R2) OR X$ > ND$(R2) THEN 7840
7800 R2 = R2 - 1
7820 GOTO 7760
7840 IF L2 > R2 THEN 7980
7860 S = S + 1
7880 T$ = ND$(L2)
7900 ND$(L2) = ND$(R2)
7920 ND$(R2) = T$
7940 L2 = L2 + 1
7960 R2 = R2 - 1
7980 IF L2 = R2 OR L2 < R2 THEN 7680
8000 IF L2 = R1 OR L2 > R1 THEN 8080
8020 S1 = S1 + 1
8040 L(S1) = L2
8060 R(S1) = R1
8080 R1 = R2
8100 IF L1 < R1 THEN 7620
8120 IF S1 > 0 THEN 7560
8140 REM SORT COMPLETED
```

Now you have access to a sorting method. The only code necessary outside this subroutine to transfer it to another program is to set:
1. The DIM of L() and R() to the number of things to be sorted.
2. The numeric variable "N" = to the number of things to be sorted.

If you have a different sort method that you like or understand better and want to include it instead, the code for your sort should replace the code between lines 7460 and 8140.

The only thing left is to display the results after sorting. I am going to present the code to display the results in their most elementary way. I will leave to those of you who want to or are able to use the flexibility in string variables to format the display in any way you desire.

```
8145 REM **--DISPLAY--**
8150 GOSUB 10000: REM PRINTER ROUTINE
8160 FOR I = 1 TO N
8180 PRINT ND$(I)
8200 PRINT
8220 NEXT I
8240 GOTO 20000: REM RETURN TO MENU ROUTINE
```

We now have an opportunity to create a sequential access file in a way that may be more powerful than in our mailing list system. The usefulness of the new file creation method depends on the programmer's knowledge of and willingness to work with string variables; i.e., LEFT$, MID$, RIGHT$, LEN, STR$, VAL, and DIM. All of the associated information with NAME$(2)—that is the address, city, state, zip code, and phone number—are all stored in the string variable ND$(1). Everything for the next name is stored in ND$(2) and so on. If you want to locate the zip code, all you need to do is use the MID$ function to determine where in the string the zip code is located. You could use the same MID$ function with our present file set up, but it might be more difficult to always locate the zip code. (For instance, some people might put the zip code on a separate line, while others would put it on the same line as the city and state. If everything is combined into one string variable, it might be easier to locate for all possible situations.) I have used a lot of conditional statements because there are a lot of possibilities, and the correct choice often depends upon many factors: the programmer's experience and preference, the value of the file being established, the necessity of backup, the amount of use the file will get, and so forth.

The code necessary to establish a separate file for our now-alphabetized information should be easy to develop. If you have trouble, the appendix has the code solution for this problem.

To finish, we need two brief subroutines used by each of our main routines: the printer subroutine and the return-to-menu subroutine.

```
10000 REM **--PRINT SUBROUTINE--**
10020 PRINT "DO YOU WANT A PAPER PRINT OUT?"
10040 INPUT "TYPE 'Y' OR 'N' ";YES$
10060 IF YES$ = "Y" THEN 10120
10080 SPEED = 150
10100 RETURN
10120 PR#1
10140 RETURN
10150 :
10160 :
20000 REM **--RETURN TO MENU SUBROUTINE--**
20020 PR#0: SPEED=255
```

```
20040 INPUT "HIT RETURN TO GO TO MENU ";L$
20060 GOTO 200: REM MENU
```

Have you saved this new MAILING LIST READER program? If not, be certain that the program is still in memory and then type the following:

SAVE MAILING LIST READER

By using the same name as our original display program, we have written over that original program and thus replaced it with this expanded version. Now the CATALOG should show:

```
A 002 HELLO
A 003 APPLE
A 007 MAILING LIST CREATOR
T 002 ADDRESS FILE
A 019 MAILING LIST READER
A 007 MAILING LIST ADDER1
A 008 MAILING LIST ADDER2
```

Notice that the only difference between this display and the display of the CATALOG at the beginning of this chapter is in the numbers before MAILING LIST READER. The READER program has grown from using two sectors on the diskette to using about nineteen sectors. (Remember that the numbers may vary slightly from computer to computer.) That is quite a difference and now makes the READER program the largest program on the diskette. In the next chapter, we will examine ways of correcting, changing, or deleting information from our file.

## QUESTIONS

1. TRUE or FALSE: DOS allows you to RUN a program from within another program.
2. What Applesoft BASIC word allows you to horizontally tab on the screen?
3. Which BASIC word is used to instruct the computer to go to a subroutine?
4. Which Basic word is used to instruct the computer to return from a subroutine?
5. TRUE or FALSE: In programming, it is a good idea to have just one main entrance and exit point in every routine.
6. Name three public domain sorting routines.
7. What are the three main BASIC words that provide a great deal of power in working with strings?
8. What BASIC word retrieves a specified number of characters from a specified position within a string variable?
9. Name three other BASIC words that can be used in some way with string variables.
10. TRUE or FALSE: When you save a file with the same name as a file already on the disk, the first file is replaced by the second file.

## ANSWERS

1. TRUE
2. HTAB
3. GOSUB
4. RETURN
5. TRUE
6. Bubble, Quicksort, Shell-Metzner
7. LEFT$, RIGHT$, MID$
8. MID$
9. LEN, STR$, VAL
10. TRUE

## MAILING LIST MENU1

```
10 REM ***--MAILING LIST PROGRAM MENU--***
11 :
12 :
20 D$ = CHR$ (4): REM CONTROL D
25 :
26 :
30 REM **--MENU ROUTINE--**
40 HOME : VTAB 5
60 HTAB 17: PRINT "PROGRAM MENU"
80 PRINT : PRINT
100 HTAB 8: PRINT "1. FILE CREATION PROGRAM"
120 PRINT
140 HTAB 8: PRINT "2. FILE ADDITION PROGRAM"
160 PRINT
180 HTAB 8: PRINT "3. FILE DISPLAY PROGRAM"
200 PRINT
300 HTAB 8: PRINT "4. CATALOG"
320 PRINT
340 HTAB 8: PRINT "5. END"
360 PRINT : PRINT
380 HTAB 8: INPUT "WHICH PROGRAM NUMBER? ";NUMBER
400 IF NUMBER < 1 OR NUMBER > 5 THEN 380
420 IF NUMBER = 1 THEN 1000
440 IF NUMBER = 2 THEN 2000
460 IF NUMBER = 3 THEN PRINT D$;"RUN MAILING LIST
    READER"
480 IF NUMBER = 4 THEN PRINT D$;"CATALOG": INPUT"HIT
    RETURN TO GO TO MENU ";L$: GOTO 40
500 IF NUMBER = 5 THEN END
600 :
700 :
1000 REM **--FILE CREATOR PROGRAM--**
1020 PRINT : PRINT "IF THE ADDRESS FILE ALREADY EXISTS"
1040 PRINT : PRINT "DO NOT RUN THIS PROGRAM!!"
1060 PRINT : PRINT "DO YOU WANT THE FILE CREATION
     PROGRAM?"
1070 PRINT
1080 INPUT "TYPE 'YES' IF YOU DO: ";YES$
1100 IF YES$ = "YES" THEN PRINT D$;"RUN MAILING LIST
     CREATOR"
1120 GOTO 40
```

```
1140 :
1160 :
2000 REM **--FILE ADDITION PROGRAM--**
2020 PRINT : PRINT "YOU WANT TO ADD TO THE EXISTING"
2040 PRINT : PRINT "ADDRESS FILE. IS THIS CORRECT? "
2060 PRINT : INPUT "TYPE 'YES' IF IT IS. ";YES$
2080 IF YES$ = "YES" THEN PRINT D$;"RUN MAILING LIST
     ADDER2"
2100 GOTO 40
```

## MAILING LIST READER2

```
10  REM ***--MAILING LIST READER--***
11 :
12 :
20  D$ = CHR$(4) : REM CONTROL D
25 :
26 :
30  REM **--INPUT ROUTINE--**
40  PRINT D$;"OPEN ADDRESS FILE"
60  PRINT D$;"READ ADDRESS FILE"
80  INPUT K
100 DIM NAME$(K),AD$(K),ND$(K),L(K),R(K)
120 FOR I = 1 TO K
140 INPUT NAME$(I)
160 NEXT I
180 PRINT D$;"CLOSE ADDRESS FILE"
190 :
191 :
200 REM **--MENU ROUTINE--**
210 HOME : VTAB 2
220 HTAB 17: PRINT "MENU"
230 PRINT : PRINT
240 HTAB 3: PRINT "1. DISPLAY INFO--ORIG. ORDER"
250 PRINT
260 HTAB 3: PRINT "2. DISPLAY NAMES ONLY"
270 PRINT
280 HTAB 3: PRINT "3. DISPLAY INFO--NO PHONE"
290 PRINT
300 HTAB 3: PRINT "4. DISPLAY SPECIFIC NAME"
310 PRINT
320 HTAB 3: PRINT "5. DISPLAY SPECIFIC NAME--NO PHONE"
```

## CHAPTER 6 DISPLAYING SEQUENTIAL FILES

```
330  PRINT
340  HTAB 3: PRINT "6. DISPLAY INFO- -RANGE"
350  PRINT
360  HTAB 3: PRINT "7. DISPLAY INFO- -ALPHABETICAL"
365  PRINT
370  HTAB 3: PRINT "8. RETURN TO PROGRAM MENU"
375  PRINT
380  HTAB 3: INPUT "WHICH NUMBER? "; NUMBER
385  IF NUMBER < 1 OR NUMBER > 8 THEN PRINT "INCORRECT
     NUMBER": GOTO 380
410  IF NUMBER = 1 THEN 1000
420  IF NUMBER = 2 THEN 2000
430  IF NUMBER = 3 THEN 3000
440  IF NUMBER = 4 THEN 4000
450  IF NUMBER = 5 THEN 5000
460  IF NUMBER = 6 THEN 6000
470  IF NUMBER = 7 THEN 7000
480  IF NUMBER = 8 THEN PRINT D$; "RUN MENU"
600  :
700  :
1000 REM **- -ORIGINAL ORDER ROUTINE- -**
1020 GOSUB 10000: REM PRINTER ROUTINE
1040 HOME : VTAB 5
1060 FOR I = 1 TO K
1080 IF NAME$(I) = "*" THEN 1140
1100 IF NAME$(I) = "!" THEN PRINT : GOTO 1140
1120 PRINT NAME$(I)
1140 NEXT I
1160 GOTO 20000: REM RETURN TO MENU ROUTINE
1500 :
1600 :
2000 REM **- -NAME ONLY ROUTINE- -**
2020 GOSUB 10000: REM PRINTER ROUTINE
2040 HOME: VTAB 5
2060 FOR I = 1 TO K - 1
2080 IF NAME$(I) = NAME$(2) THEN PRINT I; " "; NAME$(I)
2100 IF NAME$(I) = "!" THEN PRINT I; " "; NAME$(I + 1)
2120 NEXT I
2140 GOTO 20000: REM RETURN TO MENU ROUTINE
2500 :
2600 :
3000 REM **- -NO PHONE ROUTINE- -**
3020 GOSUB 10000: REM PRINTER ROUTINE
3040 HOME : VTAB 5
```

```
3060  FOR I = 1 TO K
3080  IF NAME$(I) = "*" THEN I = I + 1: GOTO 3140
3100  IF NAME$(I) = "!" THEN PRINT : GOTO 3140
3120  PRINT NAME$(I)
3140  NEXT I
3160  GOTO 20000: REM RETURN TO MENU ROUTINE
3500 :
3600 :
4000  REM **--SEARCH ROUTINE--**
4020  HOME : VTAB 5
4040  PRINT "TYPE 'END' WHEN FINISHED"
4050  INPUT "NAME TO FIND? ";FIND$
4060  IF FIND$ = "END" THEN 4400
4070  GOSUB 10000: REM PRINTER ROUTINE
4080  PRINT
4100  FOR I = 1 TO K
4120  IF NAME$(I) = FIND$ THEN 4160
4140  GOTO 4340
4160  IF NAME$(I) = "*" THEN 4340
4180  IF NAME$(I) = "!" THEN PRINT : GOTO 4340
4200  PRINT NAME$(I)
4220  PRINT NAME$(I + 1)
4240  PRINT NAME$(I + 2)
4260  IF NAME$(I + 3) < > "*" THEN PRINT NAME$(I + 3)
4280  IF NAME$(I + 4) = "*" THEN 4320
4300  PRINT NAME$(I + 4): GOTO 4340
4320  PRINT NAME$(I + 5)
4340  NEXT I
4360  PRINT
4380  GOTO 4040
4400  GOTO 20000: REM RETURN TO MENU ROUTINE
4500 :
4600 :
5000  REM **--SEARCH ROUTINE NO PHONE--**
5020  HOME : VTAB 5
5040  PRINT "TYPE 'END' WHEN FINISHED"
5050  INPUT "NAME TO FIND? ";FIND$
5060  IF FIND$ = "END" THEN 5400
5070  GOSUB 10000: REM PRINTER ROUTINE
5080  PRINT
5100  FOR I = 1 TO K
5120  IF NAME$(I) = FIND$ THEN 5160
5140  GOTO 5340
5160  IF NAME$(I) = "*" THEN I = I + 1: GOTO 5340
```

```
5180 IF NAME$(I) = "!" THEN PRINT : GOTO 5340
5200 PRINT NAME$(I)
5220 PRINT NAME$(I + 1)
5240 PRINT NAME$(I + 2)
5260 IF NAME$(I + 3) < > "*" THEN PRINT NAME$(I + 3)
5270 IF NAME$(I + 3) = "*" THEN I = I + 1: GOTO 5340
5280 IF NAME$(I + 4) = "*" THEN I = I + 1: GOTO 5340
5300 PRINT NAME$(I + 4) : GOTO 5340
5320 PRINT NAME$(I + 5)
5340 NEXT I
5360 PRINT
5380 GOTO 5040
5400 GOTO 20000: REM RETURN TO MENU ROUTINE
5500 :
5600 :
6000 REM **--RANGE ROUTINE--**
6020 HOME : VTAB 5
6040 INPUT "TYPE BEGINNING LINE NUMBER "; BL
6060 PRINT
6080 IF BL < 2 THEN PRINT "NUMBER TOO SMALL" : GOTO 6040
6100 INPUT "TYPE ENDING LINE NUMBER "; EL
6120 PRINT
6140 IF EL > K THEN PRINT "NUMBER TOO LARGE": GOTO 6100
6160 GOSUB 10000: REM PRINTER ROUTINE
6180 FOR I = BL TO EL
6200 IF NAME$(I) = "*" THEN I = I + 1: GOTO 6260
6220 IF NAME$(I) = "!" THEN PRINT : GOTO 6260
6240 PRINT NAME$(I)
6260 NEXT I
6280 GOTO 20000: REM RETURN TO MENU ROUTINE
6500 :
6600 :
6700 :
6800 :
6900 :
7000 REM **--ALPHABETICAL ORDER ROUTINE--**
7040 HOME : VTAB 5
7060 PRINT "WORKING--PLEASE DONT TOUCH!!"
7065 :
7066 :
7070 REM GET FIRST INFO-LINE
7080 FOR I = 2 TO K - 1
7100 IF NAME$(I) = NAME$(2) THEN 7160
7120 IF NAME$(I) = "!" THEN I = I + 1: GOTO 7160
```

```
7140  GOTO 7340
7145 :
7146 :
7150  REM REVERSE ORDER
7160 LN = LEN (NAME$(I))
7180  FOR J1 = 1 TO LN: IF MID$(NAME$(I),J1,1) = " " THEN J2
     = J1
7200  NEXT J1
7210  IF J2 = 0 OR J2 > LN THEN AD$(I) = NAME$(I) : GOTO 7240
7220  AD$(I) = MID$ (NAME$(I),J2 + 1,LN - J2) + " " +
     LEFT$ (NAME$(I),J2)
7240  AD$(I) = AD$(I) + "**" + NAME$(I + 1) + "**" +
     NAME$(I + 2)
7260  IF NAME$(I + 3) < > "*" THEN AD$(I) = AD$(I) + "**"
     + NAME$(I + 3)
7280  IF NAME$(I + 4) = "*" THEN 7320
7300  AD$(I) = AD$(I) + "**" + NAME$(I + 4) : GOTO 7340
7320  AD$(I) = AD$(I) + "**" + NAME$(I + 5)
7340  NEXT I
7345 :
7346 :
7350  REM RENUMBER FOR SORT
7360  J = 1
7380  FOR I = 1 TO K
7400  IF LEN (AD$(I)) > 0 THEN ND$(J) = AD$(I) : J = J + 1
7420  NEXT I
7440 N = J - 1
7445 :
7446 :
7460  REM ***--QUICKSORT--***
7480 S1 = 1
7500  PRINT "WORKING--PLEASE DONT TOUCH!!"
7520 L(1) = 1
7540 R(1) = N
7560 L1 = L(S1)
7580 R1 = R(S1)
7600 S1 = S1 - 1
7620 L2 = L1
7640 R2 = R1
7660 X$ = ND$( INT ((L1 + R1) / 2))
7680 C = C + 1
7700  IF ND$(L2) = X$ OR ND$(L2) > X$ THEN 7760
7720 L2 = L2 + 1
7740  GOTO 7680
```

```
7760 C = C1
7780 IF X$ = ND$(R2) OR X$ > ND$(R2) THEN 7840
7800 R2 = R2 - 1
7820 GOTO 7760
7840 IF L2 > R2 THEN 7980
7860 S = S + 1
7880 T$ = ND$(L2)
7900 ND$(L2) = ND$(R2)
7920 ND$(R2) = T$
7940 L2 = L2 + 1
7960 R2 = R2 - 1
7980 IF L2 = R2 OR L2 < R2 THEN 7680
8000 IF L2 = R1 OR L2 > R1 THEN 8080
8020 S1 = S1 + 1
8040 L(S1) = L2
8060 R(S1) = R1
8080 R1 = R2
8100 IF L1 < R1 THEN 7620
8120 IF S1 > 0 THEN 7560
8140 REM SORT COMPLETED
8142 :
8143 :
8145 REM **--DISPLAY--**
8150 GOSUB 10000: REM PRINTER ROUTINE
8160 FOR I = 1 TO N
8180 PRINT ND$(I)
8200 PRINT
8220 NEXT I
8240 GOTO 20000: REM RETURN TO MENU ROUTINE
8500 :
8600 :
8700 :
8800 :
8900 :
10000 REM **--PRINT ROUTINE--**
10020 PRINT "DO YOU WANT A PAPER PRINT OUT?"
10040 INPUT "TYPE 'Y' OR 'N' ";YES$
10060 IF YES$= "Y" THEN 10120
10080 SPEED = 150
10100 RETURN
10120 PR#1
10140 RETURN
15000 :
16000 :
```

```
20000 REM **--RETURN TO MENU ROUTINE--**
20020 PR#0 : SPEED = 255
20040 INPUT "HIT RETURN TO GO TO MENU ";L$
20060 GOTO 200 : REM MENU
```

# 7
# Correcting Sequential Files

We can use the same beginning for our MAILING LIST CORRECTOR program as we have for all our MAILING LIST READER programs. The one exception is line 10 which should be changed to reflect the proper name for this program. Once again we need a menu, so our routine beginning at line 200 will be much the same also.

```
200 REM ***--MENU ROUTINE--***
220 HOME: VTAB 5
240 HTAB 19
260 PRINT "CORRECTOR MENU"
280 PRINT: PRINT
300 HTAB 8: PRINT "1. CHANGE OR CORRECT INFO"
320 PRINT
340 HTAB 8: PRINT "2. DELETE INFO"
360 PRINT
380 HTAB 8: PRINT "3. WRITE REVISED FILE"
400 PRINT
420 HTAB 8: PRINT "4. RETURN TO PROGRAM MENU"
430 PRINT: PRINT
440 HTAB 8: INPUT "WHICH NUMBER "; NB
460 IF NB < 1 OR NB > 4 THEN 440
510 IF NB = 1 THEN 1000
520 IF NB = 2 THEN 2000
530 IF NB = 3 THEN 3000
540 IF NB = 4 THEN PRINT D$; "RUN MENU"
```

By now these statements should be familiar enough that no further explanation need be given. We are going to simply display a menu or a

## CHAPTER 7  CORRECTING SEQUENTIAL FILES

number of choices on the screen. Line 460 checks to see if the user has typed a valid number. If not, control is returned to the instruction (line 410) which again asks for a number.

The correction and deletion routines presented below are only one method out of many possible methods for accomplishing the same task. Some may object to rewriting the entire file for a single correction. Later in the chapter, we will get into the R and B parameters and how they can be used to make corrections. We will discuss some of the other methods also, but for now, the method we will use is to bring the entire file into memory, make our necessary corrections or deletions, and then write the file back out to disk again.

```
1000 REM ***--CORRECTION ROUTINE--***
1020 HOME: VTAB 5
1040 PRINT "TYPE '0' WHEN FINISHED"
1060 INPUT "DISPLAY WHICH LINE "; NUMBER
1080 IF NUMBER = 0 THEN 200
1100 PRINT
1120 PRINT NUMBER; " "; NAME$(NUMBER)
1140 PRINT
1160 PRINT "IS THIS CORRECT? ";
1180 INPUT "TYPE 'Y' OR 'N' "; YES$
1200 IF YES$ = "Y" THEN 1020
1220 PRINT
1240 PRINT "TYPE IN THE CORRECT INFORMATION"
1260 PRINT
1280 PRINT NUMBER; " ";: INPUT CN$
1300 PRINT: NAME$(NUMBER) = CN$
1320 PRINT NUMBER; " "; NAME$(NUMBER)
1340 PRINT
1360 GOTO 1160
```

This routine is also simple. We ask for the line number the user believes to contain incorrect information. The line of information is displayed. If it is not correct, the individual is given an opportunity to type in the correct information. The amount of new information or corrected information is not limited except by the normal 255 character string limitation. This feature is one big advantage over other correction methods which may require that the corrected information be exactly the same number of characters as the original information. Finally, the corrected information is displayed, and the correct information question is repeated. Line 1200 checks for a positive response to the question about correct information. If the information is correct, the user is taken back to the original request concerning the line number to be displayed. Line 1080 checks for a "0" which indicates that the user wishes to return to the menu. Line 1300 is the instruction that actually exchanges the corrected information for the old information. You will notice that nothing is written to disk at this time. This may cause problems for some individuals. Under this system, it is possible to make a number of changes before the file is rewritten to the disk. It is also possible, therefore, to forget to write the corrected file back to disk. Such a system may be impractical in certain situations—for example, when a somewhat forgetful person is making the changes. But, for our purposes, we want to make all corrections and deletions before rewriting the file.

The deletion routine is more complicated than the correction routine.

```
2000 REM ***--DELETE ROUTINE--***
2020 HOME: VTAB 5
2040 PRINT "TYPE '0' WHEN FINISHED"
2060 INPUT "DELETE WHICH LINE "; LINE
2080 IF LINE = 0 THEN 200
2100 PRINT
2120 PRINT LINE; " "; NAME$(LINE)
2140 PRINT
2220 PRINT "ARE YOU SURE? TYPE 'YES' IF SURE";
2240 INPUT YES$
2260 IF YES$ = "YES" THEN 2300
2280 GOTO 2000: REM BEGIN AGAIN
2290 :
2300 J = LINE
2320 IF NAME$(J) = "!" THEN 2360
2340 J = J + 1: GOTO 2320
2360 FOR I = LINE TO J
2380 PRINT I; " "; NAME$(I)
2400 NAME$(I) = "DELETED": D = D + 1
2420 NEXT I
2440 PRINT
2460 PRINT "DELETING THIS INFORMATION"
2480 Q = 2
2500 FOR I = 2 TO K
2520 IF NAME$(I) = "DELETED" THEN 2580
2540 NAME$(Q) = NAME$(I)
2560 Q = Q + 1
2580 NEXT I
2590 :
2600 K = K - D
2620 D = 0: J = 0
2700 GOTO 200
```

There are other ways of doing the same thing we did in this routine. Some

CHAPTER 7  CORRECTING SEQUENTIAL FILES   85

of the other ways might be shorter, but this way is understandable. Several things need to be done in this deletion routine. First, the information to be deleted must be identified (lines 2000 to 2280). Second, the information following the deleted material must be renumbered so that there are no empty lines, otherwise an END OF DATA ERROR will occur when these lines are encountered (lines 2300 to 2580). Finally, the number of deleted lines must be subtracted from the original total number of lines (line 2600). Down to 2280, there is nothing new. It is essentially the same beginning as the correction routine. At 2300, we set a counter (J) equal to the line number of the name of the individual to be deleted. Next, we increase the counter by one until we have found the information-line for the end of the information associated with the individual to be deleted; i.e., the separator symbol "!". Now we know which lines to delete: the lines beginning with LINE and going through J, so now we can use a loop (2360-2420) to delete our information. We use another loop (2500-2580) to do the resequencing of the remaining information. We use two additional counters: Q to keep track of the new information line-numbers, and D to keep track of the number of deleted lines. Q is set to 2 for the beginning of the file, but it could be set to LINE, the start of the deleted material. Line 2520 is the key to the resequencing. If NAME$(I) equals the word "DELETED", then the counter Q is not increased while the counter I is increased. Remember that Q is keeping track of the new line numbers while I is the old line number. Line 2540 is resequencing the NAME$ string array. Line 2600 subtracts the number of deleted lines from the original number of lines (K). Line 2620 is necessary in case more information is to be deleted during this session.

```
3000 REM ***--FILE ROUTINE--***
3020 PRINT D$;"OPEN ADDRESS FILE BACKUP"
3040 PRINT D$;"DELETE ADDRESS FILE BACKUP"
3060 PRINT D$;"RENAME ADDRESS FILE, ADDRESS FILE
     BACKUP"
3080 PRINT D$;"OPEN ADDRESS FILE"
3100 PRINT D$;"WRITE ADDRESS FILE"
3120 PRINT K
3140 FOR I = 1 TO K
3160 PRINT NAME$(I)
3180 NEXT I
3200 PRINT D$;"CLOSE ADDRESS FILE"
3220 PRINT D$;"RUN MENU"
```

There is something different with this file routine. Where did ADDRESS FILE BACKUP come from? What is RENAME? The DOS command OPEN will create a file by that name if the file does not already exist. Therefore, if the file, ADDRESS FILE BACKUP, does not already exist, the command, OPEN ADDRESS FILE BACKUP, will create a file by that name. Next, we delete that file since it must either be an empty file or a now unnecessary backup copy. (The first time this program is used, there will not be an ADDRESS FILE BACKUP.) Line 3060 renames the file ADDRESS FILE (which now contains our uncorrected information) so that it becomes ADDRESS FILE BACKUP. Finally, we open a new ADDRESS FILE and write out our corrected information to it (lines 3080-3200). Line 3220 returns to the Program Menu. At this point, if you have not already done so, you should save this program to the diskette that contains all the other MAILING LIST SYSTEM programs.

SAVE MAILING LIST CORRECTOR

Typing CATALOG now should show:

A 002 HELLO
A 003 APPLE
A 006 MAILING LIST CREATOR
T 002 ADDRESS FILE
A 019 MAILING LIST READER
A 007 MAILING LIST ADDER1
A 008 MAILING LIST ADDER2
A 007 MAILING LIST CORRECTOR

It is now necessary to make changes in the MENU Program in order to include this MAILING LIST CORRECTOR program. Remember that space was left between lines 200 and 300. Additional lines need to be changed also. If you have trouble making the necessary changes, the final version of the program is included at the end of this chapter. Type:

LOAD MENU

Make the necessary changes. Then type:

SAVE MENU

The MAILING LIST SYSTEM should now be complete.

This general method of correcting or deleting information has the added benefit of providing us with a backup copy of our pre-corrected ADDRESS FILE. The sequence of opening and deleting a backup file, renaming the uncorrected file as the new backup, and writing out the corrected information to a new file under the original file name is a very useful routine. If you have two disk drives, you can put the backup in one drive and the new master in the other drive and have the computer switch between the two by simply attaching the appropriate drive number at the end of the various command lines (but within the quotation marks).

3020 PRINT D$; "OPEN ADDRESS FILE BACKUP, D1"
3040 PRINT D$; "DELETE ADDRESS FILE BACKUP"

CHAPTER 7   CORRECTING SEQUENTIAL FILES                87

```
3060 PRINT D$; "RENAME ADDRESS FILE, ADDRESS FILE
     BACKUP,D2"
3080 PRINT D$; "OPEN ADDRESS FILE,D1"
3100 PRINT D$; "WRITE ADDRESS FILE"
```

The rest of the program would be handled in the same way. The more drives you have, the greater your flexibility in manipulating files in this manner.

Even without two drives, the ADDRESS FILE and the ADDRESS FILE BACKUP can be put on two different diskettes. Some method of making the computer pause after line 3060 would be necessary in order to allow the user to swap diskettes. Two possibilities would be : a loop of a certain duration, or an input statement informing the user that it is time to switch diskettes.

As mentioned, there are other ways to make corrections to a sequential access file. One way involves the R and B parameters, but I do not recommend it. The R and B parameters are supposed to provide somewhat direct access to a specified record (R) or field and a specified byte (B). So theoretically, if I know that the hundredth information-line needs to be changed, I can access just that record or field and make the change. But I have to make the new information exactly the same number of characters as the old information or I will end up with a scrambled file. If my new information contains less characters than the old information, I will have an extra record composed of the leftover characters. If my corrected information is longer than the old information, part of the next record will be overwritten and therefore lost. It is for this reason that I do not recommend using the R or B parameter to make corrections to a sequential access file.

These parameters are somewhat more useful if restricted to reading file information, but even here they can prove to be troublesome. The R parameter is supposed to position the file pointer a specific number from its present location. It is therefore relative, not absolute. When a file is opened, the file pointer points to record 0 so that R100 would access the 100th record in the file. But a following R command of R10 would access the 110th record not the 10th record of the file.

The B parameter is supposed to be absolute, except when it follows an R parameter. In other words, a command of B100 will access the 100th byte of the file. A following command of B10 will access the 10th byte of the file, not the 110th byte. But an intervening R command is supposed to set the file pointer R records from its present position and then B bytes from the beginning of that record. The problem is that, besides this rather confusing set of rules, it doesn't actually work the way it should. You cannot access the Bth byte of the Rth record. The parameters seem to work fine when used separately, but when combined, some very strange results occur. Combining the R and B parameters appears to result in adding the value of R to the value of B and accessing the resultant byte value in the file. Therefore, the examples will use the parameters separately. Type in the following:

```
NEW
20 D$ = CHR$(4) : REM CONTROL D
40 HOME: VTAB 5
60 INPUT "READ WHICH RECORD ";R
80 PRINT D$;"OPEN ADDRESS FILE"
100 PRINT D$;"POSITION ADDRESS FILE,R";R
120 PRINT D$;"READ ADDRESS FILE"
140 INPUT LINE$
160 PRINT D$;"CLOSE ADDRESS FILE"
180 PRINT
200 PRINT LINE$
220 PRINT
240 GOTO 60
```

The only unfamiliar line should be line 100. This is the instruction that positions the file pointer to the desired record. Remember that this is a relative pointer. If you add the following four lines, the results will be very different.

```
145 PRINT D$;"POSITION ADDRESS FILE,R5"
150 PRINT D$;"READ ADDRESS FILE"
155 INPUT LINE2$
210 PRINT LINE2$
```

LINE2$ should be five information-lines beyond LINE$ in the ADDRESS FILE. If you want to check this, use the display program and the range selection to see if this is so.

Now change the program so that you will be reading bytes instead of records.

```
60 INPUT "READ WHICH BYTE ";B
120 PRINT D$;"READ ADDRESS FILE,B";B
DELETE THE FOLLOWING LINES:
100
145
150
155
210
```

This change should make the program able to read specific bytes within the ADDRESS FILE. If you increase the number requested in line 60 by one, you should see the result of reading specific bytes within a file.

If you want to attempt writing to a specific record or byte, you now have the necessary information, but I again repeat that I would not recommend using these parameters to make corrections to any file you value.

In the next chapter, we will take a look at some more exotic uses of sequential access files: the use of the EXEC command and the CHAIN command plus an explanation of some of the differences between Applesoft and Integer.

## QUESTIONS

1. TRUE or FALSE: Under the correction method presented in this chapter, corrected information is immediately written to the disk.
2. What happens to the original ADDRESS FILE once information in it has been changed?
3. What is the DOS command used to remove unwanted files?
4. What is the DOS command used to change the name of files?
5. What are the two parameters that can be used with sequential files?
6. Which parameter is relative?
7. Which parameter is absolute?
8. Give the DOS command used with the R parameter.

## ANSWERS

1. FALSE
2. It becomes ADDRESS FILE BACKUP
3. DELETE
4. RENAME
5. R and B
6. R
7. B
8. POSITION

## MAILING LIST CORRECTOR

```
10   REM **--MAILING LIST CORRECTOR--**
11   :
12   :
20   D$ = CHR$ (4) : REM CONTROL D
25   :
26   :
30   REM **--INPUT ROUTINE--**
40   PRINT D$; "OPEN ADDRESS FILE"
60   PRINT D$; "READ ADDRESS FILE"
80   INPUT K
100  DIM NAME$(K), LINE$(K)
120  FOR I = 1 TO K
140  INPUT NAME$(I)
160  NEXT I
180  PRINT D$; "CLOSE ADDRESS FILE"
190  :
191  :
200  REM **--MENU ROUTINE--**
220  HOME : VTAB 5
240  HTAB 19
260  PRINT "MENU"
280  PRINT : PRINT
300  HTAB 8: PRINT "1. CHANGE OR CORRECT INFO"
320  PRINT
340  HTAB 8: PRINT "2. DELETE INFO"
360  PRINT
380  HTAB 8: PRINT "3. WRITE REVISED FILE"
400  PRINT
420  HTAB 8: PRINT "4. RETURN TO PROGRAM MENU"
430  PRINT : PRINT
440  HTAB 8: INPUT "WHICH NUMBER "; NB
460  IF NB < 0 OR NB > 4 THEN 440
510  IF NB = 1 THEN 1000
520  IF NB = 2 THEN 2000
530  IF NB = 3 THEN 3000
540  IF NB = 4 THEN PRINT D$; "RUN MENU"
600  :
700  :
1000 REM **--CORRECTION ROUTINE--**
1020 HOME : VTAB 5
1040 PRINT "TYPE '0' WHEN FINISHED"
```

CHAPTER 7 CORRECTING SEQUENTIAL FILES

```
1060 INPUT "DISPLAY WHICH LINE ";NUMBER
1080 IF NUMBER = 0 THEN 200
1100 PRINT
1120 PRINT NUMBER;" ";NAME$(NUMBER)
1140 PRINT
1160 PRINT "IS THIS CORRECT? ";
1180 INPUT "TYPE 'Y' OR 'N' ";YES$
1200 IF YES$ = "Y" THEN 1020
1220 PRINT
1240 PRINT "TYPE IN THE CORRECT INFORMATION"
1260 PRINT
1280 PRINT NUMBER;" ";: INPUT CN$
1300 PRINT : NAME$(NUMBER) = CN$
1320 PRINT NUMBER;" ";NAME$(NUMBER)
1340 PRINT
1360 GOTO 1160
1500 :
1600 :
2000 REM **--DELETE ROUTINE--**
2020 HOME : VTAB 5
2040 PRINT "TYPE '0' WHEN FINISHED"
2060 INPUT "DELETE WHICH LINE ";LINE
2080 IF LINE = 0 THEN 200
2100 PRINT
2120 PRINT LINE;" ";NAME$(LINE)
2140 PRINT
2220 PRINT "ARE YOU SURE? TYPE 'YES' IF SURE";
2240 INPUT YES$
2260 IF YES$ = "YES" THEN 2300
2280 GOTO 2000
2300 J = LINE
2320 IF NAME$(J) = "!" THEN 2360
2340 J = J + 1: GOTO 2320
2360 FOR I = LINE TO J
2380 PRINT I;" ";NAME$(I)
2400 NAME$(I) = "DELETED":D = D + 1
2420 NEXT I
2440 PRINT
2460 PRINT "DELETING THIS INFORMATION"
2480 Q = 2
2500 FOR I = 2 TO K
2520 IF NAME$(I) = "DELETED" THEN 2580
2540 NAME$(Q) = NAME$(I)
2560 Q = Q + 1
```

```
2580 NEXT I
2600 K = K - D
2620 D = 0:J = 0
2700 GOTO 200
2800 :
2900 :
3000 REM **--FILE ROUTINE--**
3020 PRINT D$;"OPEN ADDRESS FILE BACKUP"
3040 PRINT D$;"DELETE ADDRESS FILE BACKUP"
3060 PRINT D$;"RENAME ADDRESS FILE, ADDRESS FILE BACKUP"
3080 PRINT D$;"OPEN ADDRESS FILE"
3100 PRINT D$;"WRITE ADDRESS FILE"
3120 PRINT K
3140 FOR I = 1 TO K
3160 PRINT NAME$(I)
3180 NEXT I
3200 PRINT D$;"CLOSE ADDRESS FILE"
3220 PRINT D$;"RUN MENU"
```

## R&B PROG1

```
20  D$ = CHR$(4) : REM CONTROL D
40  HOME : VTAB 5
60  INPUT "READ WHICH RECORD ";R
80  PRINT D$;"OPEN ADDRESS FILE"
100 PRINT D$;"POSITION ADDRESS FILE,R";R
120 PRINT D$;"READ ADDRESS FILE"
140 INPUT LINE$
160 PRINT D$;"CLOSE ADDRESS FILE"
180 PRINT
200 PRINT LINE$
220 PRINT
240 GOTO 60
```

## R&B PROG2

```
20  D$ = CHR$(4) : REM CONTROL D
40  HOME : VTAB 5
60  INPUT "READ WHICH RECORD ";R
80  PRINT D$;"OPEN ADDRESS FILE"
100 PRINT D$;"POSITION ADDRESS FILE,R";R
```

```
120  PRINT D$; "READ ADDRESS FILE"
140  INPUT LINE$
145  PRINT D$; "POSITION ADDRESS FILE, R5"
150  PRINT D$; "READ ADDRESS FILE"
155  INPUT LINE2$
160  PRINT D$; "CLOSE ADDRESS FILE"
180  PRINT
200  PRINT LINE$
210  PRINT LINE2$
220  PRINT
240  GOTO 60
```

## R&B PROG3

```
20 D$ = CHR$(4) : REM CONTROL D
40  HOME: VTAB 5
60  INPUT "READ WHICH BYTE "; B
80  PRINT D$; "OPEN ADDRESS FILE"
120  PRINT D$; "READ ADDRESS FILE, B"; B
140  INPUT LINE$
160  PRINT D$; "CLOSE ADDRESS FILE"
180  PRINT
200  PRINT LINE$
220  PRINT
240  GOTO 60
```

## MAILING LIST MENU

```
10 REM ***--MAILING LIST PROGRAM MENU--***
11:
12:
20 D$ = CHR$(4) : REM CONTROL D
25:
26:
30 REM **--MENU ROUTINE--**
40  HOME : VTAB 5
60  HTAB 17: PRINT "PROGRAM MENU"
80  PRINT : PRINT
100  HTAB 8: PRINT "1. FILE CREATION PROGRAM"
120  PRINT
140  HTAB 8: PRINT "2. FILE ADDITION PROGRAM"
```

# MAILING LIST MENU

```
160  PRINT
180  HTAB 8: PRINT "3. FILE DISPLAY PROGRAM"
200  PRINT
220  HTAB 8: PRINT "4. FILE CORRECTION PROGRAM"
240  PRINT
300  HTAB 8: PRINT "5. CATALOG"
320  PRINT
340  HTAB 8: PRINT "6. END"
360  PRINT : PRINT
380  HTAB 8: INPUT "WHICH PROGRAM NUMBER? "; NUMBER
400  IF NUMBER < 1 OR NUMBER > 6 THEN 380
420  IF NUMBER = 1 THEN 1000
440  IF NUMBER = 2 THEN 2000
460  IF NUMBER = 3 THEN PRINT D$; "RUN MAILING LIST
     READER"
470  IF NUMBER = 4 THEN PRINT D$; "RUN MAILING LIST
     CORRECTOR"
480  IF NUMBER = 5 THEN PRINT D$; "CATALOG": INPUT "HIT
     RETURN TO GO TO MENU "; L$: GOTO 40
500  IF NUMBER = 6 THEN END
600  :
700  :
1000 REM **--FILE CREATOR PROGRAM--**
1020 PRINT : PRINT "IF THE ADDRESS FILE ALREADY EXISTS"
1040 PRINT : PRINT "DO NOT RUN THIS PROGRAM!!"
1060 PRINT : PRINT "DO YOU WANT THE FILE CREATION
     PROGRAM?"
1070 PRINT
1080 INPUT "TYPE 'YES' IF YOU DO: "; YES$
1100 IF YES$ = "YES" THEN PRINT D$; "RUN MAILING LIST
     CREATOR"
1120 GOTO 40
1140 :
1160 :
2000 REM **--FILE ADDITION PROGRAM--**
2020 PRINT : PRINT "YOU WANT TO ADD TO THE EXISTING"
2040 PRINT : PRINT "ADDRESS FILE. IS THIS CORRECT? "
2060 PRINT : INPUT "TYPE 'YES' IF IT IS. "; YES$
2080 IF YES$ = "YES" THEN PRINT D$; "RUN MAILING LIST
     ADDER2"
2100 GOTO 40
```

# 8
# EXEC and CHAIN

This chapter is for those who like to try unusual things. It is not necessary in order to understand normal file handling using Applesoft BASIC, but if you want to see some rather unusual disk use, read on. Otherwise, unless you are determined to use Integer for your files, you can skip this chapter. The Disk Operating System provides two commands that will form the basis for this chapter: EXEC and CHAIN. CHAIN, as a command, is only available with Integer BASIC. Apple has provided a binary routine on their Apple System Master Diskettes (CHAIN) that can be used from within an Applesoft program to simulate the actual CHAIN command available to Integer BASIC users. We will begin with the EXEC command.

**EXEC**

The manuals say that EXEC is a way of controlling the computer as if the instructions were being typed at the keyboard. The immediate question arises as to why one would not just type the instructions. The easiest explanation of the EXEC usefulness is in the form of a demonstration of an existing program. If you were to try to show a friend what the MAILING LIST SYSTEM does, you would need to type in all the responses necessary. But if you had a program that made use of the EXEC command, you could simply run that program and sit back and watch as the EXEC command took control of the computer and operated as an automatic and continuous demonstration of the MAILING LIST SYSTEM. Such a program is very useful when a demonstration is necessary and a knowledgeable operator is not available. Another example, although probably less useful with the copy programs available, is the use of the EXEC command to load and save programs from one disk to another without operator intervention. Such use might prove beneficial if only a selected number of programs were to be transferred to a working disk and the operator did not know how to do this. Deletions of certain unwanted programs by an untrained operator is best handled by either a program or a text file under the EXEC command. Another example provided by Apple manuals is a method of transferring programs from one form of BASIC to the other. They thoughtfully remind you that transferring the program to the other BASIC will not guarantee that the program will operate without modification. It probably will not. These are a few of the uses of this very versatile command. It is really fairly unexplored as far as written material goes. It should be possible to set up a text file in such a way that by using the EXEC command, you could conduct some business while you are away from the computer.

Let's look at how you can set up and use an EXEC file. There are at least three steps to using an EXEC file. The first is the creation of a program that will write out a text file containing the instructions the computer will eventually follow. The second step is the actual text file creation, and the third is the EXECing of that text file. Thus, in the language of the manuals, you create a program that makes a text file that is then EXECed. It is not as difficult to understand and follow as it may sound. The first step is to create a program that makes a text file. Type the following:

```
10 REM ***--EXEC FILE CREATOR--***
20 D$ = CHR$(4) : REM CONTROL D
30 Q$ = CHR$(34) : REM QUOTATION MARK
40 PRINT D$;"OPEN DEMO"
60 PRINT D$;"DELETE DEMO"
80 PRINT D$;"OPEN DEMO"
100 PRINT D$;"WRITE DEMO"
```

These are the opening lines in our text file creation program. The only line that should be unfamiliar is line 30. We need to set a string variable equal to CHR$(34) in order to do some of the things we will do within this demonstration program. As previously explained, we open and delete our file before writing to it so that we do not end up with unwanted or unexpected results due to overwriting. After our initial lines, we want to write instructions to the file. These instructions are the same as if we were typing at the keyboard, but because the instructions are all included with their carriage return, we need to slow down the entire process.

```
105 PRINT "SPEED = 50"
107 PRINT "MON I,O,C"
110 PRINT "RUN MENU"
```

```
120 PRINT "1"
140 PRINT "N"
160 PRINT "2"
180 PRINT "N"
200 PRINT "3"
220 PRINT "1"
240 PRINT "N"
260 PRINT "N"
280 PRINT "2"
300 PRINT "N"
320 PRINT "N"
```

Lines 105 through 320 print the information between the quotation marks to the text file called DEMO. Line 105 sets the speed of the computer at a rate of 50 rather than its normal 255. Line 107 uses the command that allows us to see the instructions as they are passed to and from the disk. Line 110 writes the instruction that will go to the disk, load in the program MENU, and begin operation of that program. The program MENU will take command of the computer, and all following text file characters will be viewed by the computer as responses to the various input requests in the MENU program and any other program called by MENU. Therefore, line 120 prints out a "1". When used under the EXEC command, this is taken by the computer as a response to the question "WHICH PROGRAM NUMBER?" The N in line 140 indicates that we do not want to continue with the MAILING LIST CREATOR program. Lines 160 and 180 go through the same process with program number 2 in the PROGRAM MENU. In line 200, however, we have requested program number 3, the MAILING LIST READER or display program. Under EXEC control, the computer will leave the MENU program and go to the MAILING LIST READER program for its next set of instructions. The print statements from lines 220 to 660 will all be taken as answers to input questions within the MAILING LIST READER program. I have left room between line 320 and line 500 for you to fill in a name from your own mailing list. At the end of this chapter, you will find the missing lines and a fictitious name. You can use those lines as a guide if you are having trouble filling in the proper responses.

```
500 PRINT "6"
520 PRINT "50"
540 PRINT "75"
560 PRINT "N"
580 PRINT "N"
600 PRINT "7"
620 PRINT "N"
640 PRINT "N"
660 PRINT "8"
```

These lines, again, are taken as responses to questions asked in the MAILING LIST READER program. In the next section, we will move back to the MENU program and then on to the MAILING LIST CORRECTOR program. Finally, we go back to the MENU program and end.

```
680 PRINT "4"
700 PRINT "1"
720 PRINT "100"
725 PRINT "N"
730 PRINT "TEST OF CORRECTION PROGRAM"
740 PRINT "Y"
750 PRINT "0"
760 PRINT "2"
780 PRINT "135"
800 PRINT "YES"
820 PRINT "4"
840 PRINT "6"
```

These are responses to the various questions asked in either the MENU program or MAILING LIST CORRECTOR program. The final section puts things back the way they were originally.

```
860 PRINT "SPEED=255"
880 PRINT "NOMON I,O,C"
900 PRINT "PRINT"Q$"THE ADDRESS FILE WAS NOT
    CHANGED"Q$
920 PRINT "PRINT"Q$"BY THIS DEMO. "Q$
1000 PRINT D$;"CLOSE DEMO"
```

There are several new things in this section. Line 880 is the opposite of line 107 in that it turns off the display of commands to and from the disk. Line 900 is the first instance of using Q$. To understand why we use Q$, remember that we are writing this information to a text file. To write to the disk requires the use of the command PRINT with the characters that are to be written enclosed by quotation marks. One of the things that we now want to write to disk is the word PRINT itself, used as a computer command, with material also enclosed in quotes. In effect, we must have quotes within quotes. One of the ways to accomplish this is by setting Q$ equal to CHR$(34), which is the value for a quotation mark. Thus, we actually have quotation marks within quotation marks. Finally, line 1000 closes the file. The first step of the EXEC process is finished.

Once you have this program typed in and checked, save the program on the same disk as the rest of the programs from this book. Then type: RUN. The disk should come on for a short time and then the cursor will return. If you type CATALOG, you should see two new files on the disk. One file will be the program you saved before typing RUN. The other file should be DEMO and should be a text file. You have just completed the second step in using EXEC files.

The final step is using the EXEC command. All that is really necessary now is to type: EXEC DEMO. You should be able to sit back and watch the computer go through a demonstration of the MAILING LIST SYSTEM programs. Another way would be to modify the MENU program to include the option of this automatic demo. You can use the EXEC command from within a program in the same way you use other DOS commands such as RUN. The time required to complete the demonstration will depend upon the size of your mailing list. If you either have typed an error or do not have a mailing list with at least 135 lines of information, you will probably get an error message. You have now used the EXEC command in a fairly sophisticated way to control the computer without actually typing in responses. As I indicated, this is only one way to use this versatile command. I do not think that its potentialities have even been approached.

## CHAIN

CHAIN allows a programmer to link two or more programs while keeping the values created or established in the first program. In other words, the variable values remain from one program to the other. If you have a program that creates a value that could be used by a second program, one of the ways to accomplish such use is by way of CHAIN.

CHAIN is really two different things. In Integer BASIC, it can be used as a command. In Applesoft BASIC, however, there is no CHAIN command. Rather, there is a chain binary file that can be called from within the program. We will go over both methods, but I repeat that knowledge of CHAIN is not necessary in order to make effective use of the Apple's file handling capabilities.

## INTEGER BASIC CHAIN

We begin with Integer BASIC CHAIN, because it is by far the easiest to use. Theoretically, all that is necessary to use CHAIN is to include this line as the last executed line of the first program.

PRINT D$; "CHAIN name of program to follow"

For example, if you want to pass the values of your variables from program A to program B, the line would read: PRINT D$;"CHAIN B". At that point, the computer would go to the disk, load the program called B, and begin execution according to the instructions of program B. The values created or established in program A would remain in memory and not be erased as is normally the case when a new program is RUN. Program B can then chain to program C and so on. The time involved for this chain to occur is no longer than if the line read: PRINT D$;"RUN B". This fact is a definite advantage for Integer BASIC users, since the Applesoft BASIC chain method takes more time than simply RUNning a new program. CHAIN in Integer is very easy to use and might often be nice to have available. There are some things to watch out for, but they are common to both methods and will be discussed after the Applesoft BASIC method.

## APPLESOFT BASIC CHAIN

Pages 106 and 107, in both the DOS 3.2 and 3.3 manuals, give the instructions for using CHAIN. In order to be able to pass the values of variables from one program to another using a chain method, one must load the binary program CHAIN off of the Apple System Master Diskette which comes with the purchase of a disk drive and a controller card. The binary program CHAIN must be transferred to your diskette by either the FID program (for 3.3 users) or by BLOADing and BSAVEing in one of the following ways:
1. APPLESOFT ON DISKETTE:
    BLOAD CHAIN, A12296
    Remove Apple diskette, insert your diskette.
    BSAVE CHAIN, A12296, L456
2. APPLESOFT IN ROM:
    BLOAD CHAIN, A2056
    Remove Apple diskette, insert your diskette.
    BSAVE CHAIN, A2056, L456

Now you must add two lines as the last two instruction lines executed in the first program. The last line must be typed very carefully.
    PRINT D$;"BLOAD CHAIN, A520"
    CALL 520"name of program to follow"

BLOAD, BSAVE and BRUN are the commands necessary to access binary files. Both of the above lines need line numbers and these lines must be the last two instructions executed in the initial program. The line with CALL in it has to be typed without a space between the 0 and the ". For example, if you want to pass the values of your variables from program A to program B, the line would read: CALL 520"B". When executed, the computer must first load the binary file CHAIN, then protect the variable values, and finally load program B from disk and begin operation according to the instructions in program B. I found that this procedure can take quite a while, in some circumstances almost a minute. I know that one minute does not sound like a very long time, but when compared to the few seconds that same operation takes in Integer, it seems quite long.

When chaining programs together in both Integer and Applesoft, you must make sure that variables that should not overlap, do not overlap. If you are not careful, the value in chaining can be lost by obtaining inaccurate results. For example, if you try to chain two programs that were created separately, it is possible to bring from the first program a variable that is redefined to a constant or one used as a counter in the second program. Variables that require DIMensioning may also cause a problem. If you have DIMensioned a variable in program A, and also DIMensioned that same variable in program B, you will get a REDIM'D ARRAY ERROR when the chain occurs.

One way to simulate a chain without using the Applesoft CHAIN method is to write out the variable values you want to keep to a temporary sequential text file. Then, any program that you want to use those values can access them by opening the file and reading them in. This method requires extra code in both programs but, in the long run, may prove faster and safer.

The difference in handling chaining makes Integer BASIC appear to be the better BASIC, but this is just about the only time Integer has the edge over Applesoft when dealing with files. Probably the biggest problem with Integer is that it does not have string array capability. This lack can cause some real problems when trying to input string data from disk. It can be done but requires a great deal more code than Applesoft. A sample Integer program for reading in and printing out the contents of the ADDRESS FILE is included at the end of this chapter.

In the next chapter, we will take a look at other sequential access programs and different techniques for sequential file handling.

## QUESTIONS

1. What DOS command can control the computer as if instructions were being typed at the keyboard?
2. How many steps are there in using the EXEC command?
3. What does CHR$(34) equal?
4. What Applesoft BASIC word controls the display rate?
5. What function allows a programmer to link two or more programs while preserving the original value of the variables?
6. TRUE or FALSE: Applesoft can make use of the DOS command CHAIN.
7. Name the commands necessary to access Binary files.

# ANSWERS

1. EXEC
2. 3
3. A Quotation Mark
4. SPEED
5. Chain
6. FALSE
7. BLOAD, BSAVE, and BRUN

## EXEC FILE CREATOR

```
10  REM ***--EXEC FILE CREATOR--***
11 :
12 :
20  D$ = CHR$(4) : REM CONTROL D
30  Q$ = CHR$(34) : REM QUOTATION MARKS
31 :
32 :
35  REM **--FILE CREATION--**
40  PRINT D$;"OPEN DEMO"
60  PRINT D$;"DELETE DEMO"
80  PRINT D$;"OPEN DEMO"
100 PRINT D$;"WRITE DEMO"
101 :
102 :
104 REM **--SET UP--**
105 PRINT "SPEED = 50"
107 PRINT "MON I,O,C"
108 :
109 REM **--MENU--**
110 PRINT "RUN MENU"
120 PRINT "1"
140 PRINT "N"
160 PRINT "2"
180 PRINT "N"
200 PRINT "3"
205 :
206 :
210 REM **--READER--**
220 PRINT "1"
240 PRINT "N"
260 PRINT "N"
280 PRINT "2"
300 PRINT "N"
320 PRINT "N"
340 PRINT "4"
360 PRINT "RON WISE"
380 PRINT "N"
400 PRINT "END"
410 PRINT "N"
420 PRINT "5"
```

# EXEC FILE CREATOR

```
440  PRINT "RON WISE"
460  PRINT "N"
480  PRINT "END"
490  PRINT "N"
500  PRINT "6"
520  PRINT "50"
540  PRINT "75"
560  PRINT "N"
580  PRINT "N"
600  PRINT "7"
620  PRINT "N"
640  PRINT "N"
660  PRINT "8"
665 :
666 :
670  REM **--MENU--**
680  PRINT "4"
685 :
686 :
690  REM **--CORRECTOR--**
700  PRINT "1"
720  PRINT "100"
725  PRINT "N"
730  PRINT "TEST OF CORRECTION PROGRAM"
740  PRINT "Y"
750  PRINT "0"
760  PRINT "2"
780  PRINT "135"
800  PRINT "YES"
820  PRINT "4"
825 :
826 :
830  REM **--MENU--**
840  PRINT "6"
845 :
846 :
850  REM **--RESTORE--**
860  PRINT "SPEED = 255"
880  PRINT "NOMON I,O,C"
900  PRINT "PRINT"Q$"THE ADDRESS FILE WAS NOT CHANGED"Q$
920  PRINT "PRINT"Q$"BY THIS DEMO. "Q$
1000 PRINT D$; "CLOSE DEMO"
```

## MENU CHAIN

```
10  REM MAILING LIST PROGRAM MENU
20  D$ = CHR$(4) : REM CONTROL D
40  HOME : VTAB 5
60  HTAB 17: PRINT "PROGRAM MENU"
80  PRINT : PRINT
100 HTAB 8: PRINT "1. FILE CREATION PROGRAM"
120 PRINT
140 HTAB 8: PRINT "2. FILE ADDITION PROGRAM"
160 PRINT
180 HTAB 8: PRINT "3. FILE DISPLAY PROGRAM"
200 PRINT
220 HTAB 8: PRINT "4. FILE CORRECTION PROGRAM"
240 PRINT
300 HTAB 8: PRINT "5. CATALOG"
320 PRINT
340 HTAB 8: PRINT "6. END"
360 PRINT : PRINT
380 HTAB 8: INPUT "WHICH PROGRAM NUMBER? "; NUMBER
400 IF NUMBER < 1 OR NUMBER > 6 THEN 380
420 IF NUMBER = 1 THEN 1000
440 IF NUMBER = 2 THEN 2000
460 IF NUMBER = 3 THEN PRINT D$; "RUN MAILING LIST
    READER"
470 IF NUMBER = 4 THEN 4000
480 IF NUMBER = 5 THEN PRINT D$; "CATALOG" : INPUT "HIT
    RETURN TO GO TO MENU "; L$: GOTO 40
500 IF NUMBER = 6 THEN END
1000 REM FILE CREATOR PROGRAM
1020 PRINT : PRINT "IF THE ADDRESS FILE ALREADY EXISTS"
1040 PRINT : PRINT "DO NOT RUN THIS PROGRAM!!"
1060 PRINT : PRINT "DO YOU WANT THE FILE CREATION PROGRAM?"
1070 PRINT
1080 INPUT "TYPE 'YES' IF YOU DO: "; YES$
1100 IF YES$ = "YES THEN PRINT D$; RUN MAILING LIST
     CREATOR"
1120 GOTO 40
2000 REM FILE ADDITION PROGRAM
2020 PRINT : PRINT "YOU WANT TO ADD TO THE EXISTING"
2040 PRINT : PRINT "ADDRESS FILE. IS THIS CORRECT? "
2060 PRINT : INPUT "TYPE 'YES' IF IT IS. "; YES$
```

```
2080  IF YES$ = "YES THEN PRINT D$; RUN MAILING LIST
      ADDER2"
2100  GOTO 40
4000  PRINT D$; "BLOAD CHAIN, A520"
4020  CALL 520"MAILING LIST CORRECTOR"
```

## PROGRAM CAPTURE

```
10   REM ***--PROGRAM CAPTURE--***
15   D$ = CHR$(4) : REM CONTROL D
16   PRINT D$; "OPEN PROGRAM CAPTURE"
17   PRINT D$; "WRITE PROGRAM CAPTURE"
18   POKE 33,33
19   LIST 1,2100: PRINT D$; "CLOSE PROGRAM CAPTURE": TEXT
     : END
20   D$ = CHR$(4) : REM CONTROL D
40   HOME: VTAB 5
60   HTAB 17: PRINT "PROGRAM MENU"
80   PRINT : PRINT
100  HTAB 8: PRINT "1. FILE CREATION PROGRAM"
120  PRINT
140  HTAB 8: PRINT "2. FILE ADDITION PROGRAM"
160  PRINT
180  HTAB 8: PRINT "3. FILE DISPLAY PROGRAM"
200  PRINT
220  HTAB 8: PRINT "4. FILE CORRECTION PROGRAM"
240  PRINT
300  HTAB 8: PRINT "5. CATAlOG"
320  PRINT
340  HTAB 8: PRINT "6. END"
360  PRINT : PRINT
380  HTAB 8: INPUT "WHICH PROGRAM NUMBER? "; NUMBER
400  IF NUMBER < 1 OR NUMBER > 6 THEN 380
420  IF NUMBER = 1 THEN 1000
440  IF NUMBER = 2 THEN 2000
460  IF NUMBER = 3 THEN PRINT D$; "RUN MAILING LIST
     READER"
470  IF NUMBER = 4 THEN PRINT D$; "RUN MAILING LIST
     CORRECTOR"
480  IF NUMBER = 5 THEN PRINT D$; "CATALOG": INPUT "HIT
     RETURN TO GO TO MENU"; L$: GOTO 40
500  IF NUMBER = 6 THEN END
```

```
1000 REM FILE CREATOR PROGRAM
1020 PRINT : PRINT "IF THE ADDRESS FILE ALREADY EXISTS"
1040 PRINT : PRINT "DO NOT RUN THIS PROGRAM! ! "
1060 PRINT : PRINT "DO YOU WANT THE FILE CREATION PROGRAM? "
1070 PRINT
1080 INPUT "TYPE 'YES' IF YOU DO: "; YES$
1100 IF YES$ = "YES" THEN PRINT D$; "RUN MAILING LIST
     CREATOR"
1120 GOTO 40
2000 REM FILE ADDITION PROGRAM
2020 PRINT : PRINT "YOU WANT TO ADD TO THE EXISTING"
2040 PRINT : PRINT "ADDRESS FILE. IS THIS CORRECT? "
2060 PRINT : INPUT "TYPE 'YES' IF IT IS. "; YES$
2080 IF YES$ = "YES" THEN PRINT D$; "RUN MAILING LIST
     ADDER2"
2100 GOTO 40
```

## INTEGER MAILING LIST READER

```
10 REM INTEGER MAILING LIST READER
20 D$ = "" : REM CONTROL D
40 DIM NAME$ (50) , A$ (50) , BLANK$ (10)
50 REM INPUT ROUTINE
60 Q = 1
80 PRINT D$; "OPEN ADDRESS FILE"
100 PRINT D$; "READ ADDRESS FILE"
120 INPUT K
140 INPUT BLANK$
160 DIM A (K*15)
180 FOR I = 1 TO K-1
200 INPUT NAME$
220 IF NAME$ = "" THEN 380
240 LN = LEN (NAME$)
260 FOR J = 1 TO LN
280 A (Q) = ASC (NAME$ (J) ) : REM CONVERT TO NUMBER
300 Q = Q+ 1
320 NEXT J
340 A (Q) = 161
360 Q = Q+ 1
380 NEXT I
400 PRINT D$; "CLOSE ADDRESS FILE"
500 REM DISPLAY ROUTINE
520 T = Q-1
```

```
540 FOR Q = 1 TO T
560 GOSUB 5000: REM CONVERT TO CHARACTER
580 IF A$ = "!" THEN PRINT
600 IF A$ = "!" THEN 660
620 IF A$ = "*" THEN 660
640 PRINT A$;
660 NEXT Q
1000 END
```

## CHARACTER ROUTINE1

```
5000 REM CHARACTER ROUTINE
5160 IF A(Q) = 160 THEN 6160
5161 IF A(Q) = 161 THEN 6161
5162 IF A(Q) = 162 THEN 6162
5163 IF A(Q) = 163 THEN 6163
5164 IF A(Q) = 164 THEN 6164
5165 IF A(Q) = 165 THEN 6165
5166 IF A(Q) = 166 THEN 6166
5167 IF A(Q) = 167 THEN 6167
5168 IF A(Q) = 168 THEN 6168
5169 IF A(Q) = 169 THEN 6169
5170 IF A(Q) = 170 THEN 6170
5171 IF A(Q) = 171 THEN 6171
5172 IF A(Q) = 172 THEN 6172
5173 IF A(Q) = 173 THEN 6173
5174 IF A(Q) = 174 THEN 6174
5175 IF A(Q) = 175 THEN 6175
5176 IF A(Q) = 176 THEN 6176
5177 IF A(Q) = 177 THEN 6177
5178 IF A(Q) = 178 THEN 6178
5179 IF A(Q) = 179 THEN 6179
5180 IF A(Q) = 180 THEN 6180
5181 IF A(Q) = 181 THEN 6181
5182 IF A(Q) = 182 THEN 6182
5183 IF A(Q) = 183 THEN 6183
5184 IF A(Q) = 184 THEN 6184
5185 IF A(Q) = 185 THEN 6185
5186 IF A(Q) = 186 THEN 6186
5187 IF A(Q) = 187 THEN 6187
5188 IF A(Q) = 188 THEN 6188
5189 IF A(Q) = 189 THEN 6189
5190 IF A(Q) = 190 THEN 6190
```

```
5191 IF A(Q) = 191 THEN 6191
5192 IF A(Q) = 192 THEN 6192
5193 IF A(Q) = 193 THEN 6193
5194 IF A(Q) = 194 THEN 6194
5195 IF A(Q) = 195 THEN 6195
5196 IF A(Q) = 196 THEN 6196
5197 IF A(Q) = 197 THEN 6197
5198 IF A(Q) = 198 THEN 6198
5199 IF A(Q) = 199 THEN 6199
5200 IF A(Q) = 200 THEN 6200
5201 IF A(Q) = 201 THEN 6201
5202 IF A(Q) = 202 THEN 6202
5203 IF A(Q) = 203 THEN 6203
5204 IF A(Q) = 204 THEN 6204
5205 IF A(Q) = 205 THEN 6205
5206 IF A(Q) = 206 THEN 6206
5207 IF A(Q) = 207 THEN 6207
5208 IF A(Q) = 208 THEN 6208
5209 IF A(Q) = 209 THEN 6209
5210 IF A(Q) = 210 THEN 6210
5211 IF A(Q) = 211 THEN 6211
5212 IF A(Q) = 212 THEN 6212
5213 IF A(Q) = 213 THEN 6213
5214 IF A(Q) = 214 THEN 6214
5215 IF A(Q) = 215 THEN 6215
5216 IF A(Q) = 216 THEN 6216
5217 IF A(Q) = 217 THEN 6217
5218 IF A(Q) = 218 THEN 6218
6160 A$ = " " : RETURN
6161 A$ = "!" : RETURN
6162 A$ = "'" : RETURN
6163 A$ = "#" : RETURN
6164 A$ = "$" : RETURN
6165 A$ = "%" : RETURN
6166 A$ = "&" : RETURN
6167 A$ = "'" : RETURN
6168 A$ = "(" : RETURN
6169 A$ = ")" : RETURN
6170 A$ = "*" : RETURN
6171 A$ = "+" : RETURN
6172 A$ = "," : RETURN
6173 A$ = "-" : RETURN
6174 A$ = "." : RETURN
6175 A$ = "/" : RETURN
```

## CHARACTER ROUTINE1

```
6176 A$ = "0" : RETURN
6177 A$ = "1" : RETURN
6178 A$ = "2" : RETURN
6179 A$ = "3" : RETURN
6180 A$ = "4" : RETURN
6181 A$ = "5" : RETURN
6182 A$ = "6" : RETURN
6183 A$ = "7" : RETURN
6184 A$ = "8" : RETURN
6185 A$ = "9" : RETURN
6186 A$ = ":" : RETURN
6187 A$ = ";" : RETURN
6188 A$ = "<" : RETURN
6189 A$ = "=" : RETURN
6190 A$ = ">" : RETURN
6191 A$ = "?" : RETURN
6192 A$ = "@" : RETURN
6193 A$ = "A" : RETURN
6194 A$ = "B" : RETURN
6195 A$ = "C" : RETURN
6196 A$ = "D" : RETURN
6197 A$ = "E" : RETURN
6198 A$ = "F" : RETURN
6199 A$ = "G" : RETURN
6200 A$ = "H" : RETURN
6201 A$ = "I" : RETURN
6202 A$ = "J" : RETURN
6203 A$ = "K" : RETURN
6204 A$ = "L" : RETURN
6205 A$ = "M" : RETURN
6206 A$ = "N" : RETURN
6207 A$ = "O" : RETURN
6208 A$ = "P" : RETURN
6209 A$ = "Q" : RETURN
6210 A$ = "R" : RETURN
6211 A$ = "S" : RETURN
6212 A$ = "T" : RETURN
6213 A$ = "U" : RETURN
6214 A$ = "V" : RETURN
6215 A$ = "W" : RETURN
6216 A$ = "X" : RETURN
6217 A$ = "Y" : RETURN
6218 A$ = "Z" : RETURN
```

## CHARACTER ROUTINE2

```
5000 REM CHARACTER ROUTINE
5160 IF A(Q) = 160 THEN A$ = " "
5161 IF A(Q) = 161 THEN A$ = "!"
5162 IF A(Q) = 162 THEN A$ = "'"
5163 IF A(Q) = 163 THEN A$ = "#"
5164 IF A(Q) = 164 THEN A$ = "$"
5165 IF A(Q) = 165 THEN A$ = "%"
5166 IF A(Q) = 166 THEN A$ = "&"
5167 IF A(Q) = 167 THEN A$ = "'"
5168 IF A(Q) = 168 THEN A$ = "("
5169 IF A(Q) = 169 THEN A$ = ")"
5170 IF A(Q) = 170 THEN A$ = "*"
5171 IF A(Q) = 171 THEN A$ = "+"
5172 IF A(Q) = 172 THEN A$ = ","
5173 IF A(Q) = 173 THEN A$ = "-"
5174 IF A(Q) = 174 THEN A$ = "."
5175 IF A(Q) = 175 THEN A$ = "/"
5176 IF A(Q) = 176 THEN A$ = "0"
5177 IF A(Q) = 177 THEN A$ = "1"
5178 IF A(Q) = 178 THEN A$ = "2"
5179 IF A(Q) = 179 THEN A$ = "3"
5180 IF A(Q) = 180 THEN A$ = "4"
5181 IF A(Q) = 181 THEN A$ = "5"
5182 IF A(Q) = 182 THEN A$ = "6"
5183 IF A(Q) = 183 THEN A$ = "7"
5184 IF A(Q) = 184 THEN A$ = "8"
5185 IF A(Q) = 185 THEN A$ = "9"
5186 IF A(Q) = 186 THEN A$ = ":"
5187 IF A(Q) = 187 THEN A$ = ";"
5188 IF A(Q) = 188 THEN A$ = "<"
5189 IF A(Q) = 189 THEN A$ = "="
5190 IF A(Q) = 190 THEN A$ = ">"
5191 IF A(Q) = 191 THEN A$ = "?"
5192 IF A(Q) = 192 THEN A$ = "@"
5193 IF A(Q) = 193 THEN A$ = "A"
5194 IF A(Q) = 194 THEN A$ = "B"
5195 IF A(Q) = 195 THEN A$ = "C"
5196 IF A(Q) = 196 THEN A$ = "D"
5197 IF A(Q) = 197 THEN A$ = "E"
```

```
5198 IF A(Q) = 198 THEN A$ = "F"
5199 IF A(Q) = 199 THEN A$ = "G"
5200 IF A(Q) = 200 THEN A$ = "H"
5201 IF A(Q) = 201 THEN A$ = "I"
5202 IF A(Q) = 202 THEN A$ = "J"
5203 IF A(Q) = 203 THEN A$ = "K"
5204 IF A(Q) = 204 THEN A$ = "L"
5205 IF A(Q) = 205 THEN A$ = "M"
5206 IF A(Q) = 206 THEN A$ = "N"
5207 IF A(Q) = 207 THEN A$ = "O"
5208 IF A(Q) = 208 THEN A$ = "P"
5209 IF A(Q) = 209 THEN A$ = "Q"
5210 IF A(Q) = 210 THEN A$ = "R"
5211 IF A(Q) = 211 THEN A$ = "S"
5212 IF A(Q) = 212 THEN A$ = "T"
5213 IF A(Q) = 213 THEN A$ = "U"
5214 IF A(Q) = 214 THEN A$ = "V"
5215 IF A(Q) = 215 THEN A$ = "W"
5216 IF A(Q) = 216 THEN A$ = "X"
5217 IF A(Q) = 217 THEN A$ = "Y"
5218 IF A(Q) = 218 THEN A$ = "Z"
6000 RETURN
```

# 9
# Additional Sequential File Techniques

We are going to explore some other ways to work with sequential files and look at additional techniques for file handling. In this chapter, I am going to concentrate on the file routines of the various programs presented and not discuss the rest of the programming. The listings for the complete programs are included at the end of this chapter along with an explanation of the new commands used in these programs.

We will begin with a series of programs that allows an individual to practice math and keep a record of the scores achieved. These programs are essentially drill and practice and may not be the best educational use of the computer, but for the purpose of demonstrating how files can be used in a variety of ways, these drill and practice programs will be sufficient.

We again start with careful thought and preparation. We need a separate program for each mathematical operation, along with a program for the scores. This means that another program menu would be convenient. The essential difference between the operation programs is the sign of the operation—" + " for addition, "x" for multiplication, and so forth. With the exception of division, the numbers can be displayed in basically the same way. Therefore, the program presented for addition can also be used for subtraction and multiplication with changes made to only five lines: 10, 140, 220, 500, and 520. In all of those lines, the references to addition should be changed to the desired operation.

LOAD ADD

LIST 10

Line 10 is:

10 REM ***--ADDITION--***

# CHAPTER 9  ADDITIONAL SEQUENTIAL FILE TECHNIQUES

Line 10 should be:

```
10 REM ***--SUBTRACTION--***
```

or

```
10 REM ***--MULTIPLICATION--***
```

Lines 500 and 520 are the most important to change.

```
500 C = B + A
520 S$ = "+"
```

Change to:

```
500 C = B - A
520 S$ = "-"
```

or

```
500 C = B * A
520 S$ = "X"
```

Once all the changes have been made, SAVE the new program:

SAVE SUBTRACT

or

SAVE MULTIPLY

The program for division has additional code because the numbers must be formatted differently, and provisions have been made so that all problems come out even. All these programs can be included in one large program, but the flow of logic in the program would not be as easy to follow as it is with separate programs. Little would be gained by forcing everything into one program since DOS allows us to switch from one program to another.

We must carefully consider what we want to save in our scores file. There are several pieces of information that might be important to save, but a good rule is to save only what is absolutely necessary—what it would be hard or impossible to calculate from existing information. For example, we could save the total number of problems, the number correct, the number wrong, the percentage, the name of the individual, the kind of mathematical operation, the number of digits chosen, and so forth. If the programs were slightly altered, we could also save the actual problems missed, the number of tries on a particular problem, and the last question the person tried. Obviously, all this information is not necessary, although certain individuals might value and save information others would not want.

The first step is to decide what information to save. In this example, we will save four things: the type of operation, the number of digits in the operation, the number of correct answers, and the number of wrong answers. Once we decide what to save, we need only save the assigned variables for these pieces of information. The code to do this is given below.

```
2000 REM **--FILE ROUTINE--**
2020 D$ = CHR$(4) : REM CONTROL D
2040 ONERR GOTO 2180
2060 PRINT D$; "APPEND"; NAME$
2080 PRINT D$; "WRITE"; NAME$
2085 PRINT S$: REM SIGN OF OPERATION
2090 PRINT DT: REM # OF DIGITS
2100 PRINT CR: REM # CORRECT
2120 PRINT WR: REM # WRONG
2140 PRINT D$; "CLOSE"; NAME$
2160 PRINT D$; "RUN MATH MENU"
2180 POKE 216, 0
2200 PRINT D$; "OPEN"; NAME$
2220 GOTO 2080
```

Some of this should be familiar, but the sequence and a few commands may appear different. The statement in 2040, ONERR GOTO, can be a very useful feature in file manipulation. This statement allows us to combine easily in one program what it normally takes two programs to do.

Remember that in our MAILING LIST SYSTEM programs, we used one program to create the ADDRESS FILE and another program to add to it. Such a sequence is usually necessary when creating a file that will later be added to. The use of the OPEN command places the file pointer at the very beginning of the file. Then any information that you write to the file overwrites information already in the file. That is the reason for the APPEND command. But you cannot create a file with the APPEND command. However, with the above routine, the need for two such programs is eliminated.

The ONERR GOTO statement tells the computer that when it tries to APPEND information to a non-existent file, instead of halting operation and displaying an error message, control is to be transferred to the instructions beginning with line 2180. The instruction at 2180 is known as resetting the error flag. This simply means that everything is put back the way it was before the error occurred. I have found that this is usually the best idea, because some undesirable results can occur if the error flag is not immediately reset after a disk error has occurred. POKE 216,0 is not exactly descriptive or memorable, but it is necessary and is one of those "numbers" important to remember. This combination of ONERR GOTO...POKE 216,0 will be used in other file routines, and you may be surprised at how useful this combination can be. Following 2180, 2200

CHAPTER 9   ADDITIONAL SEQUENTIAL FILE TECHNIQUES   117

instructs the computer to PRINT D$;"OPEN";NAME$. So now we have the following sequence: The first time the program is run, the computer will attempt to APPEND information into a file that does not yet exist. When that error occurs, control will be transferred to the instructions that (1) reset the error flag, then (2) OPEN the file, and finally, (3) send the computer back up to the instructions that come after the APPEND command. Thus the file is created, and the first set of information written into the file. The second time (and succeeding times) the program is run, the computer APPENDs information into the file because it finds that such a file does exist. Therefore, no error occurs, and the instructions at 2180, 2200, and 2220 are not encountered. We have accomplished in one routine what would normally have taken two routines to do.

You should also notice that we are using a variable for the file name. In our MAILING LIST SYSTEM programs, we always used the constant "ADDRESS FILE". But in this situation, and most file routines, it is more convenient to assign a variable as the file name. Anytime an individual uses any of these programs, the information is kept in a file under that person's name. By using a variable for the file name, we eliminate the need for separate programs for each person who uses the ADDITION program or any of the other math operation programs.

You must be careful to type your name the same way every time you use the programs. For example, if I answer that my name is DAVID the first time I use these programs, the file will be created under the name of DAVID. If I come back later and answer that my name is DAVE, a new and separate file will be created for DAVE. As with most things, there are advantages to the use of a variable for the file name, but there are also disadvantages. The user may get tired of being required to type his/her name, but the use of a variable for the file name remains a popular programming technique. The variable must be a string variable since no file name can begin with a number. (The file name can contain a number but just cannot begin with a number.)

The file routine used in the program SCORES is very similar to the one just discussed, but instead of writing information to the disk, this routine reads information from the disk.

```
20 D$ = CHR$(4) : REM CONTROL D
40 ONERR GOTO 380
60 DIM S$(100), DT(100)
80 DIM CR(100), WR(100)
100 I = 1
120 HOME: VTAB 5
140 INPUT "STUDENT'S NAME PLEASE? ";NAME$
160 :
180 :
```

```
200 REM **--FILE ROUTINE--**
220 PRINT D$;"OPEN";NAME$
240 PRIND D$;"READ";NAME$
260 INPUT S$(I)
280 INPUT DT(I)
300 INPUT CR(I)
320 INPUT WR(I)
340 I = I + 1
360 GOTO 260
380 POKE 216,0
400 PRINT D$;"CLOSE";NAME$
```

Lines 20 to 180 are necessary to set up the file input routine in lines 200 to 400. This time, the ONERR GOTO is used to test for the end of the file or the end of the data. If the ONERR GOTO statement is not included, we have no way of telling how much information or how many records exist in the file. We did not keep track of that information by writing out a counter to the file like we did in the MAILING LIST SYSTEM. Without the ONERR GOTO statement, we would get an END OF DATA error message and the program would halt. With that ONERR GOTO statement, the computer is instructed to go to the instruction at line 380 and proceed from there. First, we reset the error flag and then close the file, since we are now certain that we have all the information the file contains. Once again, the ONERR GOTO statement has come in very handy. This time it saves both programming and disk space.

You should notice one other major difference in this routine. In most of our programs, we have used FOR...NEXT loops. But this time, we do not know how many items the file contains and therefore, do not know how large the counter eventually needs to become. It is true that we could pick an arbitrary number, but a better method is the one used in this routine. This method is still a loop since the computer is instructed to follow the instructions down to line 360 and then go back to the instruction at line 260 and do everything over again. What gets us out of this loop? The ONERR GOTO statement does when it executes as the end of the file is encountered. When this loop is finished, we should have the values we want from the file and be able to proceed to the display routine.

These math programs provide additional file handling techniques, as well as a set of useful drill and practice programs. The menu program uses the same method we have been using to display a set of choices and then run the appropriate program. Apart from file handling, the math programs also have some programming techniques that might prove interesting.

To conclude this chapter, I have added two other programs that make use of file handling and fit our purpose of demonstrating filing techniques. Both programs are presented in a rough form. Individuals may wish to add

CHAPTER 9  ADDITIONAL SEQUENTIAL FILE TECHNIQUES         119

parts to these programs or modify the format. A random access recipe file would probably provide a better method for this particular application. To make this recipe program useful, recipes should probably be grouped; i.e., desserts, appetizers, main dish, etc. with a separate file for each grouping (change recipe name to appetizer name, dessert name, and so forth).

In the final chapter on sequential access files, we will take a brief look at the possibility of a standard method for storing data so that the data can be used by a variety of commercial programs.

---

### NEW COMMANDS OR TERMS
### IN THE FOLLOWING PROGRAMS

1. ^: Raise to the power of the number following this symbol.
2. RND: Generate a random number.
3. INT: Take only the integer portion of the number in parantheses.
4. LEN: Find the length of the string in terms of the number of characters in the string.
5. STR$: Convert the specified number into a string value.
6. CHR$(95): Display the underline character.
7. DEF FN: Define a function. This useful command allows a programmer to set a single variable equal to a complete equation.
8. SGN: Produce the sine of the specified value.
9. VAL: Give the numeric value of a string.

## QUESTIONS

1. A good rule to follow in deciding what information to save is to save: (A) everything possible, (B) as little as possible, (C) only what is absolutely necessary.
2. What Applesoft command allows programmers to check for an error and then proceed to a specific set of instructions?
3. TRUE or FALSE: APPEND can open a file.
4. TRUE or FALSE: It is never possible to use a variable as a file name.
5. Which BASIC statement retrieves only the integer portion of a number?
6. Which BASIC statement converts a number into a string?
7. Which BASIC statement converts a string into a number?

# ANSWERS

1. C
2. ONERR..GOTO
3. FALSE
4. FALSE
5. INT
6. STR$
7. VAL

## MATH MENU

```
10  REM ***--MATH MENU--***
20  D$ = CHR$(4) : REM CONTROL D
40  HOME : VTAB 2
60  HTAB 14: PRINT "MATH MENU"
80  PRINT : PRINT
100 HTAB 8: PRINT "1. ADDITION"
120 PRINT
140 HTAB 8: PRINT "2. SUBTRACTION"
160 PRINT
180 HTAB 8: PRINT "3. MULTIPLICATION"
200 PRINT
220 HTAB 8: PRINT "4. DIVISION"
240 PRINT
300 HTAB 8: PRINT "5. SCORES"
320 PRINT
340 HTAB 8: PRINT "6. INFORMATION"
345 PRINT
350 HTAB 8: PRINT "7. END"
360 PRINT : PRINT
380 HTAB 8: INPUT "WHICH PROGRAM NUMBER? ";NUMBER
400 IF NUMBER 1 < OR NUMBER > 7 THEN 380
420 IF NUMBER = 1 THEN PRINT D$;"RUN ADD"
440 IF NUMBER = 2 THEN PRINT D$;"RUN SUBTRACT"
460 IF NUMBER = 3 THEN PRINT D$;"RUN MULTIPLY"
470 IF NUMBER = 4 THEN PRINT D$;"RUN DIVIDE"
480 IF NUMBER = 5 THEN PRINT D$;"RUN SCORES"
500 IF NUMBER = 6 THEN 1000
520 IF NUMBER = 7 THEN END
1000 REM **--INFORMATION--**
1020 HOME
1040 PRINT "THIS IS A SERIES OF MATH DRILL AND"
1050 PRINT
1060 PRINT "PRACTICE PROGRAMS. IT IS DESIGNED TO"
1070 PRINT
1080 PRINT "ALLOW FOR AS MUCH FLEXIBILITY AS"
1090 PRINT
1100 PRINT "POSSIBLE. THE QUESTION ABOUT THE "
1110 PRINT
1120 PRINT "NUMBER OF DIGITS MIGHT, AT FIRST, "
1130 PRINT
1140 PRINT "SEEM CONFUSING. THE QUESTION SIMPLY"
```

```
1150 PRINT
1160 PRINT "ASKS FOR THE GREATEST NUMBER OF "
1170 PRINT
1180 PRINT "DIGITS POSSIBLE IN EITHER FIGURE. "
1190 PRINT
1200 PRINT "THE NEXT TWO QUESTIONS FURTHER ALLOW"
1210 PRINT
1220 PRINT "YOU TO LIMIT THE POSSIBLE PROBLEMS. "
1230 GOSUB 5000
1240 PRINT "FOR EXAMPLE, IF YOU WANTED TO PRACTICE"
1250 PRINT
1260 PRINT "MULTIPLYING BY '5', YOU COULD CHOOSE"
1270 PRINT
1280 PRINT "THREE DIGIT NUMBERS AND THEN ANSWER"
1290 PRINT
1300 PRINT "WITH A '5' FOR EACH OF THE NEXT TWO"
1310 PRINT
1320 PRINT "QUESTIONS. YOU WOULD THEN BE GIVEN"
1330 PRINT
1340 PRINT "PROBLEMS LIKE: 345 X 5 OR 823 X 5. "
1350 GOSUB 5000
1360 PRINT "ANOTHER EXAMPLE WOULD BE TO ADD TWO"
1370 PRINT
1380 PRINT "DIGIT NUMBERS BY ANSWERING THE"
1390 PRINT
1400 PRINT "QUESTIONS IN THIS WAY: "
1410 PRINT
1420 PRINT "HOW MANY DIGITS- -2"
1430 PRINT
1440 PRINT "LARGEST NUMBER- -99"
1450 PRINT
1460 PRINT "SMALLEST NUMBER- -1"
1470 PRINT
1480 PRINT "YOU COULD THEN GET PROBLEMS LIKE: "
1490 PRINT
1500 PRINT "58 + 34 OR 87 + 9. "
1510 GOSUB 5000
1520 PRINT "TRYING THE DIFFERENT POSSIBILITIES"
1530 PRINT
1540 PRINT "WILL SOON INDICATE THE FLEXIBILITY. "
1550 PRINT
1560 PRINT "THE DIVISION SECTION WILL ONLY GIVE"
1570 PRINT
1580 PRINT "PROBLEMS THAT COME OUT EVEN. "
```

```
1590 PRINT
1600 PRINT "YOU MAY HAVE TO WAIT A SHORT TIME. "
1610 PRINT
1620 PRINT "FOR THE NEXT PROBLEM. THIS"
1630 PRINT
1640 PRINT "IS BECAUSE THE NUMBERS GENERATED"
1650 PRINT
1660 PRINT "MUST MEET CERTAIN SPECIFICATIONS. "
1670 GOSUB 5000
1680 PRINT "THIS IS NOT A PROFESSIONAL PROGRAM"
1690 PRINT
1700 PRINT "AND THEREFORE DOES NOT DO A LOT OF"
1710 PRINT
1720 PRINT "ERROR CHECKING. YOU CAN CRASH THE "
1730 PRINT
1740 PRINT "PROGRAMS WITH CONFUSING ANSWERS"
1750 PRINT
1760 PRINT "OR MISTAKES IN TYPING. TYPING A "
1770 PRINT
1780 PRINT "'CTRL' 'C' WILL END ANY PROGRAMS. "
1790 PRINT
1800 PRINT "YOU MUST THEN START OVER. "
1810 PRINT
1820 PRINT "THIS SERIES OF PROGRAMS WAS DONE"
1830 PRINT
1840 PRINT "MAINLY TO DEMONSTRATE, IN A USEFUL"
1850 PRINT
1860 PRINT "MANNER, CERTAIN FILE HANDLING"
1870 PRINT
1880 PRINT "CAPABILITIES. "
2000 GOSUB 5000
2020 GOTO 40
5000 PRINT
5020 INPUT "HIT RETURN TO CONTINUE ";L$
5040 HOME
5060 RETURN
```

## ADD

```
10 REM ***--ADDITION--***
11 :
12 :
20 REM **--VARIABLE LIST--**
21 REM A  = BOTTOM NUMBER
22 REM B  = TOP NUMBER
23 REM C  = CORRECT ANSWER
24 REM D  = STUDENT'S ANSWER
25 REM Q  = COUNTER
26 REM W  = PREVIOUS ANSWER
27 REM Z  = NUMBER OF TRIES
28 REM CR = CORRECT ANSWERS
29 REM WR = WRONG ANSWERS
30 REM DT = # OF DIGITS
31 REM LA = # OF DIGITS IN A
32 REM LB = # OF DIGITS IN B
33 REM LN = # OF DIGITS IN C
34 REM OTHER VARIABLES ARE
35 REM DESCRIPTIVE
36 :
37 :
40 HOME : VTAB 5
60 INPUT "HOW MANY DIGITS ";DIGIT
80 PRINT
100 PRINT "WHAT IS THE LARGEST FIGURE FOR THE"
120 PRINT
140 INPUT "NUMBER YOU ARE ADDING BY? ";MAX
160 PRINT
180 PRINT "WHAT IS THE SMALLEST FIGURE FOR THE"
200 PRINT
220 INPUT "NUMBER YOU ARE ADDING BY? ";MN
240 DT = DIGIT: DIGIT = 10 ^ DIGIT
260 PRINT
280 INPUT "WHAT IS YOUR NAME? ";NAME$
290 :
295 :
300 REM **--CREATE PROBLEM--**
320 HOME
340 HTAB 10: VTAB 2
360 PRINT "TYPE 'END' WHEN FINISHED"
380 VTAB 10
```

126  CHAPTER 9  ADDITIONAL SEQUENTIAL FILE TECHNIQUES

```
382 MAX$ = STR$ (MAX)
384 LM = LEN (MAX$)
386  IF DT = LM + 1 OR DT < LM + 1 THEN 400
388 LM = 10 ^ LM
390 A = INT ( RND (1) * LM)
392  IF A < MN THEN 390
394  IF A > MAX THEN 390
396  GOTO 480
400 Z = 1
420 A = INT ( RND (1) * DIGIT)
440  IF A < MN THEN 420
460  IF A > MAX THEN 420
480 B = INT ( RND (1) * DIGIT)
500 C = B + A
520 S$ = "+"
540  IF C < 0 THEN 480
560  IF C = W THEN 480
580 W = C
600 A$ = STR$ (A)
620 LA = LEN (A$)
640 B$ = STR$ (B)
660 LB = LEN (B$)
680  HTAB 22 - LB: PRINT B
700  HTAB 22 - (LA + 1) : PRINT S$; A
720 C$ = STR$ (C)
740 LN = LEN (C$)
760 Q = 1
780  IF LA < LB THEN Q = 0
800  HTAB 22 - (DT + Q) : FOR I = 1 TO DT + Q: PRINT
     CHR$ (95) ; : NEXT I
810 :
815 :
820 REM **- -GET ANSWER- -**
840  PRINT : HTAB 22 - (LN + 1) : INPUT ANSWER$
860  IF ANSWER$ = "END" THEN 1060
880 D = VAL (ANSWER$)
900  IF D = C THEN PRINT : PRINT : PRINT : HTAB 19: PRINT
     "GOOD": FOR I = 1 TO 1000: NEXT I : CR = CR + 1: GOTO 320
920  IF Z < 3 THEN PRINT : HTAB 10: PRINT "NO, PLEASE TRY
     AGAIN. " : Z = Z + 1: PRINT : WR = WR + 1: GOTO 660
940  PRINT
960  PRINT "NO, THE ANSWER IS "; C
980  PRINT : PRINT B; " "; S$; " "; A; " = "; C
1000  PRINT : Z = 1: WR = WR + 1
```

```
1020  INPUT "HIT RETURN WHEN READY TO GO ON "; L$
1040  GOTO 320
1050 :
1055 :
1060  REM **--TOTAL ROUTINE--**
1080  HOME : VTAB 5
1100  PRINT "YOU GOT " ; CR; " RIGHT!"
1120  PRINT
1140  PRINT "YOU MISSED "; WR
1160 :
1180 :
2000  REM **--FILE ROUTINE--**
2020  D$ = CHR$(4)
2040  ONERR GOTO 2180
2060  PRINT D$; "APPEND"; NAME$
2080  PRINT D$; "WRITE"; NAME$
2085  PRINT S$
2090  PRINT DT
2100  PRINT CR
2120  PRINT WR
2140  PRINT D$; "CLOSE"; NAME$
2160  PRINT D$; "RUN MATH MENU"
2180  POKE 216, 0
2200  PRINT D$; "OPEN"; NAME$
2220  GOTO 2080
```

## SUBTRACT

```
10  REM ***--SUBTRACTION--***
11 :
12 :
20  REM **--VARIABLE LIST--**
21  REM A  = BOTTOM NUMBER
22  REM B  = TOP NUMBER
23  REM C  = CORRECT ANSWER
24  REM D  = STUDENT'S ANSWER
25  REM Q  = COUNTER
26  REM W  = PREVIOUS ANSWER
27  REM Z  = NUMBER OF TRIES
28  REM CR = CORRECT ANSWERS
29  REM WR = WRONG ANSWERS
30  REM DT = # OF DIGITS
```

```
 31  REM LA = # OF DIGITS IN A
 32  REM LB = # OF DIGITS IN B
 33  REM LN = # OF DIGITS IN C
 34  REM OTHER VARIABLES ARE
 35  REM DESCRIPTIVE
 36 :
 37 :
 40  HOME : VTAB 5
 60  INPUT "HOW MANY DIGITS ";DIGIT
 80  PRINT
100  PRINT "WHAT IS THE LARGEST FIGURE FOR THE"
120  PRINT
140  INPUT "NUMBER YOU ARE SUBTRACTING BY? ";MAX
160  PRINT
180  PRINT "WHAT IS THE SMALLEST FIGURE FOR THE"
200  PRINT
220  INPUT "NUMBER YOU ARE SUBTRACTING BY? ";MN
240  DT = DIGIT:DIGIT = 10^DIGIT
260  PRINT
280  INPUT "WHAT IS YOUR NAME? ";NAME$
290 :
295 :
300  REM **--CREATE PROBLEM--**
320  HOME
340  HTAB 10:VTAB 2
360  PRINT "TYPE 'END' WHEN FINISHED"
380  VTAB 10
382  MAX$ = STR$(MAX)
384  LM = LEN (MAX$)
386  IF DT = LM + 1 OR DT < LM + 1 THEN 400
388  LM = 10^LM
390  A = INT ( RND (1) * LM)
392  IF A < MN THEN 390
394  IF A > MAX THEN 390
396  GOTO 480
400  Z = 1
420  A = INT ( RND (1) * DIGIT)
440  IF A < MN THEN 420
460  IF A > MAX THEN 420
480  B = INT ( RND (1) * DIGIT)
500  C = B - A
520  S$ = "-"
540  IF C < 0 THEN 480
560  IF C = W THEN 480
```

```
580 W = C
600 A$ = STR$ (A)
620 LA = LEN (A$)
640 B$ = STR$ (B)
660 LB = LEN (B$)
680  HTAB 22 - LB: PRINT B
700  HTAB 22 - (LA + 1) : PRINT S$; A
720 C$ = STR$ (C)
740 LN = LEN (C$)
760 Q = 1
780  IF LA < LB THEN Q = 0
800  HTAB 22 - (DT + Q) : FOR I = 1 TO DT + Q: PRINT
     CHR$ (95) ; : NEXT I
810 :
815 :
820  REM **--GET ANSWER--**
840  PRINT : HTAB 22 - (LN + 1) : INPUT ANSWER$
860  IF ANSWER$ = "END" THEN 1060
880 D = VAL (ANSWER$)
900  IF D = C THEN PRINT : PRINT : PRINT : HTAB 19: PRINT
     "GOOD ": FOR I = 1 TO 1000: NEXT I: CR = CR + 1: GOTO 320
920  IF Z < 3 THEN PRINT : HTAB 10: PRINT "NO, PLEASE TRY
     AGAIN. " : Z = Z + 1 : PRINT : WR = WR + 1 : GOTO 660
940  PRINT
960  PRINT "NO, THE ANSWER IS "; C
980  PRINT : PRINT B; " "; S$; " "; A; " = "; C
1000  PRINT : Z = 1: WR = WR + 1
1020  INPUT "HIT RETURN WHEN READY TO GO ON "; L$
1040  GOTO 320
1050 :
1055 :
1060  REM **--TOTAL ROUTINE--**
1080  HOME : VTAB 5
1100  PRINT "YOU GOT "; CR; " RIGHT! "
1120  PRINT
1140  PRINT "YOU MISSED "; WR
1160 :
1180 :
2000  REM **--FILE ROUTINE--**
2020 D$ = CHR$ (4)
2040  ONERR GOTO 2180
2060  PRINT D$; "APPEND"; NAME$
2080  PRINT D$; "WRITE"; NAME$
2085  PRINT S$
```

130  CHAPTER 9  ADDITIONAL SEQUENTIAL FILE TECHNIQUES

```
2090 PRINT DT
2100 PRINT CR
2120 PRINT WR
2140 PRINT D$;"CLOSE";NAME$
2160 PRINT D$;"RUN MATH MENU"
2180 POKE 216,0
2200 PRINT D$;"OPEN";NAME$
2220 GOTO 2080
```

## MULITIPLY

```
10 REM ***--MULTIPLICATION--***
11 :
12 :
20 REM **--VARIABLE LIST--**
21 REM A  = BOTTOM NUMBER
22 REM B  = TOP NUMBER
23 REM C  = CORRECT ANSWER
24 REM D  = STUDENT'S ANSWER
25 REM Q  = COUNTER
26 REM W  = PREVIOUS ANSWER
27 REM Z  = NUMBER OF TRIES
28 REM CR = CORRECT ANSWERS
29 REM WR = WRONG ANSWERS
30 REM DT = # OF DIGITS
31 REM LA = # OF DIGITS IN A
32 REM LB = # OF DIGITS IN B
33 REM LN = # OF DIGITS IN C
34 REM OTHER VARIABLES ARE
35 REM DESCRIPTIVE
36 :
37 :
40 HOME : VTAB 5
60 INPUT "HOW MANY DIGITS ";DIGIT
80 PRINT
100 PRINT "WHAT IS THE LARGEST FIGURE FOR THE"
120 PRINT
140 INPUT "NUMBER YOU ARE MULTIPLYING BY? ";MAX
160 PRINT
180 PRINT "WHAT IS THE SMALLEST FIGURE FOR THE"
200 PRINT
220 INPUT "NUMBER YOU ARE MULTIPLYING BY? ";MN
```

# MULITIPLY

```
240 DT = DIGIT: DIGIT = 10 ^ DIGIT
260 PRINT
280 INPUT "WHAT IS YOUR NAME? "; NAME$
290 :
295 :
300 REM **--CREATE PROBLEM--**
320 HOME
340 HTAB 10: VTAB 2
360 PRINT "TYPE 'END' WHEN FINISHED"
380 VTAB 10
382 MAX$ = STR$ (MAX)
384 LM = LEN (MAX$)
386 IF DT = LM + 1 OR DT < LM + 1 THEN 400
388 LM = 10 ^ LM
390 A = INT ( RND (1) * LM)
392 IF A < MN THEN 390
394 IF A > MAX THEN 390
396 GOTO 480
400 Z = 1
420 A = INT ( RND (1) * DIGIT)
440 IF A < MN THEN 420
460 IF A > MAX THEN 420
480 B = INT ( RND (1) * DIGIT)
500 C = B * A
520 S$ = "X"
540 IF C < 0 THEN 480
560 IF C = W THEN 480
580 W = C
600 A$ = STR$ (A)
620 LA = LEN (A$)
640 B$ = STR$ (B)
660 LB = LEN (B$)
680 HTAB 22 - LB: PRINT B
700 HTAB 22 - (LA + 1) : PRINT S$; A
720 C$ = STR$ (C)
740 LN = LEN (C$)
760 Q = 1
780 IF LA < LB THEN Q = 0
800 HTAB 22 - (DT + Q) : FOR I = 1 TO DT + Q: PRINT
    CHR$ (95) ; : NEXT I
810 :
815 :
820 REM **--GET ANSWER--**
840 PRINT : HTAB 22 - (LN + 1) : INPUT ANSWER$
```

```
860  IF ANSWER$ = "END" THEN 1060
880  D = VAL (ANSWER$)
900  IF D = C THEN PRINT : PRINT : PRINT : HTAB 19: PRINT
     "GOOD" : FOR I = 1 TO 1000: NEXT I : CR = CR + 1: GOTO 32(
920  IF Z < 3 THEN PRINT : HTAB 10: PRINT "NO, PLEASE TRY
     AGAIN. ":Z = Z + 1:PRINT :WR = WR + 1: GOTO 660
940  PRINT
960  PRINT "NO, THE ANSWER IS ";C
980  PRINT : PRINT B; " ";S$;" ";A; " = ";C
1000 PRINT : Z = 1: WR = WR + 1
1020 INPUT "HIT RETURN WHEN READY TO GO ON ";L$
1040 GOTO 320
1050 :
1055 :
1060 REM **--TOTAL ROUTINE--**
1080 HOME : VTAB 5
1100 PRINT "YOU GOT ";CR;" RIGHT! "
1120 PRINT
1140 PRINT "YOU MISSED ";WR
1160 :
1180 :
2000 REM **--FILE ROUTINE--**
2020 D$ = CHR$ (4)
2040 ONERR GOTO 2180
2060 PRINT D$;"APPEND";NAME$
2080 PRINT D$;"WRITE";NAME$
2085 PRINT S$
2090 PRINT DT
2100 PRINT CR
2120 PRINT WR
2140 PRINT D$;"CLOSE";NAME$
2160 PRINT D$;"RUN MATH MENU"
2180 POKE 216,0
2200 PRINT D$;"OPEN";NAME$
2220 GOTO 2080
```

# DIVIDE

```
10 REM ***--DIVISION--***
11 :
12 :
20 REM **--VARIABLE LIST--**
21 REM A  = DIVISOR
22 REM B  = DIVIDEND
23 REM C  = CORRECT ANSWER
24 REM D  = STUDENT'S ANSWER
25 REM Q  = COUNTER
26 REM W  = PREVIOUS ANSWER
27 REM Z  = NUMBER OF TRIES
28 REM CR = CORRECT ANSWERS
29 REM WR = WRONG ANSWERS
30 REM DT = # OF DIGITS
31 REM LA = # OF DIGITS IN A
32 REM LB = # OF DIGITS IN B
33 REM LN = # OF DIGITS IN C
34 REM OTHER VARIABLES ARE
35 REM DESCRIPTIVE
36 :
37 :
40  HOME : VTAB 5
60  INPUT "HOW MANY DIGITS ";DIGIT
80  PRINT
100 PRINT "WHAT IS THE LARGEST FIGURE FOR THE NO. "
120 PRINT
140 INPUT "YOU ARE DIVIDING BY (DIVISOR)? ";MAX
160 PRINT
180 PRINT "WHAT IS THE SMALLEST FIGURE FOR THE NO. "
200 PRINT
220 INPUT "YOU ARE DIVIDING BY (DIVISOR)? ";MN
240 DT = DIGIT:DIGIT = 10 ^ DIGIT
260 PRINT
280 INPUT "WHAT IS YOUR NAME? ";NAME$
290 :
295 :
300 REM **--CREATE PROBLEM--**
310 MAX$ = STR$ (MAX)
320 LM = LEN (MAX$)
330  IF DT = LM + 1 OR DT < LM + 1 THEN 400
340 LM = 10 ^ LM
```

# CHAPTER 9 ADDITIONAL SEQUENTIAL FILE TECHNIQUES

```
350 A = INT ( RND (1) * LM)
360  IF A < MN THEN 350
370  IF A > MAX THEN 350
380  GOTO 480
400 Z = 1
420 A = INT ( RND (1) * DIGIT)
440  IF A < MN THEN 420
460  IF A > MAX THEN 420
480 B = INT ( RND (1) * DIGIT)
485  IF B = 0 OR B < (A) THEN 480
490  DEF FN MOD (C) = INT ( (B / A - INT (B / A) ) * A + .05)
     * SGN (B / A)
500 C = INT (B) / (A)
510 C = INT (C)
520 S$ = "/"
540  IF C < 0 THEN 420
560  IF C = W THEN 420
570  IF FN MOD (RM) < > 0 THEN 480
580 W = C
600 A$ = STR$ (A)
620 LA = LEN (A$)
640 B$ = STR$ (B)
660 LB = LEN (B$)
662  HOME
664  HTAB 10: VTAB 2
666  PRINT "TYPE 'END' WHEN FINISHED"
668  VTAB 10
670  HTAB 22: FOR I = 1 TO DT + 1 : PRINT CHR $ (95) ; : NEXT I
675  PRINT
680  HTAB 22 - LA: PRINT A; ")" ; B
720 C$ = STR$ (C)
740 LN = LEN (C$)
760 Q = 1
780  IF LB < DT THEN LN = LN + (DT - LB)
810 :
815 :
820  REM **--GET ANSWER--**
830  VTAB 9
840  PRINT : HTAB (22 + DT) - (LN - 1) : INPUT ""; ANSWER$
860  IF ANSWER$ = "END" THEN 1060
880 D = VAL (ANSWER$)
900  IF D = C THEN PRINT : PRINT : PRINT : HTAB 19: PRINT
     "GOOD": FOR I = 1 TO 500: NEXT I : CR = CR + 1: GOTO 320
```

```
 920  IF Z < 3 THEN PRINT : PRINT : PRINT : HTAB 10: PRINT "NO,
      PLEASE TRY AGAIN. ": Z = Z + 1: PRINT : WR = WR + 1
      : FOR WT = 1 TO 1000: NEXT WT: VTAB 10: GOTO 660
 940  PRINT : PRINT : PRINT
 960  PRINT "NO, THE ANSWER IS "; C
 980  PRINT : PRINT B; " "; S$; " "; A; " = "; C
1000  PRINT : Z = 1: WR = WR + 1
1020  INPUT "HIT RETURN WHEN READY TO GO ON "; L$
1040  GOTO 320
1050  :
1055  :
1060  REM **--TOTAL ROUTINE--**
1080  HOME : VTAB 5
1100  PRINT "YOU GOT "; CR; " RIGHT!"
1120  PRINT
1140  PRINT "YOU MISSED "; WR
1160  :
1180  :
2000  REM **--FILE ROUTINE--**
2020  D$ = CHR$(4)
2040  ONERR GOTO 2180
2060  PRINT D$; "APPEND"; NAME$
2080  PRINT D$; "WRITE"; NAME$
2085  PRINT S$
2090  PRINT DT
2100  PRINT CR
2120  PRINT WR
2140  PRINT D$; "CLOSE"; NAME$
2160  PRINT D$; "RUN MATH MENU"
2180  POKE 216, 0
2200  PRINT D$; "OPEN"; NAME$
2220  GOTO 2080
```

## SCORES

```
10  REM ***--SCORES--***
11 :
12 :
20  D$ = CHR$(4)
40  ONERR GOTO 380
60  DIM S$(100),DT(100)
80  DIM CR(100),WR(100)
100 I = 1
120 HOME : VTAB 5
140 INPUT "STUDENT'S NAME PLEASE? ";NAME$
160 :
180 :
200 REM **--FILE ROUTINE--**
220 PRINT D$;"OPEN";NAME$
240 PRINT D$;"READ";NAME$
260 INPUT S$(I)
280 INPUT DT(I)
300 INPUT CR(I)
320 INPUT WR(I)
340 I = I + 1
360 GOTO 260
380 POKE 216,0
400 PRINT D$;"CLOSE";NAME$
420 :
440 :
460 REM **--DISPLAY ROUTINE--**
480 HOME : VTAB 1:HTAB 19:PRINT NAME$: PRINT : PRINT
500 PRINT "SESS.";
520 HTAB 7:PRINT "OPERATION";
540 HTAB 18:PRINT "DIGITS";
560 HTAB 26:PRINT "CORRECT";
580 HTAB 35:PRINT "WRONG"
590 POKE 34,4: REM SET TOP WINDOW
600 FOR K = 1 TO I - 1
620 IF S$(K) = "+" THEN S$(K) = "ADD"
640 IF S$(K) = "-" THEN S$(K) = "SUB"
660 IF S$(K) = "X" THEN S$(K) = "MLT"
680 IF S$(K) = "/" THEN S$(K) = "DIV"
700 HTAB 3:PRINT K;
720 HTAB 10:PRINT S$(K);
740 HTAB 20:PRINT DT(K);
```

```
760  IF CR(K) > 9 THEN L = - 1
780  HTAB 29 + L: PRINT CR(K);
800  L = 0
820  IF WR(K) > 9 THEN L = - 1
840  HTAB 37 + L: PRINT WR(K)
860  L = 0
880  NEXT K
900  PRINT : INPUT "HIT RETURN WHEN FINISHED";L$
910  TEXT
920  PRINT D$; "RUN MATH MENU"
```

## RECIPES

```
2   REM ***--RECIPES--***
3  :
4  :
5   REM **--VARIABLES LIST--**
6   REM RECNBR = NUMBER OF RECORDS
7   REM INGNBR = TOTAL # OF INGRED.
8   REM ING$ & ID$ = INGRED.
9   REM REC$ & RC$ = RECIPES
10  REM IG$ = CURRENT SESS. INGRED.
11  REM RZ$ = RECIPE NAMES ONLY
12 :
13 :
14  REM **--INITIALIZATION--**
15  DIM REC(100), ING$(50), IG$(100,50), RC$(100), RZ$(100),
      ID$(50)
20  D$ = CHR$(4): REM CONTROL D
25  TB = 8: REM HTAB VALUE
27   ONERR GOTO 10000
30   PRINT D$; "OPEN RECIPE NAMES"
35   PRINT D$; "READ RECIPE NAMES"
40   INPUT NUMBERS$
45   PRINT D$; "CLOSE RECIPE NAMES"
47   POKE 216,0: REM RESET ERROR FLAG
50  LR = LEN (NUMBERS$)
55  T = 1
60   IF MID$(NUMBERS$,T,1) = "*" THEN 70
65  T = T + 1: GOTO 60
70  RECNBR = VAL ( LEFT$(NUMBERS$,T - 1))
75  INGNBR = VAL ( MID$(NUMBERS$,T + 1,LR - T))
```

```
 97 :
 98 :
 99 :
100 REM **--RECIPE MENU--**
120 HOME : VTAB 5: HTAB 15
140 PRINT "RECIPE MENU"
160 PRINT : PRINT
180 HTAB TB
200 PRINT "1. ADD RECIPE TO LIST"
220 PRINT : HTAB TB
240 PRINT "2. SELECT RECIPE FROM LIST"
260 PRINT : HTAB TB
280 PRINT "3. END PROGRAM"
380 PRINT : HTAB TB
400 INPUT "WHICH NUMBER? ";NB
420 IF NB < 1 OR NB > 3 THEN PRINT "INCORRECT NUMBER! "
    : GOTO 380
510 IF NB = 1 THEN 1000
520 IF NB = 2 THEN 2000
530 IF NB = 3 THEN END
970 :
980 :
990 :
1000 REM **--ADD TO RECIPE LIST--**
1002 R = 1
1005 HOME : VTAB 5
1010 INPUT "NAME OF RECIPE ";REC$(R)
1015 I = 1
1020 PRINT : PRINT "TYPE 'END' WHEN FINISHED. " : VTAB 10
1025 PRINT "TYPE IN INGREDIENT #";I;" " BELOW THIS LINE. "
1030 INPUT ING$(I)
1035 IF ING$(I) = "END" THEN 1050
1040 I = I + 1
1045 HOME : VTAB 6: GOTO 1020
1050 HOME : VTAB 5: PRINT REC$(R) : PRINT : PRINT
1055 FOR J = 1 TO I - 1
1060 PRINT J;" ";ING$(J)
1065 NEXT J
1070 PRINT
1075 INPUT "IS THIS CORRECT? ";YES$
1080 IF YES$ = "Y" THEN 1110
1085 PRINT
1090 INPUT "WHICH NUMBER IS WRONG? ";WR
1095 PRINT "TYPE IN CORRECT INFO. FOR INGREDIENT #";WR
```

```
1100  INPUT ING$(WR)
1105  GOTO 1050
1110  REC$(R) = REC$(R) + "!" + STR$(INGNBR) + "*"
      + STR$(I - 1)
1111  FOR J = 1 TO I - 1
1112  IG$(R, J) = ING$(J)
1113  NEXT J
1115  INGNBR = INGNBR + I - 1
1120  R = R + 1
1125  PRINT
1130  INPUT "ADD MORE RECIPES? "; YES$
1135  IF YES$ = "Y" THEN 1005
1136  RECNBR = RECNBR + R - 1
1137  NUMBERS$ = STR$(RECNBR) + "*" + STR$(INGNBR)
1140  PRINT D$; "APPEND RECIPE NAMES"
1145  PRINT D$; "APPEND INGRED"
1146  FOR K = 1 TO R - 1
1150  PRINT D$; "WRITE RECIPE NAMES"
1155  PRINT REC$(K)
1160  LN = LEN (REC$(K))
1165  T = 1
1170  IF MID$(REC$(K), T, 1) = "*" THEN 1180
1175  T = T + 1: GOTO 1170
1180  Q = VAL ( MID$(REC$(K), T + 1, LN - T))
1185  PRINT D$; "WRITE INGRED"
1190  FOR H = 1 TO Q
1195  PRINT IG$(K, H)
1200  NEXT H
1220  NEXT K
1225  PRINT D$; "CLOSE"
1265  PRINT D$; "OPEN RECIPE NAMES"
1270  PRINT D$; "WRITE RECIPE NAMES"
1275  PRINT NUMBERS$
1280  PRINT D$; "CLOSE"
1285  GOTO 100: REM MENU
1970 :
1980 :
1990 :
2000  REM **--SELECT RECIPE--**
2002  HOME : VTAB 5
2005  PRINT D$; "OPEN RECIPE NAMES"
2007  PRINT D$; "POSITION RECIPE NAMES, R2"
2010  PRINT D$; "READ RECIPE NAMES"
2015  FOR I = 1 TO RECNBR
```

```
2020 INPUT RC$(I)
2021 T = 1
2022 IF MID$(RC$(I),T,1) = "!" THEN 2024
2023 T = T + 1: GOTO 2022
2024 RZ$(I) = LEFT$(RC$(I),T - 1)
2025 NEXT I
2030 PRINT D$;"CLOSE RECIPE NAMES"
2032 FOR I = 1 TO RECNBR
2033 PRINT I;" ";RZ$(I)
2034 NEXT I
2035 PRINT : PRINT
2040 INPUT "WHICH RECIPE? ";RC
2045 LN = LEN (RC$(RC))
2050 T = 1
2055 IF MID$(RC$(RC),T,1) = "!" THEN 2065
2060 T = T + 1: GOTO 2055
2065 T1 = T
2070 IF MID$(RC$(RC),T1,1) = "*" THEN 2080
2075 T1 = T1 + 1: GOTO 2070
2080 IGNB = VAL ( MID$ (RC$(RC),T + 1,T1 - 1)) + 1
2085 LIGNB = VAL ( MID$(RC$(RC),T1 + 1,LN - T1))
2087 HOME : PRINT RZ$(RC)
2090 PRINT D$;"OPEN INGRED"
2095 PRINT D$;"POSITION INGRED,R";IGNB
2100 PRINT D$;"READ INGRED"
2105 FOR K = 1 TO LIGNB
2110 INPUT ID$(K)
2115 PRINT ID$(K)
2120 NEXT K
2125 PRINT D$;"CLOSE INGRED"
2130 PRINT : PRINT
2135 INPUT "HIT RETURN WHEN READY ";L$
2140 INPUT "SELECT ANOTHER RECIPE? ";YES$
2145 IF YES$ = "Y" THEN 2155
2150 GOTO 100: REM MENU
2155 HOME
2160 GOTO 2032: REM SELECT ANOTHER
2970 :
2980 :
2990 :
10000 REM **--FIRST TIME--**
10002 POKE 216,0: REM RESET ERROR FLAG
10005 PRINT D$;"WRITE RECIPE NAMES"
10010 PRINT "0*0-----------"
```

```
10015  PRINT D$; "CLOSE RECIPE NAMES"
10020  PRINT D$; "OPEN INGRED"
10025  PRINT D$; "WRITE INGRED"
10030  PRINT "RECIPE INGREDENTS"
10040  PRINT D$; "CLOSE INGRED"
10045  GOTO 30
```

## CREATE Q & A

```
10  REM **--INPUT Q & A--**
11 :
12 :
20 D$ = CHR$ (4) : REM CONTROL D
40  DIM Q$(50), A$(50)
60 I = 1
70 :
75 :
100  REM **--INPUT ROUTINE--**
105  HOME : VTAB 10
110  INPUT "SUBJECT NAME "; SUB$
120  PRINT : PRINT
140  PRINT "QUESTION # "; I: INPUT Q$(I)
160  IF Q$(I) = "END" THEN 300
180  INPUT "ANSWER "; A$(I)
200  PRINT : PRINT : PRINT Q$(I)
220  PRINT : PRINT A$(I)
230  PRINT
240  INPUT "IS THIS CORRECT? "; Y$
250  PRINT
260  IF Y$ = "N" THEN 140
280  I = I + 1: GOTO 140
290 :
295 :
300  REM **--FILE ROUTINE--**
310  PRINT D$; "OPEN" + SUB$
320  PRINT D$; "WRITE" + SUB$
340  PRINT I - 1
360  FOR J = 1 TO I - 1
380  PRINT Q$(J)
400  PRINT A$(J)
420  NEXT J
440  PRINT D$; "CLOSE"
```

## DRILL Q & A

```
10 REM ***--DRILL & PRACTICE--***
11 :
12 :
20 D$ = CHR$ (4) : REM CONTROL D
40 DIM Q$(50),A$(50)
50 DIM QP(11),AP(11)
60 D$ = CHR$(4)
90 :
95 :
100 REM **--FILE ROUTINE--**
105 HOME : VTAB 10
110 INPUT "SUBJECT NAME ";SUB$
115 PRINT : PRINT
120 PRINT D$;"OPEN" + SUB$
140 PRINT D$;"READ" + SUB$
160 INPUT J
180 FOR I = 1 TO J
200 INPUT Q$(I),A$(I)
220 NEXT I
240 PRINT D$;"CLOSE"
245 :
246 :
250 REM **--GET Q & A--**
260 I = RND (1) * 10: I = INT (I)
280 IF I > J OR I < 1 THEN 260
300 PRINT Q$(I)
320 PRINT : PRINT
340 INPUT "YOUR ANSWER IS ";S$
360 IF S$ = "END" THEN 600
380 IF S$ = A$(I) THEN PRINT "CORRECT" : A = A + 1
    : GOTO 540
400 IF Z > 0 THEN 500
420 PRINT "NO, TRY ONCE MORE"
440 Z = 1
460 A2 = A2 + 1
480 GOTO 340
500 PRINT "NO, THE ANSWER IS ";A$(I)
520 M = M + 1
540 Z = 0
560 PRINT : PRINT
580 GOTO 260
```

```
590 :
595 :
600 REM **--DISPLAY SCORE--**
610 A2 = A2 - M
620 A = A - A2
640 HOME : VTAB 10
660 PRINT "YOU GOT ";A;" RIGHT ON THE FIRST TRY"
680 PRINT : PRINT
700 PRINT "YOU GOT ";A2;" RIGHT ON THE SECOND TRY"
720 PRINT : PRINT
740 PRINT "YOU MISSED ";M;" ANSWERS"
```

# 10
# DIF Files

One of the more exciting possibilities in file handling is the prospect of a standard format for transferring file information. At least one such standard is now being supported by a number of major pieces of application software. The DIF™ file format was developed by Software Arts, writers of VisiCalc.® It is important to keep in mind the intent of this standard. The standard does not suggest that all files be stored according to the DIF format. Such a requirement would place an impossible burden on too many applications to make the standard truly acceptable. Instead, the standard suggests a specific format for file transfer.

If you never expect to transfer your file information from one program to another, you have no real need to use DIF, but if you wish to have different programs share the same data, then a standard such as DIF is very valuable. For example, if you never expect to use another program with your MAILING LIST SYSTEM names and addresses, then there is no reason to store those names and addresses according to the DIF format. On the other hand, if you want to use VisiCalc with the scores obtained from the MATH SYSTEM, then the DIF format becomes important. Without the standard, it would be necessary to type all the scores into the VisiCalc program. With DIF, VisiCalc can read the scores directly from the disk. On a small file, retyping is not a big consideration, but as the file grows, it becomes a major problem. Regardless of the file

---

DIF™ is a trademark of Software Arts, Inc.
VisiCalc® is a registered trademark of VisiCorp.

size, re-keying the information for every application program that makes use of the same data is annoying, inconvenient, and unnecessary. If an application program such as VisiCalc or VisiPlot™ makes use of or supports the DIF file format, any information stored according to that standard can be used by that application program.

Most application programs supporting DIF actually offer you two methods of saving your information or data. The first or standard method is the most efficient and effective way to store information for the specific program. The second method is the DIF format. In other words, the file is saved twice, once in the normal manner according to the program needs, and the second time in a format that allows other programs to access and use the information. This two-method system is necessary because the DIF format (or any standard format) is not a very efficient way of storing and retrieving information. Let's look at the DIF format and use it to store the scores from the MATH SYSTEM so that VisiCalc can directly access those scores.

Before getting into the exact way DIF files are stored, it is necessary to understand that in order to make a standard method of saving information, the file must contain information about itself; i.e., where it starts and ends, whether the information is numeric or alphabetic, label information, or actual data. The creators of DIF decided that all DIF files must be divided into two basic parts. The first part contains information about the file itself, and the second contains the actual data. The first part is called the Header Section and the second part the Data Section. Next, since there are many ways of displaying information, they decided to group all information into two categories: Vectors and Tuples. Basically, Vectors and Tuples are just columns and rows. Finally, each piece of information must carry with it the type of information it is: numeric, alphabetic (alphanumeric), or special (descriptive). To distinguish between these types of information, they assigned the following codes: a "0" indicates numeric information, a "1" indicates alphanumeric information, and a "-1" indicates special or descriptive file information.

The only other major decision to be made was the exact organization of the file. This organizational decision is indeed more complex, but it does follow a logical pattern and can be learned with practice. The Header Section (the part of the file that carries information about itself) comes first. Obviously then, the Data Section comes after the Header Section. The beginning and ending of each of these sections must then be indicated in some way.

If you remember in the MAILING LIST SYSTEM, we used two symbols as separators, the "!" and the "*" (see Chapter 4). In much the same way, the creators of DIF have used symbols to set off the beginning and ending points of the two file sections. The word "TABLE" is used to begin the file and is the first entry in the Header Section. The characters

"EOD" (End of Data) are used as the last entry of the file and the end of the Data Section.

Finally, something must divide the two sections. The DIF creators decided that the division should occur in the Header Section and had to be the last entry in that section. That last entry, then, has to follow the pattern for the Header Section. This means that the division between the Header Section and the Data Section needs to be in the following format:

DATA
0, 0
" "

We now have the beginning and ending of the file and the division between the two sections.

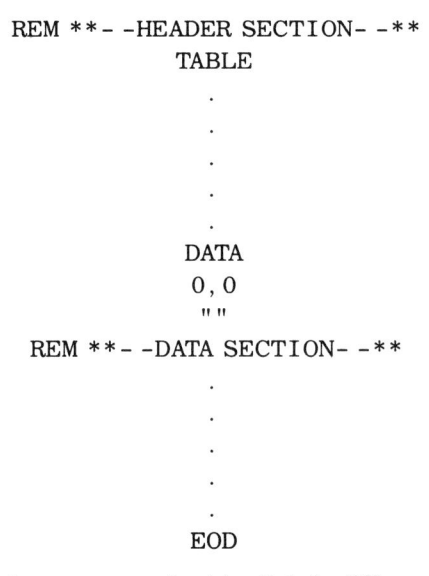

```
       REM **- -HEADER SECTION- -**
              TABLE
                .
                .
                .
                .
                .
              DATA
              0, 0
              " "
       REM **- -DATA SECTION- -**
                .
                .
                .
                .
                .
              EOD
```

The two sections are organized in slightly different ways. The Header Section requires three lines of information for each entry, while the Data Section uses two lines of information for each entry.

## HEADER SECTION

The first line in each entry of the Header Section gives the topic of the entry. TABLE, VECTORS, TUPLES, and DATA are the usual topic lines. The second line in each entry gives numeric information about each topic, such as the number of VECTORS and the number of TUPLES. The third line allows for a name for each topic if a name is necessary. A typical Header Section might look like the following:

# HEADER SECTION

```
TABLE
0, 1
"SCORES"
:
VECTORS
0, 5
" "
:
TUPLES
0, 4
" "
:
LABEL
1, 0
"SESSION #"
:
LABEL
2, 0
"OPERATION"
:
LABEL
3, 0
"DIGITS"
:
LABEL
4, 0
" # CORRECT"
:
LABEL
5, 0
" # WRONG"
:
DATA
0, 0
" "
```

(I have added the colons to separate the individual entries).

Remember that this is the way the information would look in the file and that this section contains information about the file itself. Since VECTORS and TUPLES are basically columns and rows, it is not too difficult to understand the numeric information required in the second line of information in each entry of the Header Section. The first number is the VECTOR number or column number. The second number is a specific value related to the topic of the entry.

MATH SCORES

| SESSION # | OPERATION | DIGITS | CORRECT | WRONG |
|---|---|---|---|---|
| 1 | ADD | 2 | 8 | 2 |
| 2 | MUL | 3 | 12 | 8 |
| 3 | DIV | 3 | 7 | 13 |
| 4 | SUB | 5 | 24 | 1 |

For example, in a table of five columns, the second line of information under the topic of VECTORS would be "0,5". Since the topic VECTORS is not actually in the table, it does not have a column number, (Vector name), so the "0" is first. The "5" indicates the value relating to the topic, VECTORS or five columns. Under the topic of LABEL, you can list the actual names of the columns, their relative positions and any specific value. With a LABEL, the value is usually "0" in a simple table. TUPLES, or rows, might have a second line of "0,4" indicating that the topic TUPLES was not actually in the table but had a value of 4; i.e., 4 rows. The value for the topic TABLE is the version number and must be a "1". So we see that the Header Section describes a file of information that, in our example, consists of 5 columns and 4 rows (5 VECTORS and 4 TUPLES).

## DATA SECTION

Each entry of the Data Section consists of two lines of information. The first line is numeric and gives two pieces of information: the type of information and the value associated with that information. The second line provides alphabetic information associated with the entry. For instance, if the information being stored was the number "62.5", the Data Section entry would be:

0,62.5
V

If the information being stored was the word "PERCENT", the entry would be:

1,0
"PERCENT"

In the first example, the information or data is numeric, so the first number in the first line of this entry is a "0". The value associated with this entry is the information itself, "62.5". The second line of a numeric piece of information can have one of five possibilities: (1) V for a numeric value, (2) NA for not available, the numeric value is 0, (3) ERROR when an invalid calculation has resulted in an error, the numeric value is 0, (4) TRUE for the logical value, the numeric value is 1, and (5) FALSE for the logical value, the numeric value is 0. These five possibilities lend greater flexibility to those who may have need for complex data manipulation. In simpler files, numeric information will usually have a second line of "V".

In the second example, the information being stored is alphabetic, so the first number in the first line of this entry is a "1". The value associated with alphabetic information is usually "0" so that the first line is "1,0". The second line provides alphabetic information about the entry, and since the information is alphabetic, this second line is the information itself. In other words, if the information is alphanumeric (a "1" is indicated in the first line of the entry), the second line contains that alphanumeric information.

The other possibility for an entry in the Data Section is that of a "special value." There are two special values: one for the beginning of each TUPLE and one for the end of the Data Section. Information is grouped within the Data Section by TUPLES (rows) with a special entry marking the beginning of each TUPLE. The entry for this beginning is:

-1, 0
BOT

And the entry for the end of the Data Section is:

-1, 0
EOD

The first number in the first line is the type of information (a "-1" indicating a special entry), and the second is the value associated with that entry (a "0" for special entries). The second line contains either a BOT for Beginning Of Tuple or EOD for End Of Data.

We should now be able to write out a sample TUPLE for the file using the scores from the MATH SYSTEM.

      -1, 0    (beginning of tuple)
      BOT
      :
      0, 1    (math session number)
      V
      :
      1, 0    (type of operation)
      "ADD"
      :
      0, 2    (number of digits)
      V
      :
      0, 15   (number correct)

```
            V
            :
            0, 2      (number wrong)
            V
            :
            -1, 0     (beginning of next tuple)
            BOT
```

The words in parantheses would not be included in the file. They are there to help explain each entry. Again, I have added the colons to separate each entry.

The organization of the Header Section and the Data Section allows for a large variety in file manipulation, far more variety than I have gone into with this explanation. Further information on the structure and flexibility of DIF files can be obtained from: (1) the DIF Clearinghouse, POB 527, Cambridge, MA. 02139, or (2) by reading the information contained in the VisiCalc program, or (3) by reading "DIF: A Format for Data Exchange between Applications Programs," *BYTE MAGAZINE,* November 1981, p. 174.

Now we should be able to write a simple program that will reformat our math scores file in such a way that it conforms to the DIF standard. The first part of this program reads the scores into memory. The second part does the reformatting.

```
10 REM ***--SCORES.DIF--***
11 :
12 :
20 D$ = CHR$(4) : REM CONTROL D
30 Q$ = CHR$(34) : REM QUOTATION MARK
40 ONERR GOTO 380
60 DIM S$(100),DT(100)
80 DIM CR(100),WR(100)
100 I = 1
120 HOME : VTAB 5
140 INPUT "STUDENT'S NAME PLEASE? ";NAME$
160 :
180 :
200 REM **--FILE ROUTINE--**
220 PRINT D$;"OPEN";NAME$
240 PRINT D$;"READ";NAME$
260 INPUT S$(I) : REM SIGN OF OPERATION
280 INPUT DT(I) : REM DIGITS
300 INPUT CR(I) : REM # CORRECT
320 INPUT WR(I) : REM # WRONG
330 IF S$(I) = "+" THEN S$(I) = "ADD"
```

## DATA SECTION

```
331 IF S$(I) = "-" THEN S$(I) = "SUB"
332 IF S$(I) = "X" THEN S$(I) = "MLT"
333 IF S$(I) = "/" THEN S$(I) = "DIV"
340 I = I + 1
360 GOTO 260
380 POKE 216,0: REM RESET ERROR FLAG
400 PRINT D$;"CLOSE";NAME$
420 :
440 :
```

With the exception of line 30, this is the same routine we used for the SCORES program. Line 30 has been added because we need to put quotation marks within quotation marks. This is the same procedure used in our EXEC file (see Chapter 8). The next part of the program is designed by following the necessary organization of either the Header Section or the Data Section.

```
450 REM **--DIF ROUTINE--**
460 J = I - 1: NV = 5: NT = I - 1
470 REM J = COUNTER
472 REM NV = NUMBER OF VECTORS
474 REM NT = NUMBER OF TUPLES
475 FILE$ = NAME$
480 NAME$ = NAME$ + ".DIF"
500 PRINT D$;"OPEN";NAME$
510 PRINT D$;"WRITE";NAME$
511 :
515 REM **--HEADER SECTION--**
516 :
520 PRINT "TABLE"
530 PRINT "0,1"
540 PRINT Q$FILE$Q$
545 :
550 PRINT "VECTORS"
560 PRINT "0,";NV
570 PRINT Q$Q$
575 :
580 PRINT "TUPLES"
590 PRINT "0,";NT
600 PRINT Q$Q$
605 :
610 PRINT "LABEL"
620 PRINT "1,0"
630 PRINT Q$"SESSION #"Q$
635 :
```

```
640 PRINT "LABEL"
650 PRINT "2,0"
660 PRINT Q$"OPERATION"Q$
665 :
670 PRINT "LABEL"
680 PRINT "3,0"
690 PRINT Q$"DIGITS"Q$
695 :
700 PRINT "LABEL"
710 PRINT "4,0"
720 PRINT Q$"CORRECT"Q$
725 :
730 PRINT "LABEL"
740 PRINT "5,0"
750 PRINT Q$"WRONG"Q$
755 :
760 PRINT "DATA"
770 PRINT "0,0"
780 PRINT Q$Q$
785 :
```

These lines create the Header Section. They follow the rules of the Header Section in that each entry has three lines: the topic line, the numeric line, and the title or string line. The label entries are optional. The instructions at line 460 could be handled with input variable statements instead of constants. Line 480 combines the name of the file with the suffix ".DIF" to distinguish between the two files. This suffix may be required for some application programs. The next section of code creates the Data Section.

```
880 REM **--DATA SECTION--**
805 :
810 PRINT "-1,0"
820 PRINT "BOT"
825 :
830 PRINT "1,0"
840 PRINT Q$"SESSION #"Q$
845 :
850 PRINT "1,0"
860 PRINT Q$"OPERATION"Q$
865 :
870 PRINT "1,0"
880 PRINT Q$"DIGITS"Q$
885 :
890 PRINT "1,0"
900 PRINT Q$"CORRECT"Q$
```

## DATA SECTION

```
905 :
910 PRINT "1,0"
920 PRINT Q$"WRONG"Q$
925 :
930 FOR I = 1 TO J
935 :
940 PRINT "-1,0"
950 PRINT "BOT"
955 :
960 PRINT "0,";I : REM MATH SESSION #
970 PRINT "V"
975 :
980 PRINT "1,0": REM OPERATION
990 PRINT S$(I)
995 :
1000 PRINT "0,";DT(I): REM # OF DIGITS
1010 PRINT "V"
1015 :
1020 PRINT "0,";CR(I): REM # CORRECT
1030 PRINT "V"
1035 :
1040 PRINT "0,";WR(I): REM # WRONG
1050 PRINT "V"
1055 :
1060 NEXT I
1070 PRINT "-1,0"
1080 PRINT "EOD"
1085 :
1090 PRINT D$;"CLOSE";NAME$
```

We include the labels with the Data Section so that VisiCalc will view them as data and include them in the display. (VisiCalc does not support the "LABEL" topic in the Header Section.) Once the label information has been included, we write out the actual data by printing the contents of the various arrays. We use a loop to accomplish this. When the loop is finished, the special entry EOD is written and the file closed. Now we should have a program that will create a duplicate file of an individual's math scores in the DIF file format.

We are able to create a DIF file that can be accessed by DIF supporting application programs. One other step remains. We may need to use data obtained with an application program. This requires that we create a program that reads DIF files. Reading a DIF file is simply reading a sequential file that has its information stored in a specific order. The following program will read a VisiCalc DIF file. The display portion of the program is left in its original form since each file may require a different display format.

```
10 REM ***--DIF READER--***
11 :
12 :
20 D$ = CHR$(4) : REM CONTROL D
40 DIM A$(200), S(200), N(200)
60 I = 1
80 HOME: VTAB 5
100 INPUT "FILE NAME PLEASE? ";FILE$
120 :
140 :
160 REM **--INPUT ROUTINE--**
180 PRINT D$; "OPEN";FILE$
200 PRINT D$; "READ";FILE$
220 INPUT T$: REM READ THE TOPIC NAME
240 INPUT S,N: REM READ THE VECTOR #, VALUE
260 INPUT S$: REM THE STRING VALUE
280 IF T$ = "VECTORS" THEN NV = N
300 IF T$ = "TUPLES" THEN NT = N
320 IF T$ < > "DATA" THEN 220
340 K = 1
360 INPUT S(K), N(K)
380 INPUT A$(K)
400 IF A$(K) = "EOD" THEN 440
420 K = K + 1: GOTO 360
440 PRINT D$; "CLOSE";FILE$
460 :
480 :
500 REM **--DISPLAY ROUTINE--**
520 FOR J = 1 TO K
540 PRINT S(J) ; ",";N(J)
560 PRINT A$(J)
580 NEXT J
600 END
```

The display routine is left in its elementary form since the point of the program is to show how to access files from DIF supporting application programs. All these lines should be familiar. Line 320 simply tells the computer to go back to line 220 until T$ equals "DATA". At that time, the computer is to drop down to the loop used to read and store the Data Section information (lines 340 through 420). When A$(K) reads the value "EOD", we know that the file is finished, and we need to CLOSE the file and proceed to the display routine.

As with all programs, there are other ways of writing a DIF reader program and obtaining essentially the same results. We could have read and saved all the information contained in the Header Section. We could have used a number of GOSUBS, especially in the DIF creator program, but most of these differences are stylistic differences and not substantive differences. I have included three other programs at the end of this chapter that deal with DIF files. The first reads the Header Section and writes that information back out as a sequential access file (VARIABLES). The second program reads the Header Section and the Data Section and writes the information in the Data Section back out as a random access file. The last program reads the information in both new files and displays the combined information. All three of these programs could have been combined into one, but for clarity I have used three programs. The intent is to show some of the flexibility possible with files in general and DIF files in particular and to provide a transition to the random access section of this book.

You now have the ability to read and write DIF files. That ability may not prove immediately useful, but I think you will find that this may eventually be the most valuable thing you have learned in this book. If you are not completely sure you understand the format, a second look through this chapter and additional work with DIF creator or reader programs should make you comfortable with DIF.

I have not tried to explain all the possibilities or variations of DIF. This chapter is intended only as an introduction to this file transfer standard. I firmly believe that some such standard is essential if micros are to be taken seriously.

## QUESTIONS

1. TRUE or FALSE: DIF suggests a standard way of saving all files.
2. Name the two parts of a DIF file.
3. Which part contains information about the file itself?
4. Which part contains the actual file information?
5. How many lines are associated with each entry in the Header Section?
6. How many lines are associated with each entry in the Data Section?
7. What value is used to indicate numeric information in the Data Section?
8. What value is used to indicate alphanumeric information in the Data Section?
9. What characters are used as the last entry in a DIF file?

# ANSWERS

1. FALSE
2. Header Section and Data Section
3. Header Section
4. Data Section
5. 3
6. 2
7. 0
8. 1
9. EOD

## SCORES--DIF

```
10  REM ***--SCORES.DIF--***
11 :
12 :
20  D$ = CHR$ (4) : REM CONTROL D
30  Q$ = CHR$ (34) : REM QUOTATION MARK
40  ONERR GOTO 380
60  DIM S$(100),DT(100)
80  DIM CR(100),WR(100)
100 I = 1
120 HOME : VTAB 5
140 INPUT "STUDENT'S NAME PLEASE? ";NAME$
160 :
180 :
200 REM **--FILE ROUTINE--**
220 PRINT D$;"OPEN";NAME$
240 PRINT D$;"READ";NAME$
260 INPUT S$(I): REM SIGN
280 INPUT DT(I): REM DIGITS
300 INPUT CR(I): REM # RIGHT
320 INPUT WR(I): REM # WRONG
330 IF S$(I) = "+" THEN S$(I) = "ADD"
331 IF S$(I) = "-" THEN S$(I) = "SUB"
332 IF S$(I) = "X" THEN S$(I) = "MLT"
333 IF S$(I) = "/" THEN S$(I) = "DIV"
340 I = I + 1
360 GOTO 260: REM GET ANOTHER
380 POKE 216,0: REM RESET ERR FLAG
400 PRINT D$;"CLOSE";NAME$
420 :
440 :
450 REM **--DIF ROUTINE--**
460 J = I - 1:NV = 5:NT = I - 1
470 FILE$ = NAME$
480 NAME$ = NAME$ + ".DIF"
500 PRINT D$;"OPEN";NAME$
510 PRINT D$;"WRITE";NAME$
511 :
515 REM **--HEADER SECTION--**
516 :
520 PRINT "TABLE"
530 PRINT "0,1"
```

```
540 PRINT Q$FILE$Q$
545 :
550 PRINT "VECTORS"
560 PRINT "0, ";NV
570 PRINT Q$Q$
575 :
580 PRINT "TUPLES"
590 PRINT "0, ";NT
600 PRINT Q$Q$
605 :
610 PRINT "LABEL"
620 PRINT "1,0"
630 PRINT Q$"SESSION #"Q$
635 :
640 PRINT "LABEL"
650 PRINT "2,0"
660 PRINT Q$"OPERATION"Q$
665 :
670 PRINT "LABEL"
680 PRINT "3,0"
690 PRINT Q$"DIGITS"Q$
695 :
700 PRINT "LABEL"
710 PRINT "4,0"
720 PRINT Q$"CORRECT"Q$
725 :
730 PRINT "LABEL"
740 PRINT "5,0"
750 PRINT Q$"WRONG"Q$
755 :
760 PRINT "DATA"
770 PRINT "0,0"
780 PRINT Q$Q$
785 :
800 REM **--DATA SECTION--**
805 :
810 PRINT "-1,0"
820 PRINT "BOT"
825 :
830 PRINT "1,0"
840 PRINT Q$"SESSION #"Q$
845 :
850 PRINT "1,0"
860 PRINT Q$"OPERATION"Q$
```

```
865 :
870 PRINT "1,0"
880 PRINT Q$"DIGITS"Q$
885 :
890 PRINT "1,0"
900 PRINT Q$"CORRECT"Q$
905 :
910 PRINT "1,0"
920 PRINT Q$"WRONG"Q$
925 :
930 FOR I = 1 TO J
935 :
940 PRINT "-1,0"
950 PRINT "BOT"
955 :
960 PRINT "0,";I
970 PRINT "V"
975 :
980 PRINT "1,0"
990 PRINT S$(I)
995 :
1000 PRINT "0,";DT(I)
1010 PRINT "V"
1015 :
1020 PRINT "0,";CR(I)
1030 PRINT "V"
1035 :
1040 PRINT "0,";WR(I)
1050 PRINT "V"
1055 :
1060 NEXT I
1065 :
1070 PRINT "-1,0"
1080 PRINT "EOD"
1085 :
1100 PRINT D$;"CLOSE";NAME$
```

## DIF READER

```
10 REM ***--DIF READER--***
11 :
12 :
20 D$ = CHR$(4) : REM CONTROL D
40 DIM A$(200), S(200), N(200)
60 I = 1
80 HOME : VTAB 5
100 INPUT "FILE NAME PLEASE? ";FILE$
120 :
140 :
160 REM **--INPUT ROUTINE--**
180 PRINT D$;"OPEN";FILE$
200 PRINT D$;"READ";FILE$
220 INPUT T$: REM READ THE TOPIC NAME
240 INPUT S,N: REM READ THE VECTOR #, VALUE
260 INPUT S$: REM THE STRING VALUE
280 IF T$ = "VECTORS" THEN NV = N
300 IF T$ = "TUPLES" THEN NT = N
320 IF T$ < > "DATA" THEN 220
340 K = 1
360 INPUT S(K), N(K)
380 INPUT A$(K)
400 IF A$(K) = "EOD" THEN 440
420 K = K + 1: GOTO 360
440 PRINT D$;"CLOSE";FILE$
460 :
480 :
500 REM **-DISPLAY ROUTINE-**
520 FOR J = 1 TO K
540 PRINT S(J);",";N(J)
560 PRINT A$(J)
580 NEXT J
600 END
```

## VARIABLE CREATOR

```
10 REM **-CREATE VARIABLES-**
11 :
12 :
13 REM **--VARIABLES--**
14 REM L1$ = HEAD. SECT. LINE 1
16 REM L3$ = HEAD. SECT. LINE 3
18 REM A() = HEAD. SECT. 2ND LINE
19 REM         FIRST VALUE
20 REM B() = HEAD. SECT. 2ND LINE
21 REM         SECOND VALUE
22 REM I = SET COUNTER
24 REM RL = RECORD LENGTH
26 REM LL = LABEL LENGTH
46 :
48 :
50 D$ = CHR$ (4) : REM CTRL D
60 DIM L1$(99),A(99),B(99),L3$(99)
80 HOME : VTAB 5
100 INPUT "DIF FILE NAME PLEASE! ";FILE$
110 INPUT "DO YOU WANT A PAPER PRINTOUT? ";YES$
115 IF YES$ = "Y" THEN PRINT D$;"PR#1"
120 IF RIGHT$ (FILE$,4) < > ".DIF" THEN FILE$ = FILE$ +
    ".DIF"
121 :
122 :
130 REM **--INPUT HEAD. SECT--**
140 PRINT D$;"OPEN";FILE$
160 PRINT D$;"READ";FILE$
180 I = 1
200 INPUT L1$(I)
220 INPUT A(I),B(I)
240 INPUT L3$(I)
260 IF L1$(I) = "DATA" THEN 305
280 I = I + 1
300 GOTO 200
301 :
302 :
305 REM **-DISPLAY HEAD. SECT-**
310 HOME : PRINT "    LABEL";: HTAB 23: PRINT "FIELD"
312 PRINT "     NAME";: HTAB 24: PRINT "SIZE"
315 POKE 34,3: PRINT : REM SET WINDOW
```

```
320  FOR K = 2 TO I
340  IF LEN (L3$(K)) < > 0 THEN PRINT A(K);"   ";L3$(K);:
     HTAB 25: PRINT B(K + 1)
360  LL = LL + LEN (L3$(K))
420  IF L1$(K) = "SIZE" THEN RL = RL + B(K)
440  NEXT K
445  RL = RL + B(K - 2)
450  PRINT : PRINT
460  PRINT "RECORD LENGTH = ";RL
480  PRINT "LABEL LENGTH = ";LL
490  PRINT L1$(2);" = ";B(2)
495  PRINT L1$(3);" OR NUMBER OF RECORDS = ";B(3)
496  PRINT : POKE 34,22: REM SET WINDOW
498 :
499 :
500  REM **-WRITE LABELS INFO-**
510 NF$ = FILE$ + ".UP"
512 TN = B(3)
520  PRINT D$;"OPEN VARIABLES"
540  PRINT D$;"WRITE VARIABLES"
560  PRINT NF$
580  PRINT TN
600  PRINT RL
640  PRINT B(2)
700  FOR K = 2 TO I
720  IF LEN (L3$(K)) < > 0 THEN PRINT L3$(K): PRINT B(K + 1)
740  NEXT K
1000  PRINT D$;"CLOSE"
1010  PRINT "ALL FINISHED"
1020  TEXT
1030  NORMAL
1040  PRINT D$;"PR#0"
1060  END
```

## DIF TRANSLATOR 1

```
10  REM ***- -DIF TRANSLATOR- -***
11 :
12 :
13  REM **- -VARIABLES- -**
14  REM L1$ = HEAD. SECT. LINE 1
16  REM L3$ = HEAD. SECT. LINE 3
18  REM A() = HEAD. SECT. 2ND LINE
19  REM         FIRST VALUE
20  REM B() = HEAD. SECT. 2ND LINE
21  REM         SECOND VALUE
22  REM I = SET COUNTER
24  REM RL = RECORD LENGTH
26  REM LL = LABEL LENGTH
28  REM DA$() = ACTUAL DATA VALUE
30  REM J = ARRAY COUNTER
32  REM Q = ARRAY COUNTER
34  REM A$ = DATA SECT. 1ST LINE
36  REM         VARIABLE TYPE
38  REM B$ = DATA SECT. 1ST LINE
40  REM         SECOND VALUE
42  REM C$ = DATA SECTION
44  REM         STRING VALUE
46 :
48 :
50  D$ = CHR$(4) : REM CTRL D
60  DIM L1$(99),A(99),B(99),L3$(99)
80  HOME : VTAB 5
100 INPUT "DIF FILE NAME PLEASE! ";FILE$
110 INPUT "DO YOU WANT A PAPER PRINTOUT? ";YES$
115 IF YES$ = "Y" THEN PRINT D$;"PR#1"
120 IF RIGHT$ (FILE$,4) < > ".DIF" THEN FILE$ = FILE$ +
    ".DIF"
121 :
122 :
130 REM **- -INPUT HEAD. SECT- -**
140 PRINT D$;"OPEN";FILE$
160 PRINT D$;"READ";FILE$
180 I = 1
200 INPUT L1$(I)
220 INPUT A(I),B(I)
240 INPUT L3$(I)
```

```
260  IF L1$(I) = "DATA" THEN 305
280  I = I + 1
300  GOTO 200
301 :
302 :
305  REM **-DISPLAY HEAD. SECT-**
310  HOME : PRINT "    LABEL";: HTAB 23: PRINT "FIELD"
312  PRINT "    NAME";: HTAB 24: PRINT "SIZE"
315  POKE 34,3: PRINT : REM SET WINDOW
320  FOR K = 2 TO I
340   IF LEN (L3$(K)) < > 0 THEN PRINT A(K);"   ";L3$(K);:
      HTAB 25: PRINT B(K + 1)
360  LL = LL + LEN (L3$(K))
420  IF L1$(K) = "SIZE" THEN RL = RL + B(K)
440  NEXT K
445  RL = RL + B(K - 2)
450  PRINT : PRINT
460  PRINT "RECORD LENGTH = ";RL
480  PRINT "LABEL LENGTH = ";LL
490  PRINT L1$(2);" = ";B(2)
495  PRINT L1$(3);" OR NUMBER OF RECORDS = ";B(3)
496  PRINT : POKE 34,22: REM SET WINDOW
498 :
499 :
500  REM **-INPUT DATA SECT-**
510  VTAB 24: FLASH : PRINT "READING DIF FILE--DON'T
     TOUCH!!!": NORMAL
520  DIM DA$(B(3),B(2))
540  J = 0:Q = 0
560  INPUT A$,B$
580  INPUT C$
600  IF C$ = "BOT" THEN 740
620  IF C$ = "EOD" THEN 770
625 :
630  REM IF ALPHABETICAL THEN
635  REM SAVE C$
640  IF A$ = "1" AND B$ = "0" THEN DA$(Q,J) = C$: GOTO 700
645 :
650  REM IF NUMERICAL THEN
655  REM SAVE B$
660  IF A$ = "0" AND C$ = "V" THEN DA$(Q,J) = B$: GOTO 700
665 :
670  REM IF NOT "V" THEN
675  REM SAVE BOTH
```

```
680 DA$(Q,J) = B$ + C$
685 :
700 J = J + 1
720  GOTO 560
740 J = 1: Q = Q + 1
760  GOTO 560
768 :
769 :
770 REM **--WRITE NEW FILE--**
775 VTAB 24: FLASH : PRINT "WRITING NEW FILE": NORMAL
780 PRINT D$; "OPEN"; FILE$ + ".UP"; ",L"; RL
800 FOR K = 1 TO Q
840 PRINT D$; "WRITE"; FILE$ + ".UP"; ",R"; K
850 FOR W = 1 TO J - 1
860 PRINT DA$(K,W)
880 NEXT W
900 NEXT K
920 PRINT D$; "CLOSE"; FILE$ + ".UP"
1000 PRINT D$; "CLOSE"
1010 PRINT "ALL FINISHED"
1020 TEXT
1030 NORMAL
1040 PRINT D$; "PR#0"
1060 END
```

## READ NEW FILE

```
 20 D$ = CHR$(4)
 25 PRINT D$;"OPEN VARIABLES"
 27 PRINT D$;"READ VARIABLES"
 29 INPUT NF$
 31 INPUT TN
 33 INPUT RL
 35 INPUT LABEL
 36 DIM LABEL$(LABEL),FIELDSIZE(LABEL),DA$(TN,LABEL)
 37 FOR K = 1 TO LABEL
 39 INPUT LABEL$(K)
 41 INPUT FIELDSIZE(K)
 43 NEXT K
 45 PRINT D$;"CLOSE VARIABLES"
 50 FILE$ = NF$
 60 PRINT D$;"OPEN";FILE$;",L";RL
 80 FOR I = 1 TO TN
100 PRINT D$;"READ";FILE$;",R";I
120 FOR K = 1 TO LABEL
140 INPUT DA$(I,K)
160 NEXT K
180 NEXT I
200 PRINT D$;"CLOSE"
220 FOR I = 1 TO TN
240 FOR K = 1 TO LABEL
250 PRINT LABEL$(K);: HTAB 20: PRINT FIELDSIZE(K)
260 PRINT DA$(I,K)
280 NEXT K
300 NEXT I
320 END
```

# 11
# Random File Introduction

The biggest barrier I have found in explaining random access files is fear. People are afraid that random access is too hard for them to learn. Actually, once you understand the principles behind sequential access, learning to work with random access is not that difficult. I believe that if you have followed all the examples in the previous chapters, you will be able to learn to work with random files. Don't become intimidated by the different approach random access requires.

Actually, there are two kinds of random access files: random files that consist of undivided equal length records, and random files that consist of divided equal length records. Notice the only difference is that in one kind the records are divided into parts, and in the other kind the records remain as a whole. This latter kind is the easier to explain and use, so I will discuss it first.

**UNDIVIDED RANDOM FILES**

When we use the POSITION command with sequential access files, we need the R parameter. The R parameter allows us to position the file pointer at a specific record relative to the last file pointer position. Please notice this last qualification because it points up the major difference between sequential access files used with the R parameter and random access files of the undivided kind. Remember that if you specify R25 in a sequential file immediately after opening that file, you access the 25th record, but if you follow that with a specification of R10, you access the 35th record, not the 10th record. The R parameter is relative to the file pointer position. This is not the case with random access files. The R parameter is not relative to the file pointer. It is equal to the file pointer and is absolute in its position. In the above example, a specification of R25 with random files accesses the 25th record, but a following specification of R10 does not access the 35th record in random files. Instead, it accesses the 10th record. Of course, the numbers used in this example are arbitrarily chosen and can be any record numbers. The idea is that the R parameter, when used in random files, gives access to the specified record. If you want to access the 15th record first and then the 5th record, you can do so directly.

```
20 D$ = CHR$(4) : REM CONTROL D
40 PRINT D$; "OPEN TEST FILE, L50"
60 PRINT D$; "READ TEST FILE, R15"
80 INPUT A$
100 PRINT D$; "READ TEST FILE, R5"
120 INPUT B$
140 PRINT D$; "CLOSE TEST FILE"
```

In this example, you can see how easy it is to use random files of the undivided kind. Once you open the file properly, you are able to directly read or write to any record you choose in any order you choose. But please notice the qualification about opening the file properly. Both kinds of random files must consist of equal-length records. This means that you must decide on the length of the longest record you will ever have. For instance, in our MAILING LIST SYSTEM, each line had a maximum length of 255 characters because that is the maximum number of characters allowed in a single string variable. Probably none of your lines (or records) actually had the maximum length, but that was the length possible for each record. You did not need to specify this number because, in sequential files, the next record begins immediately after the end of the last character and the record delimiter, no matter what the actual length of the record. In random files, you must specify the maximum length because the next record does not begin immediately after the last character in the previous record. It actually begins at the specified record length after the beginning of the previous record, regardless of the number of characters in that record.

In our above example, the length is given as "50". That means that each record has a maximum of 50 characters possible and that each record begins 50 characters or spaces (bytes) from the start of the previous record. If the first record begins at byte 0, the second begins at 50, the third record at 100, and so forth. You do not need to be concerned with the actual location on the disk. All that is important is to understand that since each record must be of equal length, it is very easy for the computer to calculate the starting position of each record and possible for you to specify any record in any order. You must provide the computer with that maximum length by assigning a value to the L parameter in random files. The number given after the L in an OPEN statement indicates the maximum number of characters or bytes you expect in any record in that file. It also indicates that each record will be that number of characters or bytes long.

If you have a record that is not as long as the number given after the L, then you will have a certain amount of disk space that is unused. It is, therefore, important to figure carefully and keep the number after the L as low as possible. If the number is very large and most of your records are rather small, you will be wasting a lot of disk space. A certain amount of wasted disk space is inevitable when using random files, since few files will contain information of exactly equal length. But in using random files, you are willing to waste a little disk space in order to gain the advantage of much faster disk access.

As you can see, there is not too much difference in learning to use sequential access files with the R parameter and learning to use random access files with the R parameter. Our MAILING LIST SYSTEM would be somewhat easier to work with now, but it would still be a good idea to include some kind of separator between addresses and phone numbers and also between sets of information. We would still have some difficulty picking out just the zip code or first name or city or any other part of a record if we needed just that part. It can be done with enough good programming, but an easier way is to use the second kind of random access files—the divided equal length record.

## DIVIDED RANDOM FILES

The divided random file consists of records that are broken into parts or fields of varying length. Each record is the same length, but within each record, the fields or parts of the record can be of varying lengths. In other words, a random access file that consists of records with a length of 100 characters or bytes can have each record divided into parts of equal or unequal lengths. The first field might be 25 bytes long, the second field 10 bytes, the third field 15 bytes, and the last field 50 bytes. The total number of bytes or characters equals 100, but no two fields are the same length.

In our MAILING LIST SYSTEM example, with random files that contained divided records, we could specify a certain length for the first name and other lengths for the last name, city, zip code, etc. For instance, if we decide that each line of information or each record would have no more than 30 characters in it, we could further decide that the first field of each record would exist from byte 0 to byte 10, the second from byte 11 to byte 24, and the last field from byte 25 to byte 30. The first record in each set could contain the first name in the first field, the last name in the second field, and the middle initial in the third field. The second record could contain the numerical address in the first field, the street address in the second field, and an apartment number in the third field. Finally, the third record could contain the city in the first field, the state in the second field, and the zip code in the third field. Under this set up, it would be easy to access any part of any record in any order we desired. For example, if we just wanted the zip code and first name in that order, we would have no trouble accessing just that information.

I have been using the term "byte" in connection with the word "character" so that you might get used to the idea that the length of a record is measured in bytes. Each character or number is one byte. If a file has equal length records of 50, it has 50 bytes. If the second field begins 27 characters from the first character, it starts at the 27th byte of the record. To access that byte we use another parameter---the B parameter. For example:

```
200 PRINT D$; "OPEN TEST FILE, L75"
220 PRINT D$; "READ TEST FILE, R44, B23"
240 INPUT A$
260 PRINT D$; "CLOSE"
```

This example would open the existing file called "TEST FILE", the length of which is 75 bytes for each record. Line 220 sets the file pointer to the 23rd byte of the 44th record. Line 240 brings information in starting from that 23rd byte until the delimiter is encountered or the next field is reached.

## MEDICAL RECORDS SYSTEM

With this background, we are going to go over what I hope is a useful program. The example is a file used to store personal family medical records. In these random file examples, I will not go over all the routines as I did in the sequential file examples. Instead, I will concentrate on the file routines. The complete listing for the program will be found at the end of the chapter. If you take a look at the complete listing, you will see that we begin with a menu routine and a keyboard input routine in order to obtain our original information. We are asking the user to supply: (1) the name of the individual (NAME$), (2) the date (DT$), (3) the type of record; i.e., whether it is a record of a Dr. Visit, Medication, Illness, Accident or Injury, Shot or Immunization, or X-ray (TYPE$), and (4) any miscellaneous information such as the name of the medication and frequency of use, the kind of illness, location of injury, etc. (MISC$). Once we have all our information and have verified that it is correct, we are ready to write that information out to the disk file.

```
1700 REM **--OUTPUT ROUTINE--**
1710 ONERR GOTO 1950: REM FOR FIRST USE
1720 PRINT D$; "OPEN MEDICAL FILE, L50"
1740 PRINT D$; "READ MEDICAL FILE, R0"
1760 INPUT PTR
1780 PTR = PTR + 1
1800 PRINT D$; "WRITE MEDICAL FILE, R"; PTR; ",B"; 1
1810 PRINT NAME$
1820 PRINT D$; "WRITE MEDICAL FILE, R"; PTR; ",B"; 15
1830 PRINT DT$
1840 PRINT D$; "WRITE MEDICAL FILE, R"; PTR; ",B"; 25
1850 PRINT TYPE$
1860 PRINT D$; "WRITE MEDICAL FILE; R"; PTR; ",B"; 27
1870 PRINT MISC$
1880 PRINT D$; "WRITE MEDICAL FILE, R0"
1900 PRINT PTR
1920 PRINT, D$; "CLOSE MEDICAL FILE"
1930 POKE 216, 0: REM RESET ERROR FLAG
1940 GOTO 100: REM MENU
1945 :
1946 :
1950 REM CREATE PTR FOR FIRST TIME
1955 POKE 216, 0: REM RESET ERROR FLAG
1960 PRINT D$; "WRITE MEDICAL FILE, R0"
1970 PRINT "0"
1975 GOTO 1700: REM BEGIN AGAIN
1980 :
1990 :
```

Line 1700 names the routine, and 1710 provides a check for a first use of this program. If this program has not been used before, the input statement in 1760 (INPUT PTR) will cause an error to occur. Line 1710 does not halt operation of the program upon encountering an error, rather the computer is directed to a one-time routine designed to correct the cause of the error. Therefore, upon first use, when the out of data error occurs, the computer is instructed to proceed to lines 1950 to 1975. These lines reset the error flag, and then write a " 0 " for the value of the first pointer. After the file has a value for the pointer, we can direct the computer to return to the beginning and continue as if no error has occurred.

Line 1720 opens the file, while 1740 directs the computer to get ready to read the file. Line 1760 brings in the value of the pointer and stores that value in the numeric variable PTR. Line 1780 increases the value of the pointer by one since we do not want to write over the last record in this file. Line 1800 begins the process of writing the medical information to the file.

I am going to go over every character in line 1800, so that anything that is used in this instruction will be clear. The first part of this line is the standard PRINT D$ which you should recognize as a CTRL D providing the computer with the information that it is to access the disk. After the D$, comes a " ; " which helps to make this statement easier to read. Following the semi-colon, we have an opening quotation mark and the word WRITE with the name of the file we want the information to go into. The file name is immediately followed by a comma, the letter R and the closing quotation mark. This sequence is absolutely necessary with those exact characters. After the ,R" sequence, another semi-colon is used with the variable PTR to indicate which specific record this information is to be written to. The PTR is followed by another semi-colon and another absolutely necessary sequence when using random files with the B parameter. This sequence is very much like the earlier one, except that we substitute a "B" for the previously used "R" : ",B" . This sequence informs the computer that we not only want to write information to a specific record, but to a specific location within that record. The exact location is given by the number following the B sequence---in this case the number one. That means that we want to write information to the PTRth record beginning at the first byte of that record. The actual information to be written is contained in the string variable NAME$ which is PRINTed out to the file on the disk by the instruction at line 1810.

This same procedure is used to write all of our information to this particular file. Line 1820 sets the file pointer to the 15th byte of the same record, and 1830 PRINTs the date stored in the string variable DT$. Line 1840 sets the file pointer to the 25th byte, and line 1850 writes the specific type of medical information this record contains. Finally, line 1860 establishes the file pointer at the 27th byte so that line 1870 can write out any information contained in the string variable MISC$. This part of the record has the greatest amount of storage available. MISC$ could go from the 27th byte to the 49th byte (leaving one byte for the record delimiting carriage return). Finally, we go back to record zero in line 1880 and write out our new value for PTR (line 1900). Line 1920 closes the file, and 1930 resets the error flag. We return to the main menu with the instruction at 1940.

There are several points that should be emphasized before moving on. Notice that it is not necessary to use string arrays: NAME$(). We do not have to use string arrays because of the versatility of random files. In this program, the information for a complete record is written to the disk before additional information is obtained from the user. The idea that we can use the disk without extensive use of string arrays will become more apparent with the section on reading and displaying our medical information.

Notice also that we use a variable for the value of our record or the R parameter but use a constant for all values of the byte or B parameter. It is possible to use either a variable or a constant with either or both of these parameters: the R can be a constant and the B a variable, both can be constants, or both variables. The information can go into the file in any order yet still in the same location, as specified in the above program. The computer does not care about the sequence of the instructions in this case, only the location. You may find that by changing the sequence, the process might take a little longer with some files, since the disk head will be changing direction. But with this file, little difference should be noted.

There are quite a variety of methods used to shorten the amount of programming necessary with random files. It is possible to substitute a string variable for the file name, and in that case, the closing quotation mark would come after the word WRITE with another opening quotation mark necessary before the comma-R-close-quote sequence. Line 1800 would then look like this:

1800 PRINT D$; "WRITE"; FILE$; ",R"; PTR; ",B"; 1

Another possibility is to use string variables for everything. This can be accomplished by concatenation; i.e., joining strings together.

25 A$ = D$ + "WRITE" + FILE$ + ",R"
27 B$ = " ,B "

Then 1800 would look like:

1800 PRINT A$; PTR; B$; 1

In this book, I will not use any method to shorten the necessary programming statements because I feel that the unshortened version helps the reader understand the entire process. I grant that it might be easier on the fingers if string variables are substituted for the necessary sequences, but I have found that the time and effort saved with these shortened versions is often lost when such programs must be changed or added to. Regardless of the completeness of the variable list at the beginning of the program, the statement PRINT A$ is far less self-explanatory than the statement:

PRINT D$; "WRITE MEDICAL FILE,R"; PTR; ",B"; 15

especially once the syntax is understood. But for those that like to make their statements as short as possible, I have indicated how these random file statements can be shortened.

We move now to the section of our program that allows us to see the information we have stored in the MEDICAL FILE. In this first section, we read the file and immediately display the information.

```
2000 REM **--READ ROUTINE--**
2020 PRINT D$; "OPEN MEDICAL FILE, L50"
2040 PRINT D$; "READ MEDICAL FILE, R0"
2060 INPUT PTR
2080 FOR I = 1 TO PTR
2100 PRINT D$; "READ MEDICAL FILE, R"; I; ", B"; 1
2120 INPUT NAME$
2140 PRINT D$; "READ MEDICAL FILE, R"; I; ", B"; 15
2160 INPUT DT$
2180 PRINT D$; "READ MEDICAL FILE, R"; I; ", B"; 25
2200 INPUT TYPE$
2220 PRINT D$; "READ MEDICAL FILE, R"; I; ", B"; 27
2240 INPUT MISC$
2260 TP$ = TYPE$: GOSUB 10000: TYPE$ = TP$
2280 PRINT NAME$;
2300 HTAB 10
2320 PRINT DT$;
2340 HTAB 20
2360 PRINT TYPE$
2380 HTAB 20
2400 PRINT MISC$
2420 PRINT
2440 NEXT I
2460 PRINT D$; "CLOSE MEDICAL FILE"
2480 INPUT "HIT RETURN TO GO TO MENU "; L$
2500 GOTO 100: REM MENU
2980 :
2990 :
```

The first thing that is done is to name the routine (line 2000). Next, the file is opened (line 2020) and the computer is instructed to read record zero of the file (line 2040) and bring in the value of the pointer (line 2060). Line 2080 establishes a loop that goes from the first record to the value of PTR. Lines 2100 through 2240 set the file pointer to the desired location and then read in the necessary information. Lines 2280 to 2440 display the information that has just been copied from the disk file. The instruction in line 2260 may look unusual. We first set the string variable TP$ equal to the string variable TYPE$. Control is then transferred to a subroutine located in lines 10000 to 10240. The purpose of this subroutine is to match the single character symbol with its complete corresponding TYPE name: for example, exchange "D" for "Dr. Visit ". Once the exchange has been made, control is returned to the statement immediately following the GOSUB statement. In this case, the statement immediately following the GOSUB is on the same line and makes the exchange back to TYPE$. This is one of the few times that multiple statements on the same instruction line may clarify the purpose of the instructions. The idea is to use a common variable, go to a routine that uses that common variable, return from that routine, and switch back to the original variable. We will use this procedure in other programs. Line 2480 closes the file, and lines 2500 and 2520 allow us to return to the menu portion of the program.

The only section of the program left to examine is the search routine. Lines 3000 to 3270 establish exactly what we will be searching for, and lines 3275 through 3700 conduct the actual search and display the results.

```
3275 REM **-INPUT ROUTINE-**
3280 PRINT D$;"OPEN MEDICAL FILE, L50"
3300 PRINT D$;"READ MEDICAL FILE, R0"
3320 INPUT PTR
3340 FOR I = 1 TO PTR
3360 PRINT D$;"READ MEDICAL FILE, R"; I; ",B"; BYTE
3380 INPUT FIND$
3400 IF SRCH$ < > FIND$ THEN 3640: REM NEXT RECORD
3420 FOR K = 1 TO 4
3440 IF K = 1 THEN BT = 1
3460 IF K = 2 THEN BT = 15
3480 IF K = 3 THEN BT = 25
3500 IF K = 4 THEN BT = 27
3520 PRINT D$;"READ MEDICAL FILE, R"; I; ",B"; BT
3540 INPUT A$(K)
3560 IF K = 3 THEN TP$ = A$(3) : GOSUB 10000:
     A$(3) = TP$
3580 PRINT A$(K)
3600 NEXT K
3620 PRINT
3640 NEXT I
3660 PRINT D$;"CLOSE MEDICAL FILE"
3680 INPUT "HIT RETURN WHEN READY "; L$
3700 GOTO 3000: REM SEARCH AGAIN
```

This is an elementary search and display routine. Lines 3275 to 3320 open the file and obtain the value of the pointer. Line 3340 establishes the boundaries for a loop. Within that loop, we look for just the desired part of each record. When that part is located, the rest of the information associated with that part is read and displayed (lines 3380 to 3600).

Those instruction-lines are skipped for information that does not match or equal the string variable for which we are searching. When the entire file has been searched, the file is closed, and control is transferred back to the beginning of the search routine to see if the user wishes to search for more information.

This program provides a reasonable example of the techniques involved with creating, adding to, and reading from a random access file with the B parameter. It does not get too fancy yet is a useful program. You may want to supply additional routines such as a printer routine. In the following chapter, we will use random files in a more elaborate manner to do some different things. At the end of this chapter, I have included the medical program that uses arrays. This program provides better displays on smaller files.

One additional comment needs to be made in concluding this chapter. There is another way of using random access files of the divided kind without using the B parameter. This is the method usually discussed and is the reason I chose to focus on the B parameter. This other method substitutes separate variables of varying length for the B parameter. In other words, within each random record exists a sequential access series of variables. The programmer cannot get to the third variable within each record without first reading or writing to the first two variables. An example of this type of random access file is given in the DIF TRANSLATOR program at the end of the previous chapter.

Some may argue that it is not really an inconvenience to read all the variables within a record in order to access a middle variable because of the way Apple reads and writes to the disk. But my preference is with the use of the B parameter. It provides a true random access to any byte within the file and is clearer to read than many of the programs using random files without the B parameter.

# QUESTIONS

1. Name the two kinds of random files.
2. What sequential access DOS command uses the same parameter as random access files?
3. TRUE or FALSE: Random files can contain records of different lengths.
4. In random files, the R parameter is (a) absolute or (b) relative?
5. TRUE or FALSE: In random files, the next record begins immediately after the last character in the previous record.
6. What parameter must an OPEN command have in a random file?
7. TRUE or FALSE: Random files waste disk space but have much faster disk access than do sequential files.
8. How many types of divided random files are there?
9. What is the length of each record measured in?
10. What is the necessary sequence for record identification in random files?
11. What is the necessary sequence for byte identification in random files?
12. TRUE or FALSE: Random files require greater use of string arrays than do sequential files.
13. TRUE or FALSE: The R parameter can be a variable in random files, but the B parameter must be a constant.
14. Divided random files that do not use the B parameter use what kind of access for fields within each record?

## ANSWERS

1. Divided and undivided random files.
2. POSITION
3. FALSE
4. A
5. FALSE
6. L
7. TRUE
8. 2
9. BYTES
10. " ,R "
11. " ,B "
12. FALSE
13. FALSE
14. Sequential access

# MEDICAL RECORDS

```
10  REM **- -MEDICAL RECORDS- -**
11 :
12 :
20  D$ = CHR$ (4) : REM CONTROL D
60  TB = 15: REM HTAB VALUE
80 :
90 :
100 REM **- -MENU ROUTINE- -**
110 HOME : VTAB 5
120 HTAB TB
140 PRINT "MEDICAL RECORDS"
160 PRINT : PRINT : PRINT
180 HTAB TB
200 PRINT "1. WRITE RECORD"
210 PRINT : HTAB TB
220 PRINT "2. READ RECORD"
230 PRINT : HTAB TB
240 PRINT "3. SEARCH RECORDS"
250 PRINT : HTAB TB
260 PRINT "4. END"
280 PRINT : HTAB TB
300 INPUT "WHICH NUMBER ";NUMBER
320 IF NUMBER < 1 OR NUMBER > 4 THEN 280
410 IF NUMBER = 1 THEN 1000
420 IF NUMBER = 2 THEN 2000
430 IF NUMBER = 3 THEN 3000
440 IF NUMBER = 4 THEN END
980 :
990 :
1000 REM **- -WRITE ROUTINE- -**
1020 HOME : VTAB 10
1030 HTAB TB
1040 INPUT "NAME ";NAME$
1050 IF LEN (NAME$) > 14 THEN NAME$ = LEFT$ (NAME$, 14)
1060 PRINT
1070 HTAB TB
1080 INPUT "DATE ";DT$
1100 PRINT
1120 HOME : VTAB 5
1140 HTAB TB
1160 PRINT "TYPE OF RECORD"
```

```
1180  PRINT : PRINT
1200  HTAB TB
1220  PRINT "D--DR. VISIT"
1230  PRINT : HTAB TB
1240  PRINT "M--MEDICATION"
1250  PRINT : HTAB TB
1260  PRINT "I--ILLNESS"
1270  PRINT : HTAB TB
1280  PRINT "A--ACCIDENT/INJURY"
1290  PRINT : HTAB TB
1300  PRINT "S--SHOT/IMMUNIZATION"
1310  PRINT : HTAB TB
1320  PRINT "X--X-RAY"
1330  PRINT : HTAB TB
1340  INPUT "WHICH TYPE OF RECORD ";TYPE$
1360  HOME : VTAB 5
1365  HTAB 8: PRINT "TYPE IN ANY MISC. INFO. "
1368  VTAB 10
1370  HTAB 8
1380  FOR I = 1 TO 22
1400  PRINT CHR$(95);: REM UNDERLINE
1420  NEXT I
1430  HTAB 8
1440  VTAB 10
1460  INPUT " ";MISC$: REM INPUT OVER UNDERLINE
1480  IF LEN (MISC$) > 22 THEN PRINT "TOO LONG" : PRINT :
      PRINT "DO NOT GO BEYOND THE END OF THE DASHES": FOR I = 1 '
      3000: NEXT I: GOTO 1360
1500  HOME : VTAB 5
1520  PRINT NAME$
1530  PRINT
1540  PRINT DT$
1550  PRINT
1555  TP$ = TYPE$: GOSUB 10000: REM TYPE SUBROUTINE
1560  PRINT TP$
1570  PRINT
1580  PRINT MISC$
1590  PRINT
1600  INPUT "IS THIS CORRECT? ";YES$
1620  IF YES$ < > "Y" THEN 1000: REM START OVER
1640 :
1660 :
1700  REM **--OUTPUT ROUTINE--**
1710  ONERR GOTO 1950: REM FOR FIRST USE
```

```
1720 PRINT D$;"OPEN MEDICAL FILE, L50"
1740 PRINT D$;"READ MEDICAL FILE,R0"
1760 INPUT PTR
1780 PTR = PTR + 1
1800 PRINT D$;"WRITE MEDICAL FILE,R";PTR;",B";1
1810 PRINT NAME$
1820 PRINT D$;"WRITE MEDICAL FILE,R";PTR;",B";15
1830 PRINT DT$
1840 PRINT D$;"WRITE MEDICAL FILE,R";PTR;",B";25
1850 PRINT TYPE$
1860 PRINT D$;"WRITE MEDICAL FILE,R";PTR;",B";27
1870 PRINT MISC$
1880 PRINT D$;"WRITE MEDICAL FILE,R0"
1900 PRINT PTR
1920 PRINT D$;"CLOSE MEDICAL FILE"
1930 POKE 216,0: REM RESET ERR FLAG
1940 GOTO 100: REM MENU
1945 :
1946 :
1950 REM CREATE PTR FOR FIRST TIME
1955 POKE 216,0: REM RESET ERR FLAG
1960 PRINT D$;"WRITE MEDICAL FILE,R0"
1970 PRINT "0"
1975 GOTO 1700: REM BEGIN AGAIN
1980 :
1990 :
2000 REM **--READ ROUTINE--**
2020 PRINT D$;"OPEN MEDICAL FILE, L50"
2040 PRINT D$;"READ MEDICAL FILE,R0"
2060 INPUT PTR
2080 FOR I = 1 TO PTR
2100 PRINT D$;"READ MEDICAL FILE,R";I;",B";1
2120 INPUT NAME$
2140 PRINT D$;"READ MEDICAL FILE,R";I;",B";15
2160 INPUT DT$
2180 PRINT D$;"READ MEDICAL FILE,R";I;",B";25
2200 INPUT TYPE$
2220 PRINT D$;"READ MEDICAL FILE,R";I;",B";27
2240 INPUT MISC$
2260 TP$ = TYPE$: GOSUB 10000: TYPE$ = TP$
2280 PRINT NAME$;
2300 HTAB 10
2320 PRINT DT$;
2340 HTAB 20
```

```
2360 PRINT TYPE$
2380 HTAB 20
2400 PRINT MISC$
2420 PRINT
2440 NEXT I
2460 PRINT D$; "CLOSE MEDICAL FILE"
2480 INPUT "HIT RETURN TO GO TO MENU "; L$
2500 GOTO 100: REM MENU
2980 :
2990 :
3000 REM **--SEARCH ROUTINE--**
3020 HOME : VTAB 5: HTAB TB
3030 PRINT "SEARCH FOR..."
3035 PRINT
3040 HTAB TB
3060 PRINT "1. NAME"
3070 PRINT : HTAB TB
3080 PRINT "2. DATE"
3090 PRINT : HTAB TB
3100 PRINT "3. TYPE"
3110 PRINT : HTAB TB
3120 PRINT "4. MISC"
3130 PRINT : HTAB TB
3135 PRINT "5. END SEARCH"
3137 PRINT : HTAB TB
3140 INPUT "WHICH NUMBER? "; NB
3160 IF NB < 1 OR NB > 5 THEN 3137
3180 IF NB = 1 THEN BYTE = 1: B$ = "NAME"
3200 IF NB = 2 THEN BYTE = 15: B$ = "DATE"
3220 IF NB = 3 THEN BYTE = 25: B$ = "TYPE"
3240 IF NB = 4 THEN BYTE = 27: B$ = "MISC"
3260 IF NB = 5 THEN 100: REM MENU
3265 PRINT : HTAB TB
3270 PRINT "WHICH "; B$; : INPUT "?"; SRCH$
3271 HOME : VTAB 2
3272 :
3273 :
3275 REM **--INPUT ROUTINE--**
3280 PRINT D$; "OPEN MEDICAL FILE, L50"
3300 PRINT D$; "READ MEDICAL FILE, R0"
3320 INPUT PTR
3340 FOR I = 1 TO PTR
3360 PRINT D$; "READ MEDICAL FILE, R"; I; ",B"; BYTE
3380 INPUT FIND$
```

```
3400 IF SRCH$ < > FIND$ THEN 3640: REM NEXT RECORD
3420 FOR K = 1 TO 4
3440 IF K = 1 THEN BT = 1
3460 IF K = 2 THEN BT = 15
3480 IF K = 3 THEN BT = 25
3500 IF K = 4 THEN BT = 27
3520 PRINT D$; "READ MEDICAL FILE, R"; I; ", B"; BT
3540 INPUT A$(K)
3560 IF K = 3 THEN TP$ = A$(3) : GOSUB 10000: A$(3) = TP$
3580 PRINT A$(K)
3600 NEXT K
3620 PRINT
3640 NEXT I
3660 PRINT D$; "CLOSE MEDICAL FILE"
3680 INPUT "HIT RETURN WHEN READY "; L$
3700 GOTO 3000: REM SEARCH AGAIN
9998 :
9999 :
10000 REM **--SUBROUTINES--**
10100 REM *--TYPE SUBROUTINE--*
10120 IF TP$ = "D" THEN TP$ = "DR. VISIT"
10140 IF TP$ = "M" THEN TP$ = "MEDICATION"
10160 IF TP$ = "I" THEN TP$ = "ILLNESS"
10180 IF TP$ = "A" THEN TP$ = "ACCIDENT/INJURY"
10200 IF TP$ = "S" THEN TP$ = "SHOT/IMMUNIZATION"
10220 IF TP$ = "X" THEN TP$ = "X-RAYS"
10240 RETURN
```

## MEDICAL RECORDS W/ARRAYS

```
10 REM **--MEDICAL RECORDS--**
11 :
12 :
20 D$ = CHR$(4): REM CONTROL D
40 DIM NAME$(50),DT$(50),TYPE$(50),MISC$(50)
60 TB = 15: REM HTAB VALUE
80 :
90 :
100 REM **--MENU ROUTINE--**
110 HOME : VTAB 5
120 HTAB TB
140 PRINT "MEDICAL RECORDS"
160 PRINT : PRINT : PRINT
180 HTAB TB
200 PRINT "1. WRITE RECORD"
210 PRINT : HTAB TB
220 PRINT "2. READ RECORD"
230 PRINT : HTAB TB
240 PRINT "3. SEARCH RECORDS"
250 PRINT : HTAB TB
260 PRINT "4. END"
280 PRINT : HTAB TB
300 INPUT "WHICH NUMBER ";NUMBER
320 IF NUMBER < 1 OR NUMBER > 4 THEN 280
410 IF NUMBER = 1 THEN 1000
420 IF NUMBER = 2 THEN 2000
430 IF NUMBER = 3 THEN 3000
440 IF NUMBER = 4 THEN END
980 :
990 :
1000 REM **--WRITE ROUTINE--**
1020 HOME : VTAB 10
1030 HTAB TB
1040 INPUT "NAME ";NAME$
1050 IF LEN (NAME$) > 14 THEN NAME$ = LEFT$ (NAME$,14)
1060 PRINT
1070 HTAB TB
1080 INPUT "DATE ";DT$
1100 PRINT
1120 HOME : VTAB 5
1140 HTAB TB
```

```
1160 PRINT "TYPE OF RECORD"
1180 PRINT : PRINT
1200 HTAB TB
1220 PRINT "D- -DR. VISIT"
1230 PRINT : HTAB TB
1240 PRINT "M- -MEDICATION"
1250 PRINT : HTAB TB
1260 PRINT "I- -ILLNESS"
1270 PRINT : HTAB TB
1280 PRINT "A- -ACCIDENT/INJURY"
1290 PRINT : HTAB TB
1300 PRINT "S- -SHOT/IMMUNIZATION"
1310 PRINT : HTAB TB
1320 PRINT "X- -X-RAY"
1330 PRINT : HTAB TB
1340 INPUT "WHICH TYPE OF RECORD ";TYPE$
1360 HOME : VTAB 5
1365 HTAB 8: PRINT "TYPE IN ANY MISC. INFO. "
1368 VTAB 10
1370 HTAB 8
1380 FOR I = 1 TO 22
1400 PRINT CHR$ (95);: REM UNDERLINE
1420 NEXT I
1430 HTAB 8
1440 VTAB 10
1460 INPUT " ";MISC$: REM INPUT OVER UNDERLINE
1480 IF LEN (MISC$) > 22 THEN PRINT "TOO LONG" : PRINT :
     PRINT "DO NOT GO BEYOND THE END OF THE DASHES" : FOR I = 1 TO
     3000: NEXT I: GOTO 1360
1500 HOME : VTAB 5
1520 PRINT NAME$
1530 PRINT
1540 PRINT DT$
1550 PRINT
1555 TP$ = TYPE$: GOSUB 10000: REM TYPE SUBROUTINE
1560 PRINT TP$
1570 PRINT
1580 PRINT MISC$
1590 PRINT
1600 INPUT "IS THIS CORRECT? ";YES$
1620 IF YES$ < > "Y" THEN 1000: REM START OVER
1640 :
1660 :
```

```
1700 REM **--OUTPUT ROUTINE--**
1710 ONERR GOTO 1950: REM FOR FIRST USE
1720 PRINT D$;"OPEN MEDICAL FILE, L50"
1740 PRINT D$;"READ MEDICAL FILE,R0"
1760 INPUT PTR
1780 PTR = PTR + 1
1800 PRINT D$;"WRITE MEDICAL FILE,R";PTR;",B";1
1810 PRINT NAME$
1820 PRINT D$;"WRITE MEDICAL FILE,R";PTR;",B";15
1830 PRINT DT$
1840 PRINT D$;"WRITE MEDICAL FILE,R";PTR;",B";25
1850 PRINT TYPE$
1860 PRINT D$;"WRITE MEDICAL FILE,R";PTR;",B";27
1870 PRINT MISC$
1880 PRINT D$;"WRITE MEDICAL FILE,R0"
1900 PRINT PTR
1920 PRINT D$;"CLOSE MEDICAL FILE"
1930 POKE 216,0: REM RESET ERR FLAG
1940 GOTO 100: REM MENU
1945 :
1946 :
1950 REM CREATE PTR FOR FIRST TIME
1955 POKE 216,0: REM RESET ERR FLAG
1960 PRINT D$;"WRITE MEDICAL FILE,R0"
1970 PRINT "0"
1975 GOTO 1700: REM BEGIN AGAIN
1980 :
1990 :
2000 REM **--READ ROUTINE--**
2001 :
2002 :
2010 REM **--INPUT ROUTINE--**
2020 PRINT D$;"OPEN MEDICAL FILE, L50"
2040 PRINT D$;"READ MEDICAL FILE,R0"
2060 INPUT PTR
2080 FOR I = 1 TO PTR
2100 PRINT D$;"READ MEDICAL FILE,R";I;",B";1
2120 INPUT NAME$
2140 PRINT D$;"READ MEDICAL FILE,R";I;",B";15
2160 INPUT DT$
2180 PRINT D$;"READ MEDICAL FILE,R";I;",B";25
2200 INPUT TYPE$
2220 PRINT D$;"READ MEDICAL FILE,R";I;",B";27
2240 INPUT MISC$
```

# MEDICAL RECORDS W/ARRAYS

187

```
2260 TP$ = TYPE$: GOSUB 10000: TYPE$ = TP$
2280 PRINT NAME$;
2300 HTAB 10
2320 PRINT DT$;
2340 HTAB 20
2360 PRINT TYPE$
2380 HTAB 20
2400 PRINT MISC$
2420 PRINT
2440 NEXT I
2460 PRINT D$; "CLOSE MEDICAL FILE"
2480 INPUT "HIT RETURN TO GO TO MENU "; L$
2500 GOTO 100: REM MENU
2980 :
2990 :
3000 REM **--SEARCH ROUTINE--**
3010 Q = 0
3020 HOME : VTAB 5: HTAB TB
3030 PRINT "SEARCH FOR..."
3035 PRINT
3040 HTAB TB
3060 PRINT "1. NAME"
3070 PRINT : HTAB TB
3080 PRINT "2. DATE"
3090 PRINT : HTAB TB
3100 PRINT "3. TYPE"
3110 PRINT : HTAB TB
3120 PRINT "4. MISC"
3130 PRINT : HTAB TB
3135 PRINT "5. END SEARCH"
3137 PRINT : HTAB TB
3140 INPUT "WHICH NUMBER? "; NB
3160 IF NB < 1 OR NB > 5 THEN 3137
3180 IF NB = 1 THEN BYTE = 1: B$ = "NAME"
3200 IF NB = 2 THEN BYTE = 15: B$ = "DATE"
3220 IF NB = 3 THEN BYTE = 25: B$ = "TYPE"
3240 IF NB = 4 THEN BYTE = 27: B$ = "MISC"
3260 IF NB = 5 THEN 100: REM MENU
3265 PRINT : HTAB TB
3270 PRINT "WHICH "; B$;: INPUT "?"; SRCH$
3271 :
3272 :
3275 REM **--INPUT ROUTINE--**
3280 PRINT D$; "OPEN MEDICAL FILE, L50"
```

```
3300 PRINT D$; "READ MEDICAL FILE, R0"
3320 INPUT PTR
3340 FOR I = 1 TO PTR
3360 PRINT D$; "READ MEDICAL FILE, R"; I; ", B"; BYTE
3380 INPUT FIND$
3400 IF SRCH$ < > FIND$ THEN 3600: REM NEXT RECORD
3420 K = 1
3430 FOR K = 1 TO 4
3440 IF K = 1 THEN BT = 1
3450 IF K = 2 THEN BT = 15
3460 IF K = 3 THEN BT = 25
3470 IF K = 4 THEN BT = 27
3480 PRINT D$; "READ MEDICAL FILE, R"; I; ", B"; BT
3500 INPUT A$(K)
3620 NEXT K
3540 Q = Q + 1
3560 NAME$(Q) = A$(1)
3570 DT$(Q) = A$(2)
3580 TYPE$(Q) = A$(3) : TP$ = TYPE$(Q) : GOSUB 10000: TYPE$(Q)
     = TP$
3590 MISC$(Q) = A$(4)
3600 NEXT I
3620 PRINT D$; "CLOSE MEDICAL FILE"
3621 :
3622 :
3630 REM **-DISPLAY ROUTINE-**
3640 HOME : VTAB 10
3700 FOR I = 1 to Q
3720 PRINT NAME$(I);
3730 HTAB 10
3740 PRINT DT$(I);
3750 HTAB 20
3760 PRINT TYPE$(I)
3770 HTAB 20
3780 PRINT MISC$(I)
3790 PRINT
3800 NEXT I
3820 INPUT "HIT RETURN WHEN READY"; L$
3840 GOTO 3000: REM SEARCH AGAIN
9998 :
9999 :
10000 REM **--SUBROUTINES--**
10100 REM *--TYPE SUBROUTINE--*
10120 IF TP$ = "D" THEN TP$ = "DR. VISIT"
```

```
10140 IF TP$ = "M" THEN TP$ = "MEDICATION"
10160 IF TP$ = "I" THEN TP$ = "ILLNESS"
10180 IF TP$ = "A" THEN TP$ = "ACCIDENT/INJURY"
10200 IF TP$ = "S" THEN TP$ = "SHOT/IMMUNIZATION"
10220 IF TP$ = "X" THEN TP$ = "X-RAYS"
10240 RETURN
```

# 12
# Home Inventory System

In this chapter, we are going to look at a simple, yet fairly complete, random access system for home inventory. We will examine the file handling portions of the various programs in detail with the expectation of modifying them for use with other applications. The purpose of such modification is to suggest the possibility of the development of a general purpose data base system.

There are five programs in this HOME INVENTORY SYSTEM: HOME MENU, CREATE HOME INVENTORY, DISPLAY HOME INVENTORY, SEARCH/SORT HOME INVENTORY, and CORRECT HOME INVENTORY. Each program name attempts to describe the main function of the particular program. HOME MENU is the general menu that allows the user to easily switch among the other programs. CREATE HOME INVENTORY is used to create and add to the inventory file. DISPLAY HOME INVENTORY displays the entire inventory file in the order the information was entered. SEARCH/SORT HOME INVENTORY is really the heart of the system. This program has a menu of its own with seven options. Six of these options relate to pulling specific information from the file and displaying it. The last program, CORRECT HOME INVENTORY, allows the user to change or delete information in the inventory file.

**CREATE HOME INVENTORY**

The HOME MENU program does not contain any new programming code, so it will not be discussed. The first program we will look at is the CREATE HOME INVENTORY program. The complete listing for this program is given at the end of this chapter. You will probably find it helpful to look over the program before reading this description.

There are several different things included in this program. Line 200 uses two pokes, POKE 34,7 and POKE 32,7. These two pokes set the screen window so that it begins seven lines below the usual top of the screen and seven lines to the right of the usual left hand edge. This is done to simplify input formatting and can be changed without affecting the rest of the program. In fact, these two pokes actually cause something to occur that may sound like a problem. After these two pokes have been encountered, DOS pointers apparently are affected and the disk head must recalibrate the next time the disk is accessed. Such a recalibration does not cause any harm, but the noise of the recalibration sounds like the disk has a problem. The sound is similar to that made when the disk is booted or the sound that occurs immediately after an I/O ERROR. It is nothing to worry about.

There are two other pieces of code that may be worth explaining. Each of the input sections includes an SP value. This SP value is the number of spaces that the various inputs are allowed in the file. This value is checked to see that the user does not exceed the allotted amount. Each input statement first contains empty quotation marks. Without those quotation marks, a question mark is printed for each input. I thought the format looked better without the question mark and so have included these empty quotation marks. The GOSUB routine is used to print the varying number of underline spaces; CHR$(95).

Lines 1400 to 2000 are the instructions used to check the information and allow the user to change anything before the information is written to the disk. Lines 2000 to 2700 are the file handling lines and will be discussed in detail.

```
2000 REM **--FILE ROUTINE--**
2020 TEXT
2040 ONERR GOTO 2580: REM FIRST USE ONLY
2060 PRINT D$;"OPEN INVENTORY, L100"
2080 PRINT D$;"READ INVENTORY,R0"
2100 INPUT PTR
2120 PTR = PTR + 1: POKE 216,0: REM RESET ERROR FLAG
2140 PRINT D$;"WRITE INVENTORY,R";PTR;",B";0
2160 PRINT ITEM$
2180 PRINT D$;"WRITE INVENTORY,R";PTR;",B";25
2200 PRINT SERIAL$
2220 PRINT D$;"WRITE INVENTORY,R";PTR;",B";40
2240 PRINT CST$
2260 PRINT D$;"WRITE INVENTORY,R";PTR,",B";50
2280 PRINT ROOM$
```

```
2300 PRINT D$; "WRITE INVENTORY, R"; PTR; ", B"; 70
2320 PRINT DESC$
2340 PRINT D$; "WRITE INVENTORY, R0"
2360 PRINT PTR
2380 PRINT D$; "CLOSE INVENTORY"
2400 TEXT: HOME
2420 VTAB 5
2440 PRINT "DO YOU WANT TO ADD MORE ITEMS? "
2460 PRINT
2480 INPUT "TYPE 'NO' TO STOP "; NO$
2500 IF NO$ = "NO" THEN PRINT D$; "RUN HOME MENU"
2520 GOTO 100: REM BEGIN AGAIN
```

Line 2040 is our method of checking whether or not the file has already been created. If the file exists, then no error should occur in bringing in the value of the pointer, but if this is the first time the program has been used, an error will occur. The error will occur when line 2100 tries to bring in a value for PTR, since no such value has yet been written to the disk. We do not wish the program to halt when this error happens, rather we want the problem fixed. So we use the routine located between lines 2580 and 2700 to write out a value for PTR and then return to the beginning of the FILE ROUTINE to start the process over. After use of this error routine, a value does exist on the disk, and line 2100 can input a value for PTR without an error occurring. Once we have a value for the pointer, we add one to that value. We also reset the error flag so that if some other error should happen, we will not go back to that first time routine.

Lines 2140 to 2320 instruct the computer to write out the information collected from the user to the inventory file. Each piece of information is given a certain maximum number of spaces. Most information will not take up the maximum, so some space in each field will be left blank. ITEM$ information can exist between bytes 0 and 24, SERIAL$ information between bytes 25 and 39, CST$ (cost) information between bytes 40 and 49, ROOM$ information between bytes 50 and 69, and DESC$ (description) information between bytes 70 and 99.

There is an easy way to type in these lines. Type in the first line, 2140, and then use the edit keys—ESC I, J, K, M—to move the cursor back to the beginning of 2140, the 2. Next, increase the line number by 40 so that you are now working on line 2180. Then use the right arrow key to copy over the line until you come to the zero after the ",B" sequence. Change the zero to 25 and hit return. Follow the same process for each of the remaining lines. It may sound complicated, but I have found this procedure to greatly reduce the amount of typing necessary in using random files.

When all the information has been transferred to the disk, the pointer value is placed in record zero and the inventory file closed. The user is queried about adding more information to the file and the appropriate action taken upon obtaining a response.

## DISPLAY HOME INVENTORY

The DISPLAY HOME INVENTORY program is really the reverse of the routine just covered. The word READ is substituted for the word WRITE, and the word INPUT exchanged for PRINT. Otherwise, the routines are very similar. Each field of each record is read into the computer from the disk and displayed. When all records have been read in and displayed, the total value of all items is given, and the user is transferred to the HOME MENU.

## SEARCH/SORT HOME INVENTORY

The main program of this HOME INVENTORY SYSTEM is the SEARCH/SORT HOME INVENTORY program. There are six sort or search routines and an option to return to the HOME MENU.

1. SEARCH FOR ITEM
2. SEARCH FOR SERIAL #
3. SEARCH FOR COST
4. SEARCH FOR ROOM ITEMS
5. SORT ITEMS ALPHABETICALLY
6. SORT ITEMS BY SERIAL #
7. RETURN TO MAIN MENU

The first two and selection number four use a common search subroutine. The two sort options (numbers 5 and 6) use a common sort subroutine, the Shell-Metzner sort. Option number 3 uses its own search routines for both parts of this selection. We will cover the common search subroutine first.

```
10000 REM **--SEARCH SUBROUTINE--**
10020 PRINT D$;"READ INVENTORY,R";I;",B";BYTE
10040 INPUT FIND$
10060 IF FIND$ = "D" THEN 10100
10080 IF SRCH$ = FIND$ THEN 10200
10100 I = I + 1
10120 IF < PTR or I = PTR THEN 10000
```

## CHAPTER 12 HOME INVENTORY SYSTEM

```
10140 PRINT : HTAB TB
10160 PRINT "SEARCH COMPLETED! ": FOR K = 1 TO 1000:
      NEXT K
10180 RETURN
10200 PRINT D$; "READ INVENTORY, R"; I; ",B"; 0
10220 INPUT ITEM$
10240 PRINT D$; "READ INVENTORY, R"; I; ",B"; 25
10260 INPUT SERIAL$
10280 PRINT D$; "READ INVENTORY, R"; I; ",B"; 40
10300 INPUT CST$
10320 PRINT D$; "READ INVENTORY, R"; I; ",B"; 50
10340 INPUT ROOM$
10360 PRINT D$; "READ INVENTORY, R"; I; ",B"; 70
10380 INPUT DESC$
10400 I = I + 1
10420 PRINT D$: REM CANCEL INPUT FROM DISK
10440 RETURN
```

This subroutine is common to the first two options and to the room search option. Each of the option-routines that uses this subroutine establishes the necessary conditions prior to entering the subroutine. The values of FIND$ and BYTE are determined prior to the GOSUB statement in each of the option-routines. Once these values are known, the specified part of the file can be searched for any match (line 10080). If a match occurs, control passes to the instructions at lines 10200 to 10440. These instructions read in the information associated with the item being searched for. The RETURN statement in line 10440 returns control to the instruction following the GOSUB statement in the original option-routine---1, 2, or 4. When a match does not occur, the record counter (I) is first checked to see that its value does not exceed the value of the total number of records (PTR), then the record counter is incremented by one and the process is repeated.

One additional instruction-line needs comment. Line 10420 prints a control D. The control D is printed to cancel input from the disk and allow input from the keyboard. A control D will cancel either the READ or WRITE mode and return control to the keyboard or screen. This cancellation is necessary in order to allow the user to answer the questions posed in the option routines after the search has been completed. Without this control D, the program would attempt to input more information from the disk whenever another INPUT statement occurred.

The next section of code discussed is part one of the Search For Cost option. In lines 3000 to 3210, a decision is made by the user: whether to search for items above a certain cost or items below a certain cost. The appropriate part of this option-routine is then given control. The following code is for items above a specific value.

# SEARCH/SORT HOME INVENTORY

```
3220 REM **--ITEMS ABOVE $ AMOUNT--**
3230 HTAB TB
3240 INPUT "ABOVE WHICH AMOUNT? ";AMT
3250 HOME: VTAB 2: HTAB 14
3260 PRINT "ITEMS ABOVE $";AMT
3270 FOR I = 1 TO PTR
3280 PRINT D$;"READ INVENTORY,R";I;",B";BYTE
3290 INPUT FIND$
3300 IF FIND$ = "D" THEN 3360
3310 IF AMT > VAL (FIND$) THEN 3360
3320 PRINT D$;"READ INVENTORY,R";I;",B";0
3330 INPUT ITEM$
3340 TTLAMT = TTLAMT + VAL (FIND$)
3350 PRINT ITEM$;: HTAB 30: PRINT FIND$
3360 NEXT I
3370 PRINT
3380 PRINT "TOTAL VALUE = $";TTLAMT
3390 PRINT : GOSUB 9000: REM HOUSEKEEPING
3400 GOTO 40: REM MENU
```

The items that are valued above a certain amount are searched for in line 3310. The amount was previously determined in line 3240 and displayed in 3260. Line 3270 begins a loop that extends through 3360. Each record beginning at the 40th byte is searched for costs that exceed the specified amount. Line 3310 says that if the specified amount exceeds the cost of the record being examined, then control is passed to the line that increases the record count (line 3360). This is done to skip over lines 3320 through 3350. Those lines are to be exercised only when the specified amount is less than the cost of the record being examined. When such an item has been found: (1) the item name is read (lines 3320 and 3330), (2) a running total is kept of the cumulative value of these items, and (3) the item and its value are displayed on the screen. After all the records have been examined, the total value of all items above the specific amount is given, and control is transferred to the housekeeping subroutine. Finally, control is shifted back to the menu for further instructions.

The routine to find items below a certain value is virtually the same as that just given. The only significant difference occurs in line 3590 where the sign is reversed. We are looking for items whose value is less than the specified amount. Those items whose value is greater than the specified amount are passed over.

We have looked briefly at the first four options, the search options. The next two options are sort options and use a common sort subroutine, the Shell-Metzner sort. I will explain only the procedures involved in setting up and using a sort subroutine with Apple disk files. We will look first at the alphabetizing routine.

196   CHAPTER 12   HOME INVENTORY SYSTEM

```
5000 REM **--SORT ALPHABETICALLY--**
5020 HOME : VTAB 5
5040 HTAB TB
5060 INVERSE : PRINT "WORKING--PLEASE DON'T TOUCH!!"
     : NORMAL
5080 Q = 1: REM VALID RECORD COUNTER
5100 FOR I = 1 TO PTR
5120 PRINT D$; "READ INVENTORY,R"; I; ",B"; 0
5140 INPUT C$
5160 IF C$ = "D" THEN 5220
5180 C$(Q) = C$
5200 Q = Q + 1
5220 NEXT I
5240 N = Q - 1
5260 PRINT : PRINT: HTAB TB
5280 INVERSE: PRINT "STILL WORKING--PLEASE WAIT!"
     : NORMAL
5300 GOSUB 20000: REM SORT SUBROUTINE
5320 REM DISPLAY RESULTS
5340 HOME: VTAB 5
5360 SPEED = 150
5380 FOR I = 1 TO Q - 1
5400 PRINT I; " "; C$(I)
5420 NEXT I
5440 PRINT
5460 GOSUB 9000: REM HOUSEKEEPING
5480 GOTO 40: REM MENU
```

The key to this routine is (1) reading in only the item names, (2) storing them in a string array, (3) sorting them with the sort subroutine located between 20000 and 20300, and (4) displaying them in their now alphabetized order.

Line 5060 prints a warning on the screen for the user to observe. The word INVERSE is used to instruct the computer to print the warning in inverse characters. Immediately following the warning, the computer is instructed to return to NORMAL characters. The word FLASH could have been used in place of INVERSE and then the warning would have "blinked" or flashed on and off. Regardless of which mode is used, a return to NORMAL characters is essential before accessing the disk. If NORMAL is not restored before the disk is accessed, some type of error occurs, usually an OUT OF DATA ERROR.

A separate record counter is used (line 5080) to keep track of the valid records since there may be some records that have been deleted and now contain the value "D". If there are such records, they are skipped and the loop (I) is increased by one. But the valid record counter (Q) is not increased. If the record is not valid (i.e., it contains a "D"), it is also not included in the string array of valid records to be sorted. Once the loop is completed, the string array C$() should contain all the valid item names. A new warning message is displayed (line 5280), and control is transferred to the sort subroutine. When the sorting has been completed, the results are displayed through another loop (lines 5380 to 5420).

The instruction at line 5360 tells the computer to set the speed of the character output to the screen to a rate of 150 instead of the normal or default value of 255. The effect is a visible slowing of the display of information on the screen and greater readability. The value of 150 is arbitrary and can be any value between 0 and 255.

The last two lines in this routine (5460 and 5480) are common to all the routines and simply "clean up" various conditions that may have been "set" during execution of the routine. A good example is the SPEED setting in this routine. One of the instructions in the housekeeping subroutine restores the speed setting to 255. If you look closely at the instructions in this housekeeping subroutine, I have included an instruction that restores something that no instruction in this program has changed. I have done so purposely for the following reasons: (1) for the reader to figure out which instruction does not really need to be included in this housekeeping routine, (2) to suggest to the reader a possible use for this apparently useless instruction, and (3) to inspire some of you to modify this program so that this now useless housekeeping instruction becomes worthwhile.

The last of the options, sort by serial number, again makes use of the LEFT$ and MID$ string array commands. It is also the longest of the routines. The routine sorts by serial number and then displays the resulting list in serial number order, along with the associated item name. It is conceivable that an individual or insurance company would need all of the associated information instead of just the item name. Therefore, if you are interested in developing a completely useful HOME INVENTORY SYSTEM, you might wish to add the code necessary to display all related information in both serial number order and alphabetical order.

```
6000 REM **--SORT BY SERIAL #--**
6020 HOME: VTAB 5
6040 HTAB TB
6060 INVERSE: PRINT "WORKING--PLEASE DON'T TOUCH!!":
     NORMAL
6080 Q = 1: REM VALID RECORD COUNTER
6100 FOR I = 1 TO PTR
6120 PRINT D$; "READ INVENTORY,R"; I; ",B"; 25
6140 INPUT C$
6160 IF C$ = "D" THEN 6280
6180 C$(Q) = C$
6200 PRINT D$; "READ INVENTORY,R"; I; ",B"; 0
```

```
6220 INPUT ITEM$
6240 C$(Q) = C$(Q) + "*" + ITEM$
6260 Q = Q + 1
6280 NEXT I
6300 N = Q - 1
6320 PRINT : PRINT : HTAB TB
6340 INVERSE: PRINT "STILL WORKING--PLEASE WAIT!":
     NORMAL
6360 GOSUB 20000: REM SORT ROUTINE
6380 REM DISPLAY RESULTS
6400 HOME: VTAB 5
6420 J = 1
6440 FOR I = 1 TO Q - 1
6460 LN = LEN (C$(I))
6480 PRINT I;" ";
6500 IF MID$(C$(I),J,1) = "*" THEN PRINT LEFT$(C$(I),
     J - 1);: HTAB 20: PRINT MID$(C$(I),J + 1, LN):
     GOTO 6540
6520 J = J + 1: GOTO 6500
6540 J = 1
6560 NEXT I
6580 PRINT
6600 GOSUB 9000: REM HOUSEKEEPING
6620 GOTO 40: REM MENU
```

Line 6120 and 6140 bring in the serial number of each item. If the serial number has been deleted (i.e., contains just the letter "D"), the record is skipped as in the previous routine. In fact, the two sort routines have nearly identical beginnings. The main difference occurs in line 6120 when different byte locations are specified and a different part of the file is brought in by the INPUT C$ statement. The only other major difference before the GOSUB statement occurs when the file is read a second time, this time to bring in and concatenate (join) the item name (lines 6200 to 6240). Line 6240 combines: (1) the existing value of C$(Q) (the serial number), (2) the current value of C$ (the item name), and (3) a separator (the asterisk) into one new string array value---C$(Q).

Once the entire file is read and the correct number of valid records determined, control is passed to the sort subroutine (line 6360). Lines 6380 to 6560 are used to display the results of the sort. Here again, we need to make use of the power of the LEFT$, MID$, and LEN functions. The numeric variable LN is set to equal the length of each of the string arrays (line 6460). The MID$ function is used to determine where in the string the asterisk is located (first part of 6500). The LEFT$ and MID$ functions are used to print out the desired parts of the string in an acceptable format (the rest of line 6500). Line 6500 is an IF...THEN statement that does not execute the last part if the first part is not true. The print part of 6500 is not reached until the asterisk is found. This sequence is repeated until all valid records have been displayed in serial number order. The end of this routine is the same as the end of the other five routines.

This concludes the discussion of the SEARCH/SORT HOME INVENTORY program. There are a number of other points that could be discussed, but those points relate mainly to different techniques of programming in Applesoft rather than techniques for working with Apple files. By now, if you have worked through all of the programs, you should be able to "read" a program and recognize some of the different techniques used.

## CORRECT HOME INVENTORY

The last program in this HOME INVENTORY SYSTEM provides the ability to change or delete information in the INVENTORY file. Both parts of this program make use of two subroutines: a READ FILE SUBROUTINE, and a WRITE FILE SUBROUTINE. These two subroutines have been used in our other programs in this system. The CORRECT RECORD routine (lines 1000 to 1620) is essentially the same as the correction routine in the CREATE HOME INVENTORY program (lines 1400 to 1900). The difference is that in the CREATE HOME INVENTORY program, the information being checked for accuracy comes from the keyboard. In the CORRECT HOME INVENTORY program, the information comes from the disk. That is the reason for line 1080. This line transfers control to the READ FILE SUBROUTINE which inputs from the specified record on the disk the values for ITEM $, SERIAL$, CST$, ROOM$, and DESC$. These values are then displayed and a check is made to see if they are correct.

At this point, one other new line of code is encountered (line 1040). Lines 1040, 1290, 1300, and 1610 are all related. All deal with a string variable called FLAG$. Line 1040 sets the original value of FLAG$ equal to the word "NO". This indicates that no information has yet been changed. Lines 1290 and 1300 check the value of FLAG$ and direct the computer accordingly. If the information is correct and no change has been made, the value of FLAG$ is still "NO", and the computer is directed to start this routine over again. If the information has been changed, the value of FLAG$ will have been changed by line 1610 to "YES" indicating altered information. If the information is correct and has been changed, we are now ready to write that information back out to the file on the disk (the WRITE FILE SUBROUTINE). This technique allows the user to scan through the records if he/she is not sure of the record number of the incorrect information.

The deletion routine is a relatively uncomplicated routine. The suspected record is brought in from disk (line 2160) and displayed (lines 2180 to 2300). A request is made of the user to see if this is the information to be deleted. If it is not, the deletion routine starts again. If the information is to be deleted, the user is required to type the word "YES" rather than just the "Y". If "YES" is typed, all string variables are given the value of a single character "D", and control is passed to the WRITE FILE SUBROUTINE where "D" replaces the now deleted information. Notice that the entire file does not need to be resequenced and rewritten to the disk. Instead, only the information requiring change is affected.

The change and delete routines for random access files are considerably easier than similar routines for sequential access files. This ease is one of the major strengths of random files. Access is direct to any part of the file desired. In fact, in a very large inventory system, it is possible to read from disk and check only the desired part of the record, rather than the entire record. Programming can often be simpler and easier to read. There is less need of string arrays and therefore less need of large amounts of internal computer memory. The disk can be used as an extension of the internal memory with random files since the same principles are involved. The major difference is in the time involved, disk access being much slower than internal memory access.

At the end of this chapter, I have included another system, a BACK ORDER SYSTEM, created by modifying this current HOME INVENTORY SYSTEM. The modification is not extensive. The main reason for including the BACK ORDER SYSTEM is to suggest the possibility of a general purpose data base program. All our systems have included some method for: (1) creating and adding to a file, (2) displaying information from that file in various ways, and (3) editing the file. These are the essential characteristics in any data base system. It should be possible to create a general purpose data base system that would request certain necessary information from the user. Based on the supplied information, this general data base system would create a file and set up the procedures to display and edit information in that file. This is the exact procedure used by programs like DB MASTER,™ THE DATA REPORTER,© THE GENERAL MANAGER,© and PFS.™ They are general purpose data base systems.

---

DB Master ™ is a trademark of DB Master Associates; The Data Reporter © is a software package copyrighted by Synergistic Software; The General Manager© is a software package copyrighted by ON-LINE SYSTEMS, INC.; PFS ™ is a trademark of Software Publishing Corp.

The better data base programs have expanded on the essential characteristics. They have added "features" that some users may need but others will never use. One feature that I feel is essential is transportability of file information. If a data base system does not allow some method of universal access to the files created under its system, I believe that system is severely limited in its usefulness to anyone other than the casual user. *Files created under a general data base system must be able to be accessed by other commercial application programs:* Without such access, the user must re-enter data in each application program used with the file information. This is the reason DIF is so important (see Chapter 10). Some of the general purpose data base systems do support DIF while others at least make their files available through normal DOS file structure.

In the Preface, I said that "Reading this book will not make you capable of creating complete data base programs...," but at this point, you should have an appreciation of the effort that goes into creating a good general purpose data base system. For your individual use, you may find that you can create a semi-general purpose data base system, a system that can serve your needs but would not be universal in meeting the needs of everyone. This is the reason for including the BACK ORDER SYSTEM as a modification of the HOME INVENTORY SYSTEM. Structured carefully, with enough user-supplied variables, this series of programs can form the basis for such a personal data base system.

The next chapter will deal with the planning necessary in creating the programs for any file system. The example will be a STOCK MARKET SYSTEM for keeping track of the price and volume changes of certain issues.

## QUESTIONS

1. What does POKE 34,n and POKE 32,n affect?
2. TRUE or FALSE: It is not possible to display the underline character on the screen.
3. What character is used to cancel input from the disk and allow input from the keyboard?
4. TRUE or FALSE: It is necessary to return to NORMAL characters as soon as possible after using either INVERSE or FLASH. Otherwise a disk error may occur.
5. Name the three essential characteristics in any data base system.

## ANSWERS

1. The TEXT window
2. FALSE
3. CONTROL D
4. TRUE
5. a) Creating and adding to a file
   b) Displaying information from that file
   c) Editing the file

# HOME MENU

```
10  REM **--HOME INVENTORY SYSTEM--**
11 :
12 :
20  D$ = CHR$ (4): REM CONTROL D
40  TB = 8: REM HTAB VALUE
60 :
80 :
100 REM **--MENU ROUTINE--**
120 HOME : VTAB 5
140 HTAB TB
160 PRINT "HOME INVENTORY SYSTEM"
180 PRINT : PRINT : PRINT
200 HTAB TB
220 PRINT "1. WRITE RECORD"
240 PRINT : HTAB TB
260 PRINT "2. READ RECORD"
280 PRINT : HTAB TB
300 PRINT "3. SEARCH RECORDS"
320 PRINT : HTAB TB
340 PRINT "4. CORRECT RECORD"
360 PRINT : HTAB TB
380 PRINT "5. END"
400 PRINT : HTAB TB
420 INPUT "WHICH NUMBER ";NUMBER
440 IF NUMBER < 1 OR NUMBER > 5 THEN 400
460 IF NUMBER = 1 THEN 1000
480 IF NUMBER = 2 THEN 2000
500 IF NUMBER = 3 THEN 3000
520 IF NUMBER = 4 THEN 4000
540 IF NUMBER = 5 THEN END
560 :
580 :
1000 REM **--WRITE RECORD--**
1020 PRINT D$;"RUN CREATE HOME INVENTORY"
1998 :
1999 :
2000 REM **--READ RECORD--**
2020 PRINT D$;"RUN DISPLAY HOME INVENTORY"
2998 :
2999 :
```

```
3000 REM **--SEARCH RECORDS--**
3020 PRINT D$;"RUN SEARCH HOME INVENTORY"
3998 :
3999 :
4000 REM **--CORRECT RECORDS--**
4020 PRINT D$;"RUN CORRECT HOME INVENTORY"
```

## CREATE HOME INVENTORY

```
10  REM **--CREATE HOME INVENTORY--**
11 :
12 :
20  D$ = CHR$ (4): REM CONTROL D
40  TB = 8: REM HTAB VALUE
60 :
80 :
100 REM **--INPUT ROUTINE--**
120 HOME : VTAB 5
140 HTAB TB
160 PRINT "CREATE HOME INVENTORY"
180 PRINT : PRINT : PRINT
200 POKE 34,7: POKE 32,7: REM SET WINDOW
220 :
240 :
260 HOME
280 PRINT "ITEM NAME PLEASE. "
300 PRINT : PRINT
320 SP = 25
340 GOSUB 5000: REM INPUT SUBROUTINE
360 INPUT " ";ITEM$
380 IF LEN (ITEM$) > SP THEN PRINT " ": GOTO 260:
    REM 5 CTRL G'S
400 :
420 :
440 HOME
460 PRINT "ITEM SERIAL NUMBER PLEASE. "
480 PRINT : PRINT
500 SP = 15
520 GOSUB 5000: REM INPUT SUBROUTINE
540 INPUT " ";SERIAL$
560 IF LEN (SERIAL$) > SP THEN PRINT "": GOTO 440:
    REM 5 CTRL G'S
580 :
```

```
600 :
620  HOME
640  PRINT "ITEM COST PLEASE"
660  PRINT : PRINT
680  SP = 10
700  GOSUB 5000: REM INPUT SUBROUTINE
720  INPUT " ";CST$
740  IF LEN (CST$) > SP THEN PRINT " ": GOTO 620:
     REM 5 CTRL G'S
760 :
780 :
800  HOME
820  PRINT "ROOM OF ITEM"
840  PRINT : PRINT
860  SP = 20
880  GOSUB 5000: REM INPUT SUBROUTINE
900  INPUT " ";ROOM$
920  IF LEN (ROOM$) > SP THEN PRINT " ": GOTO 800:
     REM 5 CTRL G'S
940 :
960 :
980  HOME
1000  PRINT "ITEM DESCRIPTION"
1020  PRINT : PRINT
1040  SP = 30
1060  GOSUB 5000: REM INPUT SUBROUTINE
1080  INPUT " ";DESC$
1100  IF LEN (DESC$) > SP THEN PRINT " ": GOTO 980:
      REM 5 CTRL G'S
1120 :
1140 :
1400  REM **--DISPLAY FOR CORRECTION--**
1420  HOME
1440  PRINT "1. ";ITEM$
1460  PRINT "2. ";SERIAL$
1480  PRINT "3. ";CST$
1500  PRINT "4. ";ROOM$
1520  PRINT "5. ";DESC$
1540  PRINT : PRINT
1560  INPUT "IS THIS CORRECT ('Y' OR 'N') ";YES$
1580  IF YES$ = "Y" THEN 2000: REM FILE ROUTINE
1600  INPUT "WHICH NUMBER IS WRONG ";NB
1610  IF NB < 1 OR NB > 5 THEN PRINT "INCORRECT CHOICE"
      : GOTO 1600
```

## CHAPTER 12 HOME INVENTORY SYSTEM

```
1620 IF NB = 1 THEN SP = 25
1640 IF NB = 2 THEN SP = 15
1660 IF NB = 3 THEN SP = 10
1680 IF NB = 4 THEN SP = 20
1700 IF NB = 5 THEN SP = 30
1720 PRINT
1740 PRINT "TYPE IN CORRECT INFO"
1760 INPUT CT$(NB)
1780 IF LEN (CT$(NB)) > SP THEN PRINT "TOO LONG--TRY
     AGAIN  PLEASE": GOTO 1740
1800 IF NB = 1 THEN ITEM$ = CT$(NB)
1820 IF NB = 2 THEN SERIAL$ = CT$(NB)
1840 IF NB = 3 THEN CST$ = CT$(NB)
1860 IF NB = 4 THEN ROOM$ = CT$(NB)
1880 IF NB = 5 THEN DESC$ = CT$(NB)
1900 GOTO 1400: REM CHECK AGAIN
1998 :
1999 :
2000 REM **--FILE ROUTINE--**
2020 TEXT
2040 ONERR GOTO 2580: REM FIRST USE ONLY
2060 PRINT D$;"OPEN INVENTORY, L100"
2080 PRINT D$;"READ INVENTORY,R0"
2100 INPUT PTR
2120 PTR = PTR + 1: POKE 216,0: REM RESET ERROR FLAG
2140 PRINT D$;"WRITE INVENTORY,R";PTR;",B";0
2160 PRINT ITEM$
2180 PRINT D$;"WRITE INVENTORY,R";PTR;",B";25
2200 PRINT SERIAL$
2220 PRINT D$;"WRITE INVENTORY,R";PTR;",B";40
2240 PRINT CST$
2260 PRINT D$;"WRITE INVENTORY,R";PTR;",B";50
2280 PRINT ROOM$
2300 PRINT D$;"WRITE INVENTORY,R";PTR;",B";70
2320 PRINT DESC$
2340 PRINT D$;"WRITE INVENTORY,R0"
2360 PRINT PTR
2380 PRINT D$;"CLOSE INVENTORY"
2400 TEXT : HOME
2420 VTAB 5
2440 PRINT "DO YOU WANT TO ADD MORE ITEMS?"
2460 PRINT
2480 INPUT "TYPE 'NO' TO STOP ";NO$
```

```
2500  IF NO$ = "NO" THEN PRINT D$; "RUN HOME MENU"
2520  GOTO 100: REM BEGIN AGAIN
2540 :
2560 :
2580  REM **--FIRST USE ONLY--**
2600  POKE 216,0: REM RESET ERROR FLAG
2620  PRINT D$; "WRITE INVENTORY, R0"
2640  PRINT "0"
2660  PRINT D$; "CLOSE INVENTORY"
2680  GOTO 2000: REM BEGIN FILE ROUTINE AGAIN
2700 :
2720 :
5000  REM **--SUBROUTINE--**
5040  HTAB 1
5060  FOR I = 1 TO SP
5080  PRINT CHR$ (95);: REM UNDERLINE
5100  NEXT I
5120  HTAB 1
5160  RETURN
```

## DISPLAY HOME INVENTORY

```
10  REM **--DISPLAY HOME INVENTORY--**
11 :
12 :
20  D$ = CHR$ (4): REM CONTROL D
40  PRINT D$; "OPEN INVENTORY, L100"
60  PRINT D$; "READ INVENTORY, R0"
80  INPUT PTR
90  HOME
100  FOR I = 1 TO PTR
120  PRINT D$; "READ INVENTORY, R"; I; ",B"; 0
140  INPUT ITEM$
160  PRINT D$; "READ INVENTORY, R"; I; ",B"; 25
180  INPUT SERIAL$
200  PRINT D$; "READ INVENTORY, R"; I; ",B"; 40
220  INPUT CST$
240  PRINT D$; "READ INVENTORY, R"; I; ",B"; 50
260  INPUT ROOM$
280  PRINT D$; "READ INVENTORY, R"; I; ",B"; 70
300  INPUT DESC$
320  PRINT D$
```

```
340 PRINT I;" ";ITEM$;
360 HTAB 25: PRINT SERIAL$
380 PRINT "$";CST$;
400 HTAB 15: PRINT ROOM$
420 PRINT DESC$
440 PRINT : PRINT
450 TTLCST = TTLCST + VAL (CST$)
460 NEXT I
480 PRINT D$;"CLOSE INVENTORY"
500 PRINT : PRINT : PRINT "TOTAL VALUE OF ITEMS = $";
    TTLCST
510 PRINT
520 INPUT "HIT RETURN TO RETURN TO MENU ";L$
540 PRINT D$;"RUN HOME MENU"
```

## SEARCH HOME INVENTORY

```
10  REM ***--SEARCH/SORT RECORDS--***
11 :
12 :
13 :
20  D$ = CHR$ (4): REM CONTROL D
22  PRINT D$;"OPEN INVENTORY, L100"
24  PRINT D$;"READ INVENTORY, R0"
26  INPUT PTR
28  PRINT D$: DIM C$(PTR)
30 :
32 :
40  REM **--MENU ROUTINE--**
50  HOME : VTAB 3
60  TB = 8: HTAB 12
80  PRINT "SEARCH/SORT MENU"
100 PRINT : PRINT
120 HTAB TB
140 PRINT "1. SEARCH FOR ITEM"
160 PRINT : HTAB TB
180 PRINT "2. SEARCH FOR SERIAL #"
200 PRINT : HTAB TB
220 PRINT "3. SEARCH FOR COST"
240 PRINT : HTAB TB
260 PRINT "4. SEARCH FOR ROOM ITEMS"
280 PRINT : HTAB TB
```

## SEARCH HOME INVENTORY

```
300  PRINT "5. SORT ITEMS ALPHABETICALLY"
320  PRINT : HTAB TB
340  PRINT "6. SORT ITEMS BY SERIAL #"
360  PRINT : HTAB TB
380  PRINT "7. RETURN TO MAIN MENU"
400  PRINT : HTAB TB
420  INPUT "WHICH NUMBER "; NUMBER
440  IF NUMBER < 1 OR NUMBER > 7 THEN PRINT "INCORRECT
     NUMBER!": GOTO 400
510  IF NUMBER = 1 THEN 1000
520  IF NUMBER = 2 THEN 2000
530  IF NUMBER = 3 THEN 3000
540  IF NUMBER = 4 THEN 4000
550  IF NUMBER = 5 THEN 5000
560  IF NUMBER = 6 THEN 6000
570  IF NUMBER = 7 THEN 7000
970 :
980 :
990 :
1000 REM **--SEARCH FOR ITEM--**
1020 HOME : VTAB 5
1040 HTAB TB
1060 INPUT "WHICH ITEM? "; SRCH$
1080 I = 1: BYTE = 0
1100 GOSUB 10000 : REM SEARCH ROUTINE
1120 PRINT ITEM$;: HTAB 25: PRINT SERIAL$
1140 PRINT CST$;: HTAB 15: PRINT ROOM$
1160 PRINT DESC$
1180 PRINT : ITEM$ = "" : SERIAL$ = "" : CST$ = "" : ROOM$
     = "" : DESC $ = ""
1200 IF I = PTR OR I > PTR THEN 1260
1220 INPUT "SEARCH FOR MORE? "; YES$
1240 IF YES$ = "Y" THEN GOTO 1100
1260 PRINT
1280 GOSUB 9000: REM HOUSKEEPING
1300 GOTO 40: REM MENU
1970 :
1980 :
1990 :
2000 REM **--SEARCH FOR SERIAL #--**
2020 HOME : VTAB 5
2040 HTAB TB
2060 INPUT "WHICH SERIAL # "; SRCH$
2080 I = 1: BYTE = 25
```

## CHAPTER 12  HOME INVENTORY SYSTEM

```
2100  GOSUB 10000: REM SEARCH ROUTINE
2120  PRINT SERIAL$;: HTAB 15: PRINT ITEM$
2140  PRINT
2160  GOSUB 9000: REM HOUSEKEEPING
2180  GOTO 40: REM MENU
2970 :
2980 :
2990 :
3000  REM **--SEARCH FOR COST--**
3020  HOME : VTAB 5: BYTE = 40: TTLAMT = 0: FIND$ = ""
3040  HTAB 14
3060  PRINT "SEARCH FOR ITEMS..."
3080  PRINT : HTAB TB
3100  PRINT "A...ABOVE A CERTAIN AMOUNT"
3120  PRINT : HTAB TB
3140  PRINT "B...BELOW A CERTAIN AMOUNT"
3160  PRINT : HTAB TB
3180  INPUT "WHICH LETTER 'A' OR 'B' ";LT$
3190  IF LT$ = "A" THEN 3220
3200  IF LT$ = "B" THEN 3500
3210  PRINT "INCORRECT CHOICE ": GOTO 3160
3211 :
3212 :
3220  REM **--ITEMS ABOVE $ AMOUNT--**
3230  HTAB TB
3240  INPUT "ABOVE WHICH AMOUNT? ";AMT
3250  HOME : VTAB 2: HTAB 14
3260  PRINT "ITEMS ABOVE $";AMT
3270  FOR I = 1 TO PTR
3280  PRINT D$;"READ INVENTORY,R";I;",B";BYTE
3290  INPUT FIND$
3300  IF FIND$ = "D" THEN 3360
3310  IF AMT > VAL (FIND$) THEN 3360
3320  PRINT D$;"READ INVENTORY,R";I;",B";0
3330  INPUT ITEM$
3340 TTLAMT = TTLAMT + VAL (FIND$)
3350  PRINT ITEM$;: HTAB 30: PRINT FIND$
3360  NEXT I
3370  PRINT
3380  PRINT "TOTAL VALUE = $";TTLAMT
3390  PRINT : GOSUB 9000: REM HOUSEKEEPING
3400  GOTO 40: REM MENU
3496 :
3497 :
```

```
3500  REM **--ITEMS BELOW $ AMOUNT--**
3510  HTAB TB
3520  INPUT "BELOW WHICH AMOUNT ";AMT
3530  HOME : VTAB 2: HTAB 14
3540  PRINT "ITEMS BELOW $";AMT
3550  FOR I = 1 TO PTR
3560  PRINT D$;"READ INVENTORY,R";I;",B";BYTE
3570  INPUT FIND$
3580  IF FIND$ = "D" THEN 3640
3590  IF AMT < VAL (FIND$) THEN 3640
3600  PRINT D$;"READ INVENTORY,R";I;",B";0
3610  INPUT ITEM$
3620 TTLAMT = TTLAMT + VAL (FIND$)
3630  PRINT ITEM$;: HTAB 30: PRINT FIND$
3640  NEXT I
3650  PRINT
3660  PRINT "TOTAL VALUE = $";TTLAMT
3670  PRINT
3680  GOSUB 9000: REM HOUSEKEEPING
3690  GOTO 40: REM MENU
3970 :
3980 :
3990 :
4000  REM **--SEARCH FOR ROOM ITEMS--**
4020  HOME : VTAB 5: TLROOM = 0
4040  HTAB TB
4060  INPUT "WHICH ROOM ";SRCH$
4080  I = 1: BYTE = 50: HOME : VTAB 5
4100  HTAB 14: PRINT SRCH$: PRINT : PRINT
4120  GOSUB 10000: REM SEARCH ROUTINE
4140  PRINT ITEM$;: HTAB 25: PRINT SERIAL$
4160  PRINT CST$;: HTAB 11: PRINT DESC$
4180 TLROOM = TLROOM + VAL (CST$)
4200  PRINT
4220  IF I > PTR THEN 4280: REM SEARCH COMPLETED
4240 ITEM$ = "": SERIAL$ = "": CST$ = "": DESC$ = ""
4260  GOTO 4120: REM CONTINUE SEARCH
4280  PRINT
4300  PRINT "TOTAL VALUE FOR ";SRCH$;" = ";TLROOM
4320  PRINT : GOSUB 9000: REM HOUSEKEEPING
4340  GOTO 40: REM MENU
4970 :
4980 :
4990 :
```

```
5000 REM **--SORT ALPHABETICALLY--**
5020 HOME : VTAB 5
5040 HTAB TB
5060 INVERSE : PRINT "WORKING--PLEASE DONT'T TOUCH!!"
     : NORMAL
5080 Q = 1: REM VALID RECORD COUNTER
5100 FOR I = 1 TO PTR
5120 PRINT D$; "READ INVENTORY,R"; I; ",B"; 0
5140 INPUT C$
5160 IF C$ = "D" THEN 5220
5180 C$(Q) = C$
5200 Q = Q + 1
5220 NEXT I
5240 N = Q - 1
5260 PRINT: PRINT: HTAB TB
5280 INVERSE : PRINT "STILL WORKING--PLEASE WAIT!"
     : NORMAL
5300 GOSUB 20000: REM SORT ROUTINE
5320 REM DISPLAY RESULTS
5340 HOME: VTAB 5
5360 SPEED = 150
5380 FOR I = 1 TO Q - 1
5400 PRINT I; " "; C$(I)
5420 NEXT I
5440 PRINT
5460 GOSUB 9000: REM HOUSEKEEPING
5480 GOTO 40: REM MENU
5970 :
5980 :
5990 :
6000 REM **--SORT BY SERIAL #--**
6020 HOME: VTAB 5
6040 HTAB TB
6060 INVERSE: PRINT "WORKING--PLEASE DON'T TOUCH!!"
     : NORMAL
6080 Q = 1: REM VALID RECORD COUNTER
6100 FOR I = 1 TO PTR
6120 PRINT D$; "READ INVENTORY,R"; I; ",B"; 25
6140 INPUT C$
6160 IF C$ = "D" THEN 6280
6180 C$(Q) = C$
6200 PRINT D$; "READ INVENTORY,R"; I; ",B"; 0
6220 INPUT ITEM$
6240 C$(Q) = C$(Q) + "*" + ITEM$
```

```
6260 Q = Q + 1
6280 NEXT I
6300 N = Q - 1
6320 PRINT: PRINT: HTAB TB
6340 INVERSE: PRINT "STILL WORKING- -PLEASE WAIT! "
     : NORMAL
6360 GOSUB 20000: REM SORT ROUTINE
6380 REM DISPLAY RESULTS
6400 HOME: VTAB 5
6420 J = 1
6440 FOR I = 1 TO Q - 1
6460 LN = LEN (C$(I))
6480 PRINT I; " ";
6500 IF MID$ (C$(I), J, 1) = "*" THEN PRINT LEFT$ (C$(I),
     J - 1); : HTAB 20: PRINT MID$ (C$(I), J + 1, LN): GOTO 6540
6520 J = J + 1: GOTO 6500
6540 J = 1
6560 NEXT I
6580 PRINT
6600 GOSUB 9000: REM HOUSEKEEPING
6620 GOTO 40: REM MENU
6970 :
7000 REM **- -RETURN TO HOME MENU- -**
7020 PRINT D$; "CLOSE INVENTORY"
7040 PRINT D$; "RUN HOME MENU"
7970 :
7980 :
7990 :
9000 REM **- -HOUSEKEEPING- -**
9020 ITEM$ = ""
9040 SERIAL$ = ""
9060 CST$ = ""
9080 ROOM$ = ""
9100 DESC$ = ""
9120 PRINT D$; "PR#0"
9140 SPEED = 255
9160 PRINT D$: REM CANCEL INPUT FROM DISK
9180 INPUT "HIT RETURN TO CONTINUE "; L$
9900 RETURN
9970 :
9980 :
9990 :
10000 REM **- - SEARCH SUBROUTINE- -**
10020 PRINT D$; "READ INVENTORY, R"; I; ", B"; BYTE
```

```
10040  INPUT FIND$
10060  IF FIND$ = "D" THEN 10100
10080  IF SRCH$ = FIND$ THEN 10200
10100 I = I + 1
10120  IF I < PTR OR I = PTR THEN 10000
10140  PRINT : HTAB TB
10160  PRINT "SEARCH COMPLETED! ": FOR K = 1 TO 1000:
       NEXT K
10180  RETURN
10200  PRINT D$; "READ INVENTORY,R"; I; ",B"; 0
10220  INPUT ITEM$
10240  PRINT D$; "READ INVENTORY,R"; I; ",B"; 25
10260  INPUT SERIAL$
10280  PRINT D$; "READ INVENTORY,R"; I; ",B"; 40
10300  INPUT CST$
10320  PRINT D$; "READ INVENTORY,R"; I; ",B"; 50
10340  INPUT ROOM$
10360  PRINT D$; "READ INVENTORY,R"; I; ",B"; 70
10380  INPUT DESC$
10400 I = I + 1
10420  PRINT D$: REM CANCEL INPUT FROM DISK
10440  RETURN
19970 :
19980 :
19990 :
20000  REM **--SORT SUBROUTINE--**
20020 M = N
20040 M = INT (M / 2)
20060  IF M = 0 THEN 20300
20080 J = 1: K = N - M
20100 I = J
20120 L = I + M
20140  IF C$(I) < C$(L) THEN 20240
20160 T$ = C$(I) : C$(I) = C$(L) : C$(L) = T$
20180 I = I - M
20200  IF I < 1 THEN 20240
20220  GOTO 20120
20240 J = J + 1
20260  IF J > K THEN 20040
20280  GOTO 20100
20300  RETURN
```

## CORRECT HOME INVENTORY

```
10   REM **--CORRECT HOME INVENTORY--**
11 :
12 :
20   D$ = CHR$(4) : REM CONTROL D
40   TB = 8: REM HTAB VALUE
60   PRINT D$;"OPEN INVENTORY, L100"
70   PRINT D$;"READ INVENTORY, R0"
80   INPUT PTR
90   PRINT D$: REM CANCEL INPUT FROM DISK
95 :
96 :
100  REM **--MENU ROUTINE--**
120  HOME : VTAB 5
140  HTAB 12
160  PRINT "CORRECT/DELETE MENU"
180  PRINT : PRINT
200  HTAB TB
220  PRINT "C...CORRECT INVENTORY RECORD"
240  PRINT : HTAB TB
260  PRINT "D...DELETE INVENTORY RECORD"
280  PRINT : HTAB TB
300  PRINT "R...RETURN TO HOME MENU"
320  PRINT : HTAB TB
340  INPUT "WHICH LETTER PLEASE? ";LT$
360  IF LT$ = "C" THEN 1000
380  IF LT$ = "D" THEN 2000
400  IF LT$ = "R" THEN 3000
420  PRINT : HTAB TB
440  PRINT "INCORRECT CHOICE": GOTO 320
970 :
980 :
990 :
1000 REM **--CORRECT RECORD--**
1005 HOME
1010 POKE 32,7: POKE 34,7: REM SET WINDOW
1020 HOME
1040 FLAG$ = "NO": REM INFO HAS YET TO BE CHANGED
1050 PRINT "TYPE A '0' TO RETURN TO MENU": PRINT
1060 INPUT "CORRECT WHICH RECORD? ";REC
1070 IF REC = 0 THEN TEXT : GOTO 100: REM MENU
1075 IF REC > PTR THEN PRINT "INCORRECT CHOICE": GOTO
     1060
```

```
1080 GOSUB 6000: REM READ FILE
1120 REM **--DISPLAY FOR CORRECTION--**
1140 HOME
1160 PRINT "1. "; ITEM$
1180 PRINT "2. "; SERIAL$
1200 PRINT "3. "; CST$
1220 PRINT "4. "; ROOM$
1240 PRINT "5. "; DESC$
1260 PRINT : PRINT
1280 INPUT "IS THIS CORRECT ('Y' OR 'N') "; YES$
1290 IF YES$ = "Y" AND FLAG$ = "NO" THEN 1000
1300 IF YES$ = "Y" AND FLAG$ = "YES" THEN 7000:
     REM FILE ROUTINE
1310 INPUT "WHICH NUMBER IS WRONG "; NB
1320 IF NB < 1 OR NB > 5 THEN PRINT "INCORRECT CHOICE"
     : GOTO 1320
1340 IF NB = 1 THEN SP = 25
1360 IF NB = 2 THEN SP = 15
1380 IF NB = 3 THEN SP = 10
1400 IF NB = 4 THEN SP = 20
1420 IF NB = 5 THEN SP = 30
1440 PRINT
1460 PRINT "TYPE IN CORRECT INFO"
1480 INPUT CT$(NB)
1500 IF LEN (CT$(NB)) > SP THEN PRINT "TOO LONG--TRY
     AGAIN PLEASE": GOTO 1460
1520 IF NB = 1 THEN ITEM$ = CT$(NB)
1540 IF NB = 2 THEN SERIAL$ = CT$(NB)
1560 IF NB = 3 THEN CST$ = CT$(NB)
1580 IF NB = 4 THEN ROOM$ = CT$(NB)
1600 IF NB = 5 THEN DESC$ = CT$(NB)
1610 FLAG$= "YES": REM INFO HAS BEEN CHANGED
1620 GOTO 1120: REM CHECK AGAIN
1997 :
1998 :
1999 :
2000 REM **--DELETE RECORD--**
2020 HOME
2040 POKE 32,7: POKE 34,7: REM SET WINDOW
2060 HOME
2100 PRINT "TYPE A '0' TO RETURN TO MENU": PRINT
2120 INPUT "DELETE WHICH RECORD "; REC
2140 IF REC = 0 THEN TEXT: GOTO 100: REM MENU
```

## CORRECT HOME INVENTORY

```
2150 IF REC > PTR THEN PRINT "INCORRECT CHOICE":
     GOTO 2120
2160 GOSUB 6000: REM READ RECORD
2180 HOME
2200 PRINT ITEM$
2220 PRINT SERIAL$
2240 PRINT CST$
2260 PRINT ROOM$
2280 PRINT DESC$
2300 PRINT : PRINT
2320 INPUT "DELETE THIS RECORD? ";YES$
2340 IF YES$ = "Y" THEN 2380
2360 TEXT: GOTO 2000
2380 PRINT "ARE YOU SURE?": PRINT
2390 INPUT "TYPE 'YES' TO DELETE RECORD ";YES$
2400 IF YES$ = "YES" THEN 2440
2420 TEXT: GOTO 2000
2440 ITEM$ = "D"
2460 SERIAL$ = "D"
2480 CST$ = "D"
2500 ROOM$ = "D"
2520 DESC$ = "D"
2540 GOTO 7000: REM FILE ROUTINE
2970 :
2980 :
2990 :
3000 REM **--RETURN TO HOME MENU--**
3020 TEXT: PRINT D$;"CLOSE INVENTORY"
3040 PRINT D$;"RUN HOME MENU"
3970 :
3980 :
3990 :
6000 REM **--READ FILE ROUTINE--**
6020 PRINT D$;"READ INVENTORY,R";REC;",B";0
6040 INPUT ITEM$
6060 PRINT D$;"READ INVENTORY,R";REC;",B";25
6080 INPUT SERIAL$
6100 PRINT D$;"READ INVENTORY,R";REC;",B";40
6120 INPUT CST$
6140 PRINT D$;"READ INVENTORY,R";REC;",B";50
6160 INPUT ROOM$
6180 PRINT D$;"READ INVENTORY,R";REC;",B";70
6200 INPUT DESC$
6220 PRINT D$: REM CANCEL INPUT FROM DISK
```

```
6240 RETURN
6970 :
6980 :
6990 :
7000 REM **--FILE ROUTINE--**
7020 TEXT
7040 PRINT D$;"WRITE INVENTORY,R";REC;",B";0
7060 PRINT ITEM$
7080 PRINT D$;"WRITE INVENTORY,R";REC;",B";25
7100 PRINT SERIAL$
7120 PRINT D$;"WRITE INVENTORY,R";REC;",B";40
7140 PRINT CST$
7160 PRINT D$;"WRITE INVENTORY,R";REC;",B";50
7180 PRINT ROOM$
7200 PRINT D$;"WRITE INVENTORY,R";REC;",B";70
7220 PRINT DESC$
7240 PRINT D$;"CLOSE INVENTORY"
7260 PRINT D$;"OPEN INVENTORY, L100"
7280 PRINT D$
7300 GOTO 100: REM MENU
```

## MENU

```
10 REM **--BACK ORDER SYSTEM--**
11 :
12 :
20 D$ = CHR$(4) : REM CONTROL D
40 TB = 8: REM HTAB VALUE
60 :
80 :
100 REM **--MENU ROUTINE--**
120 HOME : VTAB 5
140 HTAB TB
160 PRINT "BACK ORDER SYSTEM"
180 PRINT : PRINT : PRINT
200 HTAB TB
220 PRINT "1. WRITE RECORD"
240 PRINT : HTAB TB
260 PRINT "2. READ RECORD"
280 PRINT : HTAB TB
300 PRINT "3. SEARCH RECORDS"
320 PRINT : HTAB TB
```

```
340 PRINT "4. CORRECT RECORD"
360 PRINT : HTAB TB
380 PRINT "5. END"
400 PRINT : HTAB TB
420 INPUT "WHICH NUMBER ";NUMBER
440 IF NUMBER < 1 OR NUMBER > 5 THEN 400
460 IF NUMBER = 1 THEN 1000
480 IF NUMBER = 2 THEN 2000
500 IF NUMBER = 3 THEN 3000
520 IF NUMBER = 4 THEN 4000
540 IF NUMBER = 5 THEN END
560 :
580 :
1000 REM **--WRITE RECORD--**
1020 PRINT D$;"RUN CREATE BACK ORDER"
1998 :
1999 :
2000 REM **--READ RECORD--**
2020 PRINT D$;"RUN DISPLAY BACK ORDER"
2998 :
2999 :
3000 REM **--SEARCH RECORDS--**
3020 PRINT D$;"RUN SEARCH BACK ORDER"
3998 :
3999 :
4000 REM **--CORRECT RECORDS--**
4020 PRINT D$;"RUN CORRECT BACK ORDER"
```

## CREATE BACK ORDER

```
10  REM **--CREATE BACK ORDER--**
11 :
12 :
20  D$ = CHR$(4): REM CONTROL D
40  TB = 8: REM HTAB VALUE
60 :
80 :
100 REM **--INPUT ROUTINE--**
120 HOME: VTAB 5
140 HTAB TB
160 PRINT "CREATE BACK ORDER"
180 PRINT : PRINT
185 VTAB 20: HTAB 8
190 PRINT "CR = LAST RECORD": PRINT : HTAB 8 : PRINT "TYPE
    '-' FOR NO VALUE"
200 POKE 34,7: POKE 32,7: POKE 35,19
220 :
240 :
260 HOME
280 PRINT "ITEM NAME PLEASE. "
300 PRINT: PRINT
320 SP = 25
340 GOSUB 5000: REM INPUT SUBROUTINE
360 INPUT ""; ITEM$
380 IF LEN (ITEM$) > SP THEN PRINT " ": GOTO 260 :
    REM 5 CTRL G'S
385 IF ITEM$ = "" THEN ITEM$ = A$
390 A$ = ITEM$
400 :
420 :
440 HOME
460 PRINT "ITEM DESCRIPTION PLEASE. "
480 PRINT: PRINT
500 SP = 30
520 GOSUB 5000: REM INPUT SUBROUTINE
540 INPUT ""; DESC$
560 IF LEN (DESC$) > SP THEN PRINT "": GOTO 440 :
    REM 5 CTRL G'S
575 IF DESC$ = "" THEN DESC$= B$
576 B$ = DESC$
580 :
```

# CREATE BACK ORDER

```
600 :
620 HOME
640 PRINT "INDIVIDUAL'S NAME PLEASE."
660 PRINT: PRINT
680 SP = 20
700 GOSUB 5000: REM INPUT SUBROUTINE
720 INPUT "";NAME$
740 IF LEN (NAME$) > SP THEN PRINT "" : GOTO 620 :
    REM 5 CTRL G'S
755 IF NAME$ = "" THEN NAME$ = C$
756 C$ = NAME$
760 :
780 :
800 HOME
820 PRINT "PHONE #"
840 PRINT: PRINT
860 SP = 20
880 GOSUB 5000: REM INPUT SUBROUTINE
900 INPUT "";PHNE$
920 IF LEN (PHNE$) > SP THEN PRINT "": GOTO 800 :
    REM 5 CTRL G'S
935 IF PHNE$ = "" THEN PHNE$ = D1$
936 D1$ = PHNE$
940 :
960 :
980 HOME
1000 PRINT "DATE REQUEST WAS MADE"
1020 PRINT: PRINT
1040 SP = 10
1060 GOSUB 5000: REM INPUT SUBROUTINE
1080 INPUT "";DTE$
1100 IF LEN (DTE$) > SP THEN PRINT "": GOTO 980 :
     REM 5 CTRL G'S
1105 IF DTE$ = "" THEN DTE$ = E$
1110 E$ = DTE$
1120 :
1140 :
1160 HOME
1180 PRINT "ORDERED YET ('Y' OR 'N')"
1200 PRINT: PRINT
1220 SP = 1
1240 GOSUB 5000: REM INPUT SUBROUTINE
1260 INPUT "";OD$
1280 IF LEN (OD$) > SP THEN PRINT "" : GOTO 1160:
     REM 5 CTRL G'S
```

## CHAPTER 12 HOME INVENTORY SYSTEM

```
1285 IF OD$ = "" THEN OD$ = F$
1290 F$ = OD$
1291 :
1292 :
1300 HOME
1310 PRINT "AMOUNT DEPOSITED"
1320 PRINT : PRINT
1330 SP = 10
1340 GOSUB 5000 REM INPUT SUBROUTINE
1350 INPUT "";AMT$
1360 IF LEN (AMT$) > SP THEN PRINT "": GOTO 1300 :
     REM 5 CTRL G'S
1365 IF AMT$ = "" THEN AMT$ = G$
1370 G$ = AMT$
1391 :
1392 :
1400 REM **--DISPLAY FOR CORRECTION--**
1410 TEXT : POKE 34,7: POKE 32,7
1420 HOME
1440 PRINT "1. ";ITEM$
1460 PRINT "2. ";DESC$
1480 PRINT "3. ";NAME$
1500 PRINT "4. ";PHNE$
1520 PRINT "5. ";DTE$
1525 PRINT "6. ";OD$
1530 PRINT "7. ";AMT$
1540 PRINT : PRINT
1560 INPUT "IS THIS CORRECT ('Y' OR 'N') ";YES$
1580 IF YES$ = "Y" THEN 2000: REM FILE ROUTINE
1600 INPUT "WHICH NUMBER IS WRONG ";NB
1610 IF NB < 1 OR NB > 7 THEN PRINT "INCORRECT CHOICE"
     : GOTO 1600
1620 IF NB = 1 THEN SP = 25
1640 IF NB = 2 THEN SP = 30
1660 IF NB = 3 THEN SP = 20
1680 IF NB = 4 THEN SP = 20
1700 IF NB = 5 THEN SP = 10
1705 IF NB = 6 THEN SP = 1
1710 IF NB = 7 THEN SP = 10
1720 PRINT
1740 PRINT "TYPE IN CORRECT INFO "
1760 INPUT CT$(NB)
1780 IF LEN (CT$(NB)) > SP THEN PRINT "TOO LONG--TRY
     AGAIN PLEASE": GOTO 1740
```

```
1800  IF NB = 1 THEN ITEM$ = CT$(NB)
1820  IF NB = 2 THEN DESC$ = CT$(NB)
1840  IF NB = 3 THEN NAME$ = CT$(NB)
1860  IF NB = 4 THEN PHNE$ = CT$(NB)
1880  IF NB = 5 THEN DTE$ = CT$(NB)
1885  IF NB = 6 THEN OD$ = CT$(NB)
1890  IF NB = 7 THEN AMT$ = CT$(NB)
1900  GOTO 1400: REM CHECK AGAIN
1998 :
1999 :
2000  REM **--FILE ROUTINE--**
2020  TEXT
2040  ONERR GOTO 2580: REM FIRST USE ONLY
2060  PRINT D$;"OPEN BACKORDER,L120"
2080  PRINT D$;"READ BACKORDER,R0"
2100  INPUT PTR
2120  PTR = PTR + 1: POKE 216,0: REM RESET ERROR FLAG
2140  PRINT D$;"WRITE BACKORDER,R";PTR;",B";0
2160  PRINT ITEM$
2180  PRINT D$;"WRITE BACKORDER,R";PTR;",B";25
2200  PRINT DESC$
2220  PRINT D$;"WRITE BACKORDER,R";PTR;",B";55
2240  PRINT NAME$
2260  PRINT D$;"WRITE BACKORDER,R";PTR;",B";75
2280  PRINT PHNE$
2300  PRINT D$;"WRITE BACKORDER,R";PTR;",B";95
2320  PRINT DTE$
2325  PRINT D$;"WRITE BACKORDER,R";PTR;",B";105
2330  PRINT OD$
2335  PRINT D$;"WRITE BACKORDER,R";PTR;",B";110
2337  PRINT AMT$
2340  PRINT D$;"WRITE BACKORDER,R0"
2360  PRINT PTR
2380  PRINT D$;"CLOSE BACKORDER"
2400  TEXT: HOME
2420  VTAB 5
2440  PRINT "DO YOU WANT TO ADD MORE ITEMS?"
2460  PRINT
2480  INPUT "TYPE 'NO' TO STOP ";NO$
2500  IF NO$ = "NO" THEN PRINT D$;"RUN MENU"
2520  GOTO 100: REM BEGIN AGAIN
2540 :
2560 :
```

```
2580 REM **--FIRST USE ONLY--**
2600 POKE 216,0: REM RESET ERROR FLAG
2620 PRINT D$;"WRITE BACKORDER,R0"
2640 PRINT "0"
2660 PRINT D$;"CLOSE BACKORDER"
2680 GOTO 2000: REM BEGIN FILE ROUTINE AGAIN
2700 :
2720 :
5000 REM **--SUBROUTINE--**
5040 HTAB 1
5060 FOR I = 1 TO SP
5080 PRINT CHR$ (95);: REM UNDERLINE
5100 NEXT I
5120 HTAB 1
5160 RETURN
```

## DISPLAY BACK ORDER

```
10  REM **--DISPLAY BACK ORDER--**
11  :
12  :
20  D$ = CHR$ (4): REM CONTROL D
40  PRINT D$;"OPEN BACKORDER, L120"
60  PRINT D$;"READ BACKORDER,R0"
80  INPUT PTR
90  HOME
100 FOR I = 1 TO PTR
120 PRINT D$;"READ BACKORDER,R";I;",B";0
140 INPUT ITEM$
160 PRINT D$;"READ BACKORDER,R";I;",B";25
180 INPUT DESC$
200 PRINT D$;"READ BACKORDER,R";I;",B";55
220 INPUT NAME$
240 PRINT D$;"READ BACKORDER,R";I;",B";75
260 INPUT PHNE$
280 PRINT D$;"READ BACKORDER,R";I;",B";95
300 INPUT DTE$
305 PRINT D$;"READ BACKORDER,R";I;",B";105
310 INPUT OD$
315 PRINT D$;"READ BACKORDER,R";I;",B";110
317 INPUT AMT$
320 PRINT D$
```

```
340 PRINT I;" ";ITEM$;
360 HTAB 15: PRINT DESC$
380 PRINT NAME$;
400 HTAB 15: PRINT PHNE$
420 PRINT DTE$;: HTAB 12: PRINT OD$;: HTAB 15 : PRINT AMT$
440 PRINT: PRINT
460 NEXT I
480 PRINT D$;"CLOSE INVENTORY"
510 PRINT
520 INPUT "HIT RETURN TO RETURN TO MENU ";L$
540 PRINT D$;"RUN MENU"
```

## SEARCH BACK ORDER

```
10  REM ***--SEARCH/SORT RECORDS--***
11 :
12 :
13 :
20  D$ = CHR$ (4): REM CONTROL D
22  PRINT D$;"OPEN BACKORDER, L120"
24  PRINT D$;"READ BACKORDER, R0"
26  INPUT PTR
28  PRINT D$: DIM C$(PTR)
30 :
32 :
40  REM **--MENU ROUTINE--**
50  HOME: VTAB 3
60  TB = 8: HTAB 12
80  PRINT "SEARCH/SORT MENU"
100 PRINT: PRINT
120 HTAB TB
140 PRINT "1. SEARCH FOR ITEM"
160 PRINT: HTAB TB
180 PRINT "2. SEARCH FOR NAME"
200 PRINT : HTAB TB
220 PRINT "3. SEARCH FOR DATE"
240 PRINT : HTAB TB
260 PRINT "4. ITEMS NOT YET ORDERED"
280 PRINT : HTAB TB
300 PRINT "5. SORT ITEMS ALPHABETICALLY"
320 PRINT : HTAB TB
340 PRINT "6. SORT BY NAME"
```

```
360 PRINT : HTAB TB
380 PRINT "7. RETURN TO MAIN MENU"
400 PRINT : HTAB TB
420 INPUT "WHICH NUMBER ";NUMBER
440 IF NUMBER < 1 OR NUMBER > 7 THEN PRINT "INCORRECT
    NUMBER!": GOTO 400
510 IF NUMBER = 1 THEN 1000
520 IF NUMBER = 2 THEN 2000
530 IF NUMBER = 3 THEN 3000
540 IF NUMBER = 4 THEN 4000
550 IF NUMBER = 5 THEN 6000
560 IF NUMBER = 6 THEN 6000
570 IF NUMBER = 7 THEN 7000
970 :
980 :
990 :
1000 REM **--SEARCH FOR ITEM--**
1020 HOME : VTAB 5
1040 HTAB TB
1060 INPUT "WHICH ITEM? ";SRCH$
1080 I = 1: BYTE = 0
1090 GOSUB 8000: REM PRINT ROUTINE
1100 GOSUB 10000: REM SEARCH ROUTINE
1120 PRINT ITEM$;: HTAB 25: PRINT NAME$
1140 PRINT DTE$;: HTAB 10: PRINT PHNE$;:HTAB 30: PRINT
     OD$
1160 PRINT DESC$
1180 PRINT : ITEM$ = "":NAME$ = "":DTE$= "" :PHNE$
     = "":DESC$ = "":OD$ = "":AMT$ = ""
1200 IF I > PTR THEN 1260
1220 INPUT "SEARCH FOR MORE? ";YES$
1240 IF YES$ = "Y" THEN GOTO 1100
1260 PRINT
1280 GOSUB 9000: REM HOUSKEEPING
1300 GOTO 40: REM MENU
1970 :
1980 :
1990 :
2000 REM **--SEARCH FOR NAME--**
2020 HOME : VTAB 5
2040 HTAB TB
2060 INPUT "WHICH NAME? ";SRCH$
2070 GOSUB 8000 : REM PRINT ROUTINE
2080 I = 1:BYTE = 55
```

```
2100  GOSUB 10000: REM SEARCH ROUTINE
2120  PRINT ITEM$;: HTAB 25: PRINT NAME$
2140  PRINT DTE$;: HTAB 10: PRINT PHNE$;: HTAB 30 :
      PRINT OD$
2160  PRINT DESC$
2180  PRINT : ITEM$ = "": NAME$ = "" : DTE$ = "": PHNE$
      = "": DESC$ = "": OD$ = "": AMT$ = ""
2200  IF I = PTR OR I > PTR THEN 2260
2220  INPUT "SEARCH FOR MORE? ";YES$
2240  IF YES$ = "Y" THEN GOTO 1100
2260  PRINT
2280  GOSUB 9000: REM HOUSKEEPING
2300  GOTO 40: REM MENU
2970  :
2980  :
2990  :
3000  REM **--SEARCH FOR DATE--**
3020  HOME : VTAB 5: BYTE = 95 : FIND$ = ""
3040  HTAB 14
3060  PRINT "SEARCH FOR ITEMS..."
3080  PRINT : HTAB TB
3100  PRINT "A...AFTER A CERTAIN DATE"
3120  PRINT : HTAB TB
3140  PRINT "B...BEFORE A CERTAIN DATE"
3160  PRINT : HTAB TB
3180  INPUT "WHICH LETTER 'A' OR 'B' ";LT$
3190  IF LT$ = "A" THEN 3220
3200  IF LT$ = "B" THEN 3500
3210  PRINT "INCORRECT CHOICE ": GOTO 3160
3211  :
3212  :
3220  REM **--ITEMS AFTER DATE--**
3230  HTAB TB
3240  INPUT "AFTER WHICH DATE? ";DS$
3245  GOSUB 8000 : REM PRINT ROUTINE
3250  HOME : VTAB 2: HTAB 14
3260  PRINT "ITEMS AFTER ";DS$
3270  FOR I = 1 TO PTR
3280  PRINT D$; "READ BACKORDER, R"; I; ",B";BYTE
3290  INPUT FIND$
3300  IF FIND$ = "D" THEN 3360
3310  IF DS$ > FIND$ THEN 3360
3320  PRINT D$; "READ BACKORDER, R" ; I; ",B"; 0
3330  INPUT ITEM$
```

```
3340 PRINT D$; "READ BACKORDER, R"; I; ", B"; 55
3345 INPUT NAME$
3350 PRINT ITEM$;: HTAB 30: PRINT FIND$
3355 PRINT NAME$
3357 PRINT
3360 NEXT I
3370 PRINT
3380 REM
3390 PRINT : GOSUB 9000: REM HOUSEKEEPING
3400 GOTO 40: REM MENU
3496 :
3497 :
3500 REM **--ITEMS BEFORE DATE--**
3510 HTAB TB
3520 INPUT "BEFORE WHICH DATE "; DS$
3525 GOSUB 8000 : REM PRINT ROUTINE
3530 HOME : VTAB 2: HTAB 14
3540 PRINT "ITEMS BEFORE "; DS$
3550 FOR I = 1 TO PTR
3560 PRINT D$; "READ BACKORDER, R"; I; ", B"; BYTE
3570 INPUT FIND$
3580 IF FIND$ = "D" THEN 3640
3590 IF VAL (DS$) < VAL (FIND$) THEN 3640
3600 PRINT D$; "READ BACKORDER, R"; I; ", B"; 0
3610 INPUT ITEM$
3615 PRINT D$; "READ BACKORDER, R"; I; ", B"; 55
3620 INPUT NAME$
3630 PRINT ITEM$;: HTAB 30: PRINT FIND$
3635 PRINT NAME$
3637 PRINT
3640 NEXT I
3650 PRINT
3660 REM
3670 PRINT
3680 GOSUB 9000: REM HOUSEKEEPING
3690 GOTO 40: REM MENU
3970 :
3980 :
3990 :
4000 REM **--SEARCH FOR ROOM ITEMS--**
4020 HOME : VTAB 5
4040 HTAB TB
4060 SRCH$ = "N"
4070 GOSUB 8000 : REM PRINT ROUTINE
```

## SEARCH BACK ORDER

```
4080 I = 1: BYTE = 105: HOME : VTAB 5
4100 HTAB 10: PRINT "ITEMS NOT ORDERED YET" : PRINT: PRINT
4120 GOSUB 10000: REM SEARCH ROUTINE
4140 PRINT ITEM$;: HTAB 25: PRINT NAME$
4150 PRINT PHNE$;: HTAB 20: PRINT DTE$; : HTAB 30:
     PRINT AMT$
4160 PRINT DESC$
4180 TLROOM = TLROOM + VAL (CST$)
4200 PRINT
4220 IF I > PTR THEN 4280: REM SEARCH COMPLETED
4240 ITEM$ = "": DESC$ = "": NAME$ = "" : PHNE$ = "":
     DTE$ = "": AMT$ = ""
4260 GOTO 4120: REM CONTINUE SEARCH
4280 PRINT
4300 REM
4320 PRINT : GOSUB 9000: REM HOUSEKEEPING
4340 GOTO 40: REM MENU
4970 :
4980 :
4990 :
5000 REM **--SORT ALPHABETICALLY--**
5020 HOME : VTAB 5
5030 IF NUMBER = 5 THEN BYTE = 0
5035 IF NUMBER = 6 THEN BYTE = 55
5040 HTAB TB
5060 INVERSE : PRINT "WORKING--PLEASE DON'T
     TOUCH!!": NORMAL
5080 Q = 1: REM VALID RECORD COUNTER
5100 FOR I = 1 TO PTR
5120 PRINT D$;"READ BACK ORDER,R";I;",B";BYTE
5140 INPUT C$
5160 IF C$ = "D" THEN 5220
5180 C$(Q) = C$
5200 Q = Q + 1
5220 NEXT I
5240 N = Q - 1
5260 PRINT : PRINT : HTAB TB
5280 INVERSE : PRINT "STILL WORKING--PLEASE WAIT!"
     : NORMAL
5300 GOSUB 20000: REM SORT ROUTINE
5320 REM DISPLAY RESULTS
5340 HOME : VTAB 5
5360 SPEED = 150
5380 FOR I = 1 TO Q - 1
```

```
5400  PRINT I;" ";C$(I)
5420  NEXT I
5440  PRINT
5460  GOSUB 9000: REM HOUSEKEEPING
5480  GOTO 40: REM MENU
5970 :
5980 :
5990 :
6000  REM **--SORT BY SERIAL#--**
6010  GOSUB 8000 : REM PRINT ROUTINE
6020  HOME : VTAB 5
6030  IF NUMBER = 5 THEN BYTE = 0
6035  IF NUMBER = 6 THEN BYTE = 55
6040  HTAB TB
6060  INVERSE : PRINT "WORKING--PLEASE DON'T
      TOUCH!!": NORMAL
6080  Q = 1: REM VALID RECORD COUNTER
6100  FOR I = 1 TO PTR
6120  PRINT D$;"READ BACKORDER,R";I;",B";BYTE
6140  INPUT C$
6160  IF C$ = "D" THEN 6280
6180  C$(Q) = C$
6200  REM
6220  REM
6240  C$(Q) = C$(Q) + "*" + STR$(I)
6260  Q = Q + 1
6280  NEXT I
6300  N = Q - 1
6320  PRINT : PRINT : HTAB TB
6340  INVERSE : PRINT "STILL WORKING--PLEASE WAIT!"
      : NORMAL
6360  GOSUB 20000: REM SORT ROUTINE
6380  REM DISPLAY RESULTS
6400  HOME : VTAB 5
6420  J = 1
6440  FOR I = 1 TO Q - 1
6460  LN = LEN (C$(I))
6480  PRINT I;" ";
6500  IF MID$ (C$(I),J,1) = "*" THEN PRINT LEFT$ (C$(I),J
      - 1): K$ = MID$ (C$(I),J + 1,LN): GOTO 6540
6520  J = J + 1: GOTO 6500
6540  K = VAL (K$)
6560  PRINT D$;"READ BACKORDER,R";K;",B";0
6570  INPUT ITEM$
```

## SEARCH BACK ORDER

```
6580 PRINT D$; "READ BACKORDER, R"; K; ", B"; 25
6590 INPUT DESC$
6600 PRINT D$; "READ BACKORDER, R"; K; ", B"; 55
6610 INPUT NAME$
6620 PRINT D$; "READ BACKORDER, R"; K; ", B"; 75
6630 INPUT PHNE$
6640 PRINT D$; "READ BACKORDER, R"; K; ", B"; 95
6650 INPUT DTE$
6660 PRINT D$; "READ BACKORDER, R"; K; ", B"; 105
6670 INPUT OD$
6680 PRINT D$; "READ BACKORDER, R"; K; ", B"; 110
6690 INPUT AMT$
6700 PRINT D$: REM CANCEL INPUT FROM DISK
6710 PRINT ITEM$
6720 PRINT DESC$
6730 PRINT NAME$
6740 PRINT PHNE$
6750 PRINT DTE$
6760 PRINT OD$
6770 PRINT AMT$
6780 PRINT : PRINT
6790 J = 1
6800 NEXT I
6820 PRINT
6840 GOSUB 9000: REM HOUSEKEEPING
6860 GOTO 40: REM MENU
6970 :
6980 :
6990 :
7000 REM **--RETURN TO HOME MENU--**
7020 PRINT D$; "CLOSE BACKORDER"
7040 PRINT D$; "RUN MENU"
7970 :
7980 :
7990 :
8000 REM **--PRINT ROUTINE--**
8010 PRINT : HTAB TB
8020 INPUT "DO YOU WANT A PRINTOUT? "; YES$
8040 IF YES$ = "Y" THEN 8080
8050 PRINT
8060 RETURN
8080 PRINT D$; "PR#1"
8090 PRINT
8100 RETURN
```

## CHAPTER 12  HOME INVENTORY SYSTEM

```
8980 :
8990 :
8995 :
9000 REM **--HOUSEKEEPING--**
9020 ITEM$ = ""
9040 NAME$ = ""
9050 PHNE$ = ""
9060 DTE$ = ""
9070 OD$ = ""
9080 AMT$ = ""
9100 DESC$ = ""
9120 PRINT D$; "PR#0"
9140 SPEED = 255
9160 PRINT D$: REM CANCEL INPUT FROM DISK
9180 INPUT "HIT RETURN TO CONTINUE "; L$
9900 RETURN
9970 :
9980 :
9990 :
10000 REM **-- SEARCH SUBROUTINE--**
10020 PRINT D$; "READ BACKORDER, R"; I; ",B"; BYTE
10040 INPUT FIND$
10060 IF FIND$ = "D" THEN 10100
10080 IF SRCH$ = FIND$ THEN 10200
10100 I = I + 1
10120 IF I < PTR OR I = PTR THEN 10000
10140 PRINT : HTAB TB
10160 PRINT "SEARCH COMPLETED! ": FOR K = 1 TO 1000:
      NEXT K
10180 RETURN
10200 PRINT D$; "READ BACKORDER, R"; I; ",B"; 0
10220 INPUT ITEM$
10240 PRINT D$; "READ BACKORDER, R"; I; ",B"; 25
10260 INPUT DESC$
10280 PRINT D$; "READ BACKORDER, R"; I; ",B"; 55
10300 INPUT NAME$
10320 PRINT D$; "READ BACKORDER, R"; I; ",B"; 75
10340 INPUT PHNE$
10360 PRINT D$; "READ BACKORDER, R"; I; ",B"; 95
10380 INPUT DTE$
10385 PRINT D$; "READ BACKORDER, R"; I; ",B"; 105
10390 INPUT OD$
10395 PRINT D$; "READ BACKORDER, R"; I; ",B"; 110
10397 INPUT AMT$
```

```
10400 I = I + 1
10420 PRINT D$ : REM CANCEL INPUT FROM DISK
10440 RETURN
19970 :
19980 :
20000 REM **--SORT SUBROUTINE--**
20020 M = N
20040 M = INT (M / 2)
20060 IF M = 0 THEN 20300
20080 J = 1: K = N - M
20100 I = J
20120 L = I + M
20140 IF C$(I) < C$(L) THEN 20240
20160 T$ = C$(I): C$(I) = C$(L): C$(L) = T$
20180 I = I - M
20200 IF I < 1 THEN 20240
20220 GOTO 20120
20240 J = J + 1
20260 IF J > K THEN 20040
20280 GOTO 20100
20300 RETURN
```

## CORRECT BACK ORDER

```
10  REM **--CORRECT BACK ORDER--**
11 :
12 :
20  D$ = CHR$(4): REM CONTROL D
40  TB = 8: REM HTAB VALUE
60  PRINT D$; "OPEN BACKORDER, L120"
70  PRINT D$; "READ BACKORDER, R0"
80  INPUT PTR
90  PRINT D$: REM CANCEL INPUT FROM DISK
95 :
96 :
100  REM **--MENU ROUTINE--**
120  HOME : VTAB 5
140  HTAB 12
160  PRINT "CORRECT/DELETE MENU"
180  PRINT : PRINT
200  HTAB TB
220  PRINT "C... CORRECT BACKORDER RECORD"
```

CHAPTER 12 HOME INVENTORY SYSTEM

```
240  PRINT : HTAB TB
260  PRINT "D...DELETE BACKORDER RECORD"
280  PRINT : HTAB TB
300  PRINT "R...RETURN TO MENU"
320  PRINT : HTAB TB
340  INPUT "WHICH LETTER PLEASE? ";LT$
360  IF LT$ = "C" THEN 1000
380  IF LT$ = "D" THEN 2000
400  IF LT$ = "R" THEN 3000
420  PRINT : HTAB TB
440  PRINT "INCORRECT CHOICE": GOTO 320
970 :
980 :
990 :
1000 REM **--CORRECT RECORD--**
1005 HOME
1010 POKE 32,7: POKE 34,7: REM SET WINDOW
1020 HOME
1040 FLAG$ = "NO": REM INFO HAS YET TO BE CHANGED
1050 PRINT "TYPE A '0' TO RETURN TO MENU": PRINT
1060 INPUT "CORRECT WHICH RECORD? ";REC
1070 IF REC = 0 THEN TEXT : GOTO 100: REM MENU
1075 IF REC > PTR THEN PRINT "INCORRECT CHOICE": GOTO
     1060
1080 GOSUB 6000: REM READ FILE
1120 REM **--DISPLAY FOR CORRECTION--**
1140 HOME
1160 PRINT "1.  ";ITEM$
1180 PRINT "2.  ";DESC$
1200 PRINT "3.  ";NAME$
1220 PRINT "4.  ";PHNE$
1240 PRINT "5.  ";DTE$
1245 PRINT "6.  ";OD$
1250 PRINT "7.  ";AMT$
1260 PRINT : PRINT
1280 INPUT "IS THIS CORRECT ('Y' OR 'N') ";YES$
1290 IF YES$ = "Y" AND FLAG$ = "NO" THEN 1000
1300 IF YES$ = "Y" AND FLAG$ = "YES" THEN 7000:
     REM FILE ROUTINE
1320 INPUT "WHICH NUMBER IS WRONG ";NB
1330 IF NB < 1 OR NB > 7 THEN PRINT "INCORRECT CHOICE"
     : GOTO 1320
1340 IF NB = 1 THEN SP = 25
1360 IF NB = 2 THEN SP = 30
```

```
1380  IF NB = 3 THEN SP = 20
1400  IF NB = 4 THEN SP = 20
1420  IF NB = 5 THEN SP = 10
1425  IF NB = 6 THEN SP = 1
1430  IF NB = 7 THEN SP = 10
1440  PRINT
1460  PRINT "TYPE IN CORRECT INFO "
1480  INPUT CT$(NB)
1500   IF LEN (CT$(NB)) > SP THEN PRINT "TOO LONG--TRY
      AGAIN PLEASE": GOTO 1460
1520  IF NB = 1 THEN ITEM$ = CT$(NB)
1540  IF NB = 2 THEN DESC$ = CT$(NB)
1560  IF NB = 3 THEN NAME$ = CT$(NB)
1580  IF NB = 4 THEN PHNE$ = CT$(NB)
1600  IF NB = 5 THEN DTE$ = CT$(NB)
1603  IF NB = 6 THEN OD$ = CT$(NB)
1606  IF NB = 7 THEN AMT$ = CT$(NB)
1610  FLAG$ = "YES": REM INFO HAS BEEN CHANGED
1620  GOTO 1120: REM CHECK AGAIN
1997 :
1998 :
1999 :
2000  REM **--DELETE RECORD--**
2020  HOME
2040  POKE 32,7: POKE 34,7: REM SET WINDOW
2060  HOME
2100  PRINT "TYPE A '0' TO RETURN TO MENU" : PRINT
2120  INPUT "DELETE WHICH RECORD ";REC
2140  IF REC = 0 THEN TEXT : GOTO 100: REM MENU
2150  IF REC > PTR THEN PRINT "INCORRECT CHOICE": GOTO
      2120
2160  GOSUB 6000: REM READ RECORD
2180  HOME
2200  PRINT ITEM$
2220  PRINT DESC$
2240  PRINT NAME$
2260  PRINT PHNE$
2280  PRINT DTE$
2285  PRINT OD$
2290  PRINT AMT$
2300  PRINT : PRINT
2320  INPUT "DELETE THIS RECORD? ";YES$
2340  IF YES$ = "Y" THEN 2380
2360  TEXT: GOTO 2000
```

```
2380 PRINT "ARE YOU SURE?": PRINT
2390 INPUT "TYPE 'YES' TO DELETE RECORD ";YES$
2400 IF YES$ = "YES" THEN 2440
2420 TEXT: GOTO 2000
2440 ITEM$ = "D"
2460 NAME$ = "D"
2480 DTE$ = "D"
2500 PHNE$ = "D"
2520 DESC$ = "D"
2525 OD$ = "D"
2530 AMT$ = "D"
2540 GOTO 7000: REM FILE ROUTINE
2970 :
2980 :
2990 :
3000 REM **--RETURN TO HOME MENU--**
3020 TEXT: PRINT D$;"CLOSE BACKORDER"
3040 PRINT D$;"RUN MENU"
3970 :
3980 :
3990 :
6000 REM **--READ FILE ROUTINE--**
6020 PRINT D$;"READ BACKORDER,R";REC;",B";0
6040 INPUT ITEM$
6060 PRINT D$;"READ BACKORDER,R";REC;",B";25
6080 INPUT DESC$
6100 PRINT D$;"READ BACKORDER,R";REC;",B";55
6120 INPUT NAME$
6140 PRINT D$;"READ BACKORDER,R";REC;",B";75
6160 INPUT PHNE$
6180 PRINT D$;"READ BACKORDER,R";REC;",B";95
6200 INPUT DTE$
6205 PRINT D$;"READ BACKORDER,R";REC;",B";105
6210 INPUT OD$
6215 PRINT D$;"READ BACKORDER,R";REC;",B";110
6217 INPUT AMT$
6220 PRINT D$: REM CANCEL INPUT FROM DISK
6240 RETURN
6970 :
6980 :
6990 :
7000 REM **--WRITE FILE ROUTINE--**
7020 TEXT
7040 PRINT D$;"WRITE BACKORDER,R";REC;",B";0
```

## CORRECT BACK ORDER

```
7060 PRINT ITEM$
7080 PRINT D$; "WRITE BACKORDER, R"; REC; ",B"; 25
7100 PRINT DESC$
7120 PRINT D$; "WRITE BACKORDER, R"; REC; ",B"; 55
7140 PRINT NAME$
7160 PRINT D$; "WRITE BACKORDER, R"; REC; ",B"; 75
7180 PRINT PHNE$
7200 PRINT D$; "WRITE BACKORDER, R"; REC; ",B"; 95
7220 PRINT DTE$
7225 PRINT D$; "WRITE BACKORDER, R"; REC; ",B"; 105
7230 PRINT OD$
7235 PRINT D$; "WRITE BACKORDER, R"; REC; ",B"; 110
7237 PRINT AMT$
7240 PRINT D$; "CLOSE BACKORDER"
7260 PRINT D$; "OPEN BACKORDER, L120"
7280 PRINT D$
7300 GOTO 100: REM MENU
```

# 13
# Planning a File System

Rather than present another chapter explaining the programming of yet another system, this chapter will present the procedures involved in conceiving and creating a file system. I will use a STOCK MARKET SYSTEM as the example. Although shorter than previous chapters, this chapter is no less important than preceding ones.

There are five main steps involved in conceiving and creating a specific data base system:

1. Know your subject.
2. Plan carefully and organize your thoughts.
3. Make preliminary decisions on the number of main variables, the length of each record, and if necessary, the lengths of fields within each record.
4. Roughly plan the sequence of the data base operation and the code for each part of that operation.
5. Begin on the code for the first part of the system.

Some programmers will argue with either the sequence of the steps or the steps themselves, and some may say that such an outline is too limited. All may be right. I am merely trying to give a limited amount of guidance in the development of a specific file system. Some systems analysts get carried away with the pre-code procedures, but one thing is clear: A certain amount of planning before coding is absolutely necessary!

## STOCK MARKET SYSTEM

Results of the previous day's trading activity are printed in most daily newspapers. Normally, these results include such things as: 52 week high and low stock price; stock symbol; latest dividend; a yield figure; P/E ratio; sales; daily high, low, and closing price; and, possibly, the amount of price change from the previous day. That information is available for active issues on the New York and American Stock Exchanges. Less information is given for Over The Counter or NASDAQ issues, Option, Commodity, and Bond trading. There are figures for the various averages: NYSE INDEX, DOW, STANDARD & POORS, AMERICAN INDEX, NASDAQ COMPOSITE and/or INDUSTRIALS, to name just a few. There are other key items to watch: gold, interest rates, the value of the dollar overseas, the money supply, the President's daily intake of vitamins, and so forth. As you might guess, the list is limitless.

Although an investment record may contain more variables than information maintained on your own library, the principle is the same. You must thoroughly know your subject in order to be able to make decisions concerning the information to be saved. The first step in planning your data base should be deciding which information is of value; i.e., what information to save.

In our STOCK MARKET SYSTEM, I am going to severely limit the amount of information saved. Individuals may wish to keep additional information they believe to be important. For each issue, we will save the daily high, low and closing price, plus the day's volume and P/E ratio. In addition, we will save any price that makes either a new high or new low for that issue.

Steps two and three somewhat blend together at this point. In the planning, decisions are made. Most stock prices are under $1000 per share, so I will allow a maximum of three places before the decimal point. Prices are usually given in terms of eighths of a dollar; i. e., 3/8 or 1/2. With a little extra planning and coding, significant disk space can be saved on each issue. If the extreme right figure is always viewed as the decimal portion of the stock price, then four digits will represent all stock prices up to $999 and 7/8 per share. (This price would be saved on the disk as 9997.) Saving the high, low, and close each day already means 15 bytes per issue per day---4 bytes for the number plus one byte for the delimiter for a total of 5 bytes for each high, low, and closing price.

Volume can be handled in somewhat the same way. Most papers indicate the sales volume only in hundreds of shares sold per day. A volume of 2000 shares would be displayed as 20. Since virtually every stock trades under 9,999,900 shares in one day, we can limit the number of places to six---99999 plus one for the carriage return delimiter. Finally, all P/E ratios are under 999 for any issue. This necessitates another four bytes for each issue, which brings the total number of bytes for each issue to 25: 4 for the P/E ratio, 6 for the volume, 5 for the high, 5 for the low, and 5 for the close. We will save any new high or new low price in a separate sequential file.

Next, we must decide on the number of issues to follow on a daily basis. This is an individual choice and often depends upon the time available for following the market closely. A reasonable figure to start out with is ten issues. This means that we can calculate the maximum number of bytes or the length of each record (since each day's transactions will represent one record). Ten stocks, each requiring 25 bytes, mean a length of 250 bytes per record. Add an additional ten bytes for each day's date, and we have a total of 260 bytes. Based on approximately 125,000 available bytes on each diskette, we can store just under 500 days' trading information, which is more than a year and a half of stock market activity for ten issues on a single data diskette.

Step four is a rough plan of the sequence of programs and the code within each program. Following the procedures we have used in our previous examples, we need programs that will create the necessary files and add to those files daily. Second, we need programs to display, in a variety of ways, the information either stored in the file or derived from the information stored. Finally, we must have correction programs.

I am going to introduce another method of creating random files. This program does nothing but open the file, write a pointer value to record zero, and close the file. It is similar to our first MAILING LIST CREATOR program except that this program does not have any data.

```
20 D$ = CHR$(4) : REM CONTROL D
40 PRINT D$;"OPEN STOCKS,L260"
60 PRINT D$;"WRITE STOCKS,R0"
80 PRINT "0"
100 PRINT D$;"CLOSE STOCKS"
```

The sole purpose of this program is to create the file with a pointer value of zero so that the first time we use the file addition program, we will not get an END OF DATA error. With this program, we do not need to use the ONERR statement in the file addition program. This program will also come in handy if the pointer ever needs to be reset.

Within the file addition program, the sequence of operation is fairly standard. We need to read in: (1) the value of the file pointer (add one to that value), (2) the symbols for the various stocks, and (3) their current high and low prices. Next, the previous day's figures must be entered for each issue and checked for accuracy. The previous day's figures must also be compared to the current high and low, and if they exceed those figures, they need to replace the appropriate figure. Finally, all the new information must be written to the disk.

The editing programs should follow a similar pattern but with a few exceptions. The pointer should be determined by an input from the keyboard. The high, low, close, volume, and P/E should come from the disk instead of the keyboard, with corrections coming from the keyboard. Finally, a routine should be included that re-formats the current file, allowing for expansion of the number of issues. This routine would read in values from the old file and write them back out in a new format to the new file.

The display programs are more difficult to structure in any absolute manner. The only certain structure is that information comes from the disk and is displayed either on the screen or through the printer. In between, a variety of steps can take place. The information can be used to calculate certain values that may be used to evaluate a particular situation and project the price movement of the stock, or the disk information may simply be formatted for display on either the screen or the printer. The information may be used to graph the price movement of the stock or compare its movement against that of another stock or average. The display portion of the STOCK MARKET SYSTEM is the core of the system and is usually not a fixed set of programs. User needs change and require that the display programs change. In the system presented, the display programs will be limited in their scope. We will display individual stock histories, along with some calculated figures, averages, price and volume changes, etc. We will not get into graphic representation of stock performance in this book. A future book will deal with graphic representation of data base information, because it is such a broad topic that it requires a book of its own.

The final step is coding the programs according to the plans established. At the end of this chapter, you will find minimum programs designed according to the structure outlined in this chapter. You are encouraged to take these minimum programs and expand upon them to fit your own interest, or alter them to cover a topic of your own design. It is only by practical experience that you will learn to create Apple files.

## QUESTIONS

1. TRUE or FALSE: Good programmers can just sit down and start writing code.
2. What is the first step in planning a data base system?
3. Give the three main parts to any data base system.
4. Which part must be flexible as user needs change?

# ANSWERS

1. FALSE
2. Deciding which information is of value.
3. Creation and addition, Display, Correction.
4. Display

# STOCK MENU

```
10  REM ***--STOCK MENU--**
11 :
12 :
20  D$ = CHR$ (4): REM CONTROL D
40  TB = 8
100  HOME : VTAB 3
120  HTAB 15
140  PRINT "STOCK MENU"
160  PRINT : PRINT
180  HTAB TB
200  PRINT "1. ADD STOCK INFO. "
220  PRINT
240  HTAB TB
260  PRINT "2. DISPLAY STOCK INFO. "
280  PRINT
300  HTAB TB
320  PRINT "3. DISPLAY HI/LOW VALUES"
340  PRINT
360  HTAB TB
380  PRINT "4. CREATE/CORRECT HI/LOW"
400  PRINT
420  HTAB TB
440  PRINT "5. CORRECT DATA"
460  PRINT
480  HTAB TB
500  PRINT "6. CATALOG"
520  PRINT
540  HTAB TB
560  PRINT "7. END"
580  PRINT
600  HTAB TB
620  INPUT "WHICH NUMBER ";NB
640  IF NB < 1 OR NB > 7 THEN PRINT : HTAB TB: PRINT
     "INCORRECT CHOICE" : GOTO 600
660  IF NB = 1 THEN PRINT D$;"RUN ADD STOCKS"
680  IF NB = 2 THEN PRINT D$;"RUN DISPLAY STOCKS"
700  IF NB = 3 THEN PRINT D$;"RUN DISPLAY HI/LOW"
720  IF NB = 4 THEN PRINT D$;"RUN CREATE HI/LOW"
740  IF NB = 5 THEN PRINT D$;"RUN CORRECT STOCKS"
760  IF NB = 6 THEN PRINT D$;"CATALOG"
780  IF NB = 7 THEN END
```

```
800  PRINT
820  INPUT "HIT RETURN TO CONTINUE "; L$
840  GOTO 100
```

## ADD STOCKS

```
10   REM ***--ADD STOCK INFO--***
11   :
12   :
13   REM **--VARIABLES LIST--**
14   REM STK$  = STOCK SYMBOL
15   REM HI$   = CURRENT HI PRICE
16   REM LOW$  = CURRENT LOW PRICE
17   REM PE    = P/E RATIO
18   REM VOL   = SALES VOLUME
19   REM H     = DAILY HIGH PRICE
20   REM L     = DAILY LOW PRICE
21   REM C     = DAILY CLOSING PRICE
22   REM CR    = CORRECTED FIGURE
27   :
28   :
29   REM **--INITIALIZATION--**
30   D$ = CHR$ (4): REM CONTROL D
40   PRINT D$; "OPEN STOCKS, L260"
60   PRINT D$; "READ STOCKS, R0"
80   INPUT PTR
100  PRINT D$; "CLOSE STOCKS"
120  PTR = PTR + 1
140  PRINT D$; "OPEN STOCKS HI/LOW"
160  PRINT D$; "READ STOCKS HI/LOW"
180  FOR I = 0 TO 9
200  INPUT STK$(I)
220  INPUT HI$(I)
240  INPUT LOW$(I)
260  NEXT I
280  PRINT D$; "CLOSE STOCKS HI/LOW"
297  :
298  :
299  REM **--KEYBOARD INPUT--**
300  I = 0
320  HOME : VTAB 5
330  INPUT "TODAY'S DATE "; DT$
```

# ADD STOCKS

```
335  HOME : VTAB 5: PRINT DT$: PRINT : PRINT
340  PRINT STK$(I)
350  PRINT : PRINT
360  INPUT "TODAY'S P/E RATIO "; PE
370  PRINT
380  INPUT "TODAY'S VOLUME "; VOL
390  GOSUB 7000
400  INPUT "TODAY'S HIGH "; H
410  GOSUB 7000
420  INPUT "TODAY'S LOW "; L
430  GOSUB 7000
440  INPUT "TODAY'S CLOSE "; C
457 :
458 :
459  REM **--CORRECTION ROUTINE--**
460  HOME : VTAB 3
470  HTAB 10
480  PRINT STK$(I)
490  PRINT : PRINT
500  PRINT "1. TODAY'S P/E RATIO-- "; PE
510  PRINT
520  PRINT "2. TODAY'S VOLUME----- "; VOL
530  PRINT
540  PRINT "3. TODAY'S HIGH------- "; H
550  PRINT
560  PRINT "4. TODAY'S LOW-------- "; L
570  PRINT
580  PRINT "5. TODAY'S CLOSE------ "; C
600  PRINT
620  INPUT "ARE THESE FIGURES CORRECT? "; YES$
640  IF YES$ = "N" THEN 680
660  GOTO 900
680  PRINT
700  INPUT "WHICH NUMBER IS WRONG "; NB
720  IF NB < 1 OR NB > 5 THEN PRINT "INCORRECT CHOICE"
     : GOTO 680
740  INPUT "THE CORRECT FIGURE = "; CR
760  IF NB = 1 THEN PE = CR
780  IF NB = 2 THEN VOL = CR
800  IF NB = 3 THEN H = CR
820  IF NB = 4 THEN L = CR
840  IF NB = 5 THEN C = CR
860  GOTO 460
897 :
```

```
898 :
899 REM **--EXCHANGE HI/LOW--**
900 IF H > VAL (HI$(I)) THEN HI$(I) = STR$ (H)
920 IF L < VAL (LOW$(I)) THEN LOW$(I) = STR$ (L)
940 IF VAL (LOW$(I)) = 0 THEN LOW$(I) = STR$ (L)
957 :
958 :
959 REM **--FILE UPDATE--**
960 PRINT D$; "OPEN STOCKS,L260"
970 PRINT D$; "WRITE STOCKS,R"; PTR; ",B"; 0
975 PRINT DT$
980 PRINT D$; "WRITE STOCKS,R"; PTR; ",B"; (I * 25) + 10 + 0
1000 PRINT PE
1020 PRINT D$; "WRITE STOCKS,R"; PTR; ",B"; (I * 25) + 10 + 4
1040 PRINT VOL
1060 PRINT D$; "WRITE STOCKS,R"; PTR; ",B"; (I * 25) + 10 + 10
1080 PRINT H
1100 PRINT D$; "WRITE STOCKS,R"; PTR; ",B"; (I * 25) + 10 + 15
1120 PRINT L
1140 PRINT D$; "WRITE STOCKS,R"; PTR; ",B"; (I * 25) + 10 + 20
1160 PRINT C
1180 PRINT D$; "CLOSE STOCKS"
1200 I = I + 1
1220 IF I < 10 THEN 335
1237 :
1238 :
1239 REM **--NEW HI/LOW FILE--**
1240 PRINT D$; "OPEN STOCKS HI/LOW"
1260 PRINT D$; "WRITE STOCKS HI/LOW"
1280 FOR K = 0 TO 9
1300 PRINT STK$(K)
1320 PRINT HI$(K)
1340 PRINT LOW$(K)
1360 NEXT K
1380 PRINT D$; "CLOSE STOCKS HI/LOW"
1397 :
1398 :
1399 REM **--FILE POINTER--**
1400 PRINT D$; "OPEN STOCKS,L260"
1420 PRINT D$; "WRITE STOCKS,R0"
1440 PRINT PTR
1460 PRINT D$; "CLOSE STOCKS"
1998 :
```

```
1999 :
5000 REM **--RETURN TO STOCK MENU"
5020 PRINT D$;"RUN STOCK MENU"
6997 :
6998 :
6999 REM **--REMINDER SUBROUTINE--**
7000 HOME : VTAB 3
7020 PRINT "***--REMEMBER--***"
7040 PRINT "YOU MUST ADD THE FRACTION"
7060 PRINT "AS THE LAST DIGIT"
7080 PRINT "1/8------ = 1"
7100 PRINT "1/4------ = 2"
7120 PRINT "3/8------ = 3"
7140 PRINT "1/2------ = 4"
7160 PRINT "5/8------ = 5"
7180 PRINT "3/4------ = 6"
7200 PRINT "7/8------ = 7"
7220 PRINT "EVEN----- = 0"
7222 PRINT : PRINT
7225 PRINT "****---IMPORTANT---****"
7230 PRINT
7235 PRINT "IF THE NUMBER HAS NO"
7240 PRINT
7245 PRINT "FRACTION, PLEASE ENTER"
7250 PRINT
7255 PRINT "A '0' AFTER THE NUMBER. "
7260 PRINT
7280 RETURN
```

## DISPLAY STOCKS

```
10 REM ***--DISPLAY STOCK HISTORY--***
11 :
12 :
13 REM **--VARIABLES LIST--**
14 REM STK $  = STOCK SYMBOL
15 REM HI $   = CURRENT HI PRICE
16 REM LOW $  = CURRENT LOW PRICE
17 REM PE     = P/E RATIO
18 REM VOL    = SALES VOLUME
19 REM H      = DAILY HIGH PRICE
20 REM L      = DAILY LOW PRICE
21 REM C      = DAILY CLOSING PRICE
22 REM DT $   = DATE
23 REM F $    = FRACTION
24 REM AV     = AVERAGE VOLUME
25 REM AP     = AVERAGE PRICE
26 REM CV     = CLOSING PRICE W/O CONV.
27 REM C1     = 1ST CLOSE PRICE
28 REM C2     = LAST CLOSE PRICE
29 REM CD     = DIFF. BETWEEN C2
30 REM M      = COMMON VAR. CONV
31 REM L1     = COMMON VAR. CONV
32 REM W      = TEMP. STOCK #
46 :
47 :
48 :
49 REM **--INITIALIZATION--**
50 D$ = CHR$ (4): REM CONTROL D
55 PRINT D$;"OPEN STOCKS,L260"
60 PRINT D$;"READ STOCKS,R0"
80 INPUT PTR
100 PRINT D$;"CLOSE STOCKS"
116 :
117 :
118 :
119 REM **--SET UP--**
120 HOME : VTAB 5
140 PRINT D$;"OPEN STOCKS HI/LOW"
160 PRINT D$;"READ STOCKS HI/LOW"
180 FOR I = 0 TO 9
200 INPUT STK$(I)
```

## DISPLAY STOCKS

```
220  INPUT HI$(I)
230  M = VAL (HI$(I)): GOSUB 8000:HI$(I) = STR$ (M) + " "
     + F$
240  INPUT LOW$(I)
245  M = VAL (LOW$(I)): GOSUB 8000:LOW$(I) = STR$ (M)
     + " " + F$
250  PRINT I + 1;".";: HTAB 5: PRINT STK$(I)
260  NEXT I
270  STK$(10) = "STOCK MENU"
275  PRINT "11.";: HTAB 5: PRINT STK$(10)
280  PRINT D$;"CLOSE STOCKS HI/LOW"
285  PRINT
290  INPUT "WHICH STOCK ";W
291  IF W < 1 OR W > 11 THEN PRINT "INCORRECT CHOICE"
     : GOTO 290
292  IF W = 11 THEN PRINT D$;"RUN STOCK MENU"
293  I = W - 1
294  :
295  REM ***--TITLES--***
296  HOME : VTAB 5: HTAB 18: PRINT STK$(I): PRINT : PRINT
297  PRINT "DATE";: HTAB 10: PRINT "VOL";: HTAB 15: PRINT
     "HI";: HTAB 23: PRINT "LOW";: HTAB 31: PRINT "CLOSE"
298  :
299  :
300  REM **--DISK INPUT ROUTINE--**
320  PRINT D$;"OPEN STOCKS,L260"
330  FOR K = 1 TO PTR
335  PRINT D$;"READ STOCKS,R";K;",B";0
337  INPUT DT$
340  PRINT D$;"READ STOCKS,R";K;",B";(I * 25) + 10
360  INPUT PE
380  PRINT D$;"READ STOCKS,R";K;",B";(I * 25) + 10 + 4
400  INPUT VOL
420  PRINT D$;"READ STOCKS,R";K;",B";(I * 25) + 10 + 10
440  INPUT H
460  PRINT D$;"READ STOCKS,R";K;",B";(I * 25) + 10 + 15
480  INPUT L
500  PRINT D$;"READ STOCKS,R";K;",B";(I * 25) + 10 + 20
520  INPUT C
536  :
537  :
538  :
539  REM **--DISPLAY ROUTINE--**
540  PRINT DT$;
```

```
600  HTAB 10
620  PRINT VOL;
740  HTAB 15
750  M = H: GOSUB 8000: H = M
760  PRINT H; " "; F$;
780  HTAB 23
790  M = L: GOSUB 8000: L = M
800  PRINT L; " "; F$;
820  HTAB 31
825  IF K = 1 THEN C1 = C: V1 = VOL
827  IF K = PTR THEN C2 = C
828  CV = C: L1 = CV: GOSUB 9000: CV = L1
830  M = C: GOSUB 8000: C = M
840  PRINT C; " "; F$
850  AV = AV + VOL
860  AP = AP + CV
880  NEXT K
884  PRINT D$
885  PRINT : INPUT "HIT RETURN TO CONTINUE "; L$
886  :
887  :
888  :
889  REM**--DISPLAY SECOND PAGE--**
890  HOME : VTAB 5
895  HTAB 18 : PRINT STK$(I)
899  PRINT : PRINT
900  PRINT "CURRENT P/E RATIO = "; PE
920  PRINT
940  PRINT "CURRENT HIGH = "; HI$(I)
945  AV = AV / (K - 1)
950  AP = AP / (K - 1)
955  PRINT
960  PRINT "CURRENT LOW = "; LOW$(I)
962  PRINT
965  PRINT "AVERAGE VOL. = "; AV
967  PRINT
970  PRINT "AVERAGE PRICE = "; AP
972  PRINT
975  L1 = C2 : GOSUB 9000 : C2 = L1
976  L1 = C1 : GOSUB 9000 : C1 = L1
977  CD = C2 - C1
980  PRINT "PRICE DIFF. FROM 1ST REC. = "; CD
985  PRINT
990  PRINT "LAST PRICE = "; C; " "; F$
```

```
996 :
997 :
998 :
999  REM **--ANOTHER STOCK--**
1000  PRINT D$
1010  PRINT
1020  INPUT "HIT RETURN TO CONTINUE ";L$
1040  HOME : VTAB 5
1050 AV = 0:AP = 0
1060  FOR I = 1 TO 11
1080  PRINT I;".";: HTAB 5: PRINT STK$(I - 1)
1100  NEXT I
1120  PRINT
1140  GOTO 290
1997 :
1998 :
1999 :
5000  REM **--RETURN TO STOCK MENU"
5020  PRINT D$;"RUN STOCK MENU"
6996 :
6997 :
6998 :
8000  REM **--CONVERT TO FRACTION--**
8005 F = M - INT (M / 10) * 10
8010 M = INT (M / 10)
8020  IF F = 0 THEN F$ = ""
8040  IF F = 1 THEN F$ = "1/8"
8060  IF F = 2 THEN F$ = "1/4"
8080  IF F = 3 THEN F$ = "3/8"
8100  IF F = 4 THEN F$ = "1/2"
8120  IF F = 5 THEN F$ = "5/8"
8140  IF F = 6 THEN F$ = "3/4"
8160  IF F = 7 THEN F$ = "7/8"
8200  RETURN
8997 :
8998 :
8999 :
9000  REM **--CONVERT TO DECIMAL--**
9010 L1 = L1 / 10:S1 = INT (L1):D1 = L1 - S1
9020 D1 = (D1 * 10) / 8:L1 = S1 + D1:L1 = INT
      (L1 * 1000 + .5) / 1000
9040  RETURN
```

## DISPLAY HI/LOW

```
10 REM **--READ HI/LOW FILE--**
11 :
12 :
20 D$ = CHR$ (4) : REM CONTROL D
100 REM **--DISK INPUT--**
120 PRINT D$; "OPEN STOCKS HI/LOW"
140 PRINT D$; "READ STOCKS HI/LOW"
160 FOR I = 1 TO 10
180 INPUT STK$(I)
200 INPUT HI$(I)
220 INPUT LOW$(I)
240 NEXT I
260 PRINT D$; "CLOSE STOCKS HI/LOW"
297 :
298 :
300 REM **--DISPLAY ROUTINE--**
305 HOME : VTAB 5
310 PRINT "STOCK SYMBOL";: HTAB 18: PRINT "HI";: HTAB 28
    : PRINT "LOW"
315 PRINT
320 FOR I = 1 TO 10
330 PRINT I;".";
335 HTAB 5
340 PRINT STK$(I);
360 HTAB 18
370 M = VAL (HI$(I)): GOSUB 8000: HI$(I) = STR$ (M)
380 PRINT HI$(I);" ";F$;
390 M = VAL (LOW$(I)): GOSUB 8000: LOW$(I) = STR$ (M)
400 HTAB 28
420 PRINT LOW$(I);" ";F$
460 NEXT I
480 PRINT : INPUT "HIT RETURN TO CONTINUE ";L$
497 :
498 :
500 REM **--RETURN TO STOCK MENU--**
520 PRINT D$; "RUN STOCK MENU"
8000 REM
8005 F = M - INT (M / 10) * 10
8010 M = INT (M / 10)
```

```
8020  IF F = 0 THEN F$ = ""
8040  IF F = 1 THEN F$ = "1/8"
8060  IF F = 2 THEN F$ = "1/4"
8080  IF F = 3 THEN F$ = "3/8"
8100  IF F = 4 THEN F$ = "1/2"
8120  IF F = 5 THEN F$ = "5/8"
8140  IF F = 6 THEN F$ = "3/4"
8160  IF F = 7 THEN F$ = "7/8"
8200  RETURN
```

## CREATE STOCKS HI/LOW

```
10 REM ***--STOCKS HI/LOW--***
11 :
12 :
13 REM **--VARIABLES LIST--**
14 REM STK$ = STOCK SYMBOL
15 REM HI$  = CURRENT HIGH PRICE
16 REM LOW$ = CURRENT LOW PRICE
18 :
19 :
20 D$ = CHR$ (4): REM CONTROL D
27 :
28 :
30 REM **--KEYBOARD INPUT--**
40 FOR I = 0 TO 9
60 HOME : VTAB 5
80 INPUT "STOCK SYMBOL ";STK$(I)
85 PRINT
90 PRINT "IF YOU ARE NOT SURE OF THE"
92 PRINT
94 PRINT "HI OR LOW, ENTER A '0'."
95 PRINT
100 INPUT "HI VALUE ";HI$(I)
110 PRINT
120 INPUT "LOW VALUE ";LOW$(I)
128 :
129 :
130 REM **--CORRECTION ROUTINE--**
140 HOME
160 VTAB 5
180 PRINT "1. ";STK$(I)
```

```
200  PRINT "2.  ";HI$(I)
220  PRINT "3.  ";LOW$(I)
240  PRINT
260  INPUT "IS THIS CORRECT? ";YES$
280  IF YES$ = "N" THEN 320
300  GOTO 500
320  PRINT
340  INPUT "WHICH NUMBER IS WRONG? ";NB
360  IF NB < 1 OR NB > 3 THEN PRINT "INCORRECT CHOICE"
     :GOTO 320
380  IF NB = 1 THEN INPUT "CORRECT STOCK NAME
     PLEASE ";STK$(I)
400  IF NB = 2 THEN INPUT "CORRECT HI VALUE PLEASE ";HI$(I)
420  IF NB = 3 THEN INPUT "CORRECT LOW VALUE
     PLEASE ";LOW$(I)
440  GOTO 140
500  NEXT I
1198 :
1199 :
1200 REM **--CREATE HI/LOW FILE--**
1240 PRINT D$;"OPEN STOCKS HI/LOW"
1260 PRINT D$;"WRITE STOCKS HI/LOW"
1280 FOR K = 0 TO 9
1300 PRINT STK$(K)
1320 PRINT HI$(K)
1340 PRINT LOW$(K)
1360 NEXT K
1380 PRINT D$;"CLOSE STOCKS HI/LOW"
1398 :
1399 :
1400 REM **--RETURN TO STOCK MENU--**
1420 PRINT D$;"RUN STOCK MENU"
```

## STOCK CORRECTION

```
10  REM ***--STOCK CORRECTION--**
11  :
12  :
13  REM **--VARIABLES LIST--**
14  REM STK$   = STOCK SYMBOL
15  REM HI$    = CURRENT HI PRICE
16  REM LOW$   = CURRENT LOW PRICE
17  REM PE     = P/E RATIO
18  REM VOL    = SALES VOLUME
19  REM H      = DAILY HIGH PRICE
20  REM L      = DAILY LOW PRICE
21  REM C      = DAILY CLOSING PRICE
22  REM DT$    = DATE
46  :
47  :
48  :
49  REM **--INITIALIZATION--**
50  D$ = CHR$ (4): REM CONTROL D
55  PRINT D$;"OPEN STOCKS,L260"
60  PRINT D$;"READ STOCKS,R0"
80   INPUT PTR
100  PRINT D$;"CLOSE STOCKS"
116 :
117 :
118 :
119  REM **--SET UP--**
120  HOME : VTAB 5
140  PRINT D$;"OPEN STOCKS HI/LOW"
160  PRINT D$;"READ STOCKS HI/LOW"
180  FOR I = 0 TO 9
200  INPUT STK$(I)
220  INPUT HI$(I)
240  INPUT LOW$(I)
250  PRINT I + 1;".";: HTAB 5: PRINT STK$(I)
260  NEXT I
270  STK$(10) = "STOCK MENU"
275  PRINT "11.";: HTAB 5: PRINT STK$(10)
280  PRINT D$;"CLOSE STOCKS HI/LOW"
285  PRINT
290  INPUT "WHICH STOCK ";W: PRINT : PRINT "WHICH RECORD?
     1 TO " ;PTR;" ";: INPUT K: IF K > PTR THEN 290
```

## CHAPTER 13 PLANNING A FILE SYSTEM

```
291 IF W < 1 OR W > 11 THEN PRINT "INCORRECT CHOICE"
    : GOTO 290
292 IF W = 11 THEN 5000
293 I = W - 1
294 :
295 REM ***--TITLES--***
296 HOME : VTAB 5: HTAB 18: PRINT STK$(I) : PRINT : PRINT
298 :
299 :
300 REM **--DISK INPUT ROUTINE--**
320 PRINT D$; "OPEN STOCKS,L260"
335 PRINT D$; "READ STOCKS,R";K;",B";0
337 INPUT DT$
340 PRINT D$; "READ STOCKS,R";K;",B";(I * 25) + 10
360 INPUT PE
380 PRINT D$; "READ STOCKS,R";K;",B";(I * 25) + 10 + 4
400 INPUT VOL
420 PRINT D$; "READ STOCKS,R";K;",B";(I * 25) + 10 + 10
440 INPUT H
460 PRINT D$; "READ STOCKS,R";K;",B";(I * 25) + 10 + 15
480 INPUT L
500 PRINT D$; "READ STOCKS,R";K;",B";(I * 25) + 10 + 20
520 INPUT C
536 :
537 :
538 :
539 REM **--CORRECTION ROUTINE--**
540 PRINT "1. DATE   = ";DT$
545 PRINT
550 PRINT "2. P/E    = ";PE
555 PRINT
560 PRINT "3. VOL.   = ";VOL
565 PRINT
570 PRINT "4. HIGH   = ";H
575 PRINT
580 PRINT "5. LOW    = ";L
585 PRINT
590 PRINT "6. CLOSE  = ";C
592 PRINT
594 PRINT "0. ALL CORRECT"
595 PRINT : PRINT D$
600 INPUT "WHICH NUMBER IS WRONG ";NB
610 IF NB > 6 THEN PRINT "INCORRECT CHOICE": GOTO 600
615 IF NB = 0 THEN 710
```

```
620  INPUT "CORRECT INFORMATION = ";CR$
630  IF NB = 1 THEN DT$ = CR$
640  IF NB = 2 THEN PE = VAL (CR$)
650  IF NB = 3 THEN VOL = VAL (CR$)
660  IF NB = 4 THEN H = VAL (CR$)
670  IF NB = 5 THEN L = VAL (CR$)
680  IF NB = 6 THEN C = VAL (CR$)
690  HOME : VTAB 5: HTAB 18: PRINT STK$(I)
695  PRINT : PRINT
700  GOTO 540
707 :
708 :
709  REM **--WRITE CORRECTED FILE--**
710  PRINT D$;"WRITE STOCKS,R";K;",B";0
720  PRINT DT$
730  PRINT D$;"WRITE STOCKS,R";K;",B"; (I * 25) + 10
740  PRINT PE
750  PRINT D$;"WRITE STOCKS,R";K;",B"; (I * 25) + 10 + 4
760  PRINT VOL
770  PRINT D$;"WRITE STOCKS,R";K;",B"; (I * 25) + 10 + 10
780  PRINT H
790  PRINT D$;"WRITE STOCKS,R";K;",B"; (I * 25) + 10 + 15
800  PRINT L
810  PRINT D$;"WRITE STOCKS,R";K;",B"; (I * 25) + 10 + 20
820  PRINT C
996 :
997 :
998 :
999  REM **--ANOTHER STOCK--**
1000 PRINT D$
1010 PRINT
1020 INPUT "HIT RETURN TO CONTINUE ";L$
1040 HOME : VTAB 5
1060 FOR I = 1 TO 11
1080 PRINT I;".";: HTAB 5: PRINT STK$(I - 1)
1100 NEXT I
1120 PRINT
1140 GOTO 290
1997 :
1998 :
1999 :
5000 REM **--RETURN TO STOCK MENU"
5010 PRINT D$;"CLOSE STOCKS"
5020 PRINT D$;"RUN STOCK MENU"
```

# 14
# Binary Files

Binary files are much more complex than any of the other three file types. By their very nature, they are the least sophisticated and hardest of all the file types to work with. At the same time, however, they are the most versatile. First, Binary files may contain either instructions for the computer or lists of information, and sometimes they may contain both. Second, the information in Binary files is displayed as hexadecimal numbers (base 16) when viewed by either the computer or programmers. Third, the method of accessing those files may depend on the actual information in them.

To use a Binary file, one needs to put a "B" in front of the three main commands used with Applesoft or Integer files: BRUN, BLOAD, and BSAVE. These three commands also need the specific name of the file. BSAVE needs other information. This command must also include the actual memory location in the computer: where the Binary file starts and the exact length of the file. The address and length can be given in either decimal or hexadecimal numbers. The letters "A" and "L" precede their respective numbers. An example of saving a Binary file is:

BSAVE MAILING LIST, A768, L2000

or

BSAVE MAILING LIST, A$300, L$1DBF

The $ in the second example indicates that the addresses are given in hexadecimal numbers rather than decimal numbers as in the first example. Both examples give the same instructions to the computer: save the contents of a certain amount of the computer's memory to the disk under the given file name.

## CHAPTER 14  BINARY FILES

You can also load (or run) a Binary file from disk to a specific memory location in the computer using these same parameters.

BLOAD MAILING LIST, A768

or

BLOAD MAILING LIST, A$300

The obvious requirement is that you know something about the computer's storage.

Binary files can contain just about any type of information in addition to the computer-instruction files we have looked at. Often, a Binary file will have the information necessary to create some kind of graphic on the Hi-Resolution screens of the Apple. This information is a form of computer instruction since it gives specific information to the computer to place information in certain locations. But it cannot really be considered a program since most of these files do not contain information necessary to allow them to be BRUN. Instead, these files are used by other programs: Applesoft programs, Integer programs, even Binary programs. This use is true regardless of the type of information in the Binary file; i.e., graphic information, inventory information, name and address information, etc. By themselves, these files are often useless. They require a program to be able to read the information on the disk and decide what to do with that information.

Programmers who work extensively with Binary files must know other computer languages. For this reason, our examination of Binary files will be limited. I will not go through the same process I did with text files, because these other languages are more complicated to learn than any form of BASIC, but there are a number of things that can be of use to the BASIC programmer.

Binary files can be used from within BASIC programs in much the same manner as other BASIC programs. In other words, you can BLOAD, BRUN, or BSAVE a Binary file from within a BASIC program. You must know what is in that Binary file and the exact purpose for using that file within your BASIC program. One example would be that of BLOADing a file as a picture to the Hi-Res screens. Another example might be that of a very large list of information that needs to be sorted quickly. There are Binary files that do sorting much quicker than any BASIC sort method. In general, Binary files or programs will operate at a much faster speed than BASIC programs. The reason for this lies with the nature of the computer.

The computer operates according to on/off switches. At the lowest levels, the programmer learns to operate the computer by manually setting these on/off switches. Above this level, there is a progression of increasingly sophisticated languages. The individual must learn to program the computer according to one or more of these languages. At the level above setting switches are programs with ones and zeros (base 2). On the next level is hexadecimal programming (base 16) or machine language, and just below the higher level languages is assembly language. Binary files are involved in all of these levels of programming, in addition to storing lists of information or graphic information. At the point of higher level languages, a programmer learns to instruct the computer with a vocabulary much like English. The various languages are really a link between the computer and humans. They allow people to work with a sequence of letters and numbers that are much easier to understand than on/off switches. However, when these higher level languages are interpreted through a number of levels down to the on/off switches, the process takes more time than if the program did not require all that interpretation. That is why programs written in any of the lower level languages run much faster than programs written in BASIC.

I said that Binary files are displayed as hexadecimal numbers. This means that instead of counting to nine before adding the next digit, we count to 15.

| Decimal | Hexadecimal |
| --- | --- |
| 0 | 0 |
| 1 | 1 |
| 2 | 2 |
| 3 | 3 |
| 4 | 4 |
| 5 | 5 |
| 6 | 6 |
| 7 | 7 |
| 8 | 8 |
| 9 | 9 |
| 10 | A |
| 11 | B |
| 12 | C |
| 13 | D |
| 14 | E |
| 15 | F |
| 16 | 10 |
| 17 | 11 |
| . | . |
| . | . |
| . | . |

Let's try some examples to see how the Binary DOS commands work. It is important to follow the examples by actually typing them into the

## CHAPTER 14 BINARY FILES

computer. Make sure you have a blank diskette (a blank diskette is necessary if you are going to try the material presented in the section on RWTS in the appendix), and do the following:

Type:

```
NEW (to erase anything currently in the computer's
memory)
20 D$ = CHR$(4) : REM CONTROL D
40 PRINT D$; "CATALOG"
LIST (to see our program)
INIT HELLO
```

Remember that this erases anything currently on the diskette, places DOS on the diskette, and saves the above program as the first file on this diskette. Type:

CATALOG

You should see:

HELLO

Now type:

CALL -151

You should see:

```
* and the cursor.
```

The "*" means that you are now in the MONITOR and must work with machine language (hexadecimal numbers) rather than BASIC. Type the following exactly!

```
300: 20 58 FC A2 00 BD 11 03
```

Do not hit the RETURN key until after the last character; i.e., the 3. Type the next rows in the same way, and press the RETURN key at the end of the lines.

```
308: 20 F0 FD E8 E0 32 D0 F5
310: 60 C8 C5 CC CC CF 8D CD
318: D9 A0 CE C1 CD C5 A0 C9
320: D3 A0 C1 D0 D0 CC C5 A0
328: C9 C9 8D C9 A0 C1 CD A0
330: C1 A0 D3 CD C1 D2 D4 A0
338: C3 CF CD D0 D5 D4 C5 D2
340: 8D 8D 8D
```

We now have a machine language program in the computer's memory.

We use hexadecimal numbers to enter this program. Once we have a program in the computer's memory, we need to save it onto something permanent. Type the next steps carefully.

3D0G

This returns you to BASIC.

BSAVE APPLE7, A$300, L$43

or

BSAVE APPLE7, A768, L67

The disk comes on and the machine language program is saved to the diskette.

Now type:

BRUN APPLE7

Again the disk comes on briefly and you should see the screen clear. In the upper left corner of the screen, the following message should appear:

HELLO
MY NAME IS APPLE II

I AM A SMART COMPUTER

The machine language program has duplicated our first BASIC program.

Type:

BLOAD APPLE7

CALL 768

Again the same message appears.
After a BLOAD, a CALL to the starting decimal address will run the program.

Finally type:

CALL-151

300G

Once more the message appears.
When you are working in the MONITOR, a G after the starting hexadecimal address runs the program.

To see all that you entered, type:

300.350

This command lists the program. You should see an exact copy of the

## CHAPTER 14 BINARY FILES 263

information you entered above. You can list in another way.
Type:

300L

Now you have the Apple's mini-Assembler version of the program complete with mnemonics (LDA, STA, RTS, JSR, etc.).

When you see a listing in a magazine of an assembly code, it is not necessary to type in all the information. Assembly listings will look like the following:

```
1B9D: 84 35        157    PUT    STY YSAV1 SAVE Y
1B9F: 48           158           PHA
1BA0: 20 0A 1C     159           JSR UPDATE HIRES
```

You can enter the above by typing:

1B9D: 84 35 48 20 0A 1C

As long as the memory locations are consecutive, you can continue to enter the hexadecimal numbers. It is not advisable to go too far before hitting the return key because of the chance of error.

Now, make the following change to our program:

309: ED

This replaces the value "F0" currently in memory location 309. Type:

300G

The message reappears and everything seems the same. Now type:

3D0G

to get back to BASIC. Then type:

BSAVE APPLE3, A$300, L$43

Now type:

BLOAD APPLE3
CALL 768

Everything is still working the same. But now type:

BRUN APPLE3

You should see some strange things happen. A double message may be printed, and the cursor will disappear. You must either turn the computer off and then re-boot the computer and disk, or hit reset to get out of this mess. By changing that one byte, we altered the program so that it does not work when used with the BRUN command. Yet it does work properly when we BLOAD and issue a CALL to the starting address. This example should give you some idea of the complexity involved with programming in machine code. For those interested, we changed the JSR instruction from the subroutine at FDF0 to the somewhat more standard subroutine at FDED. The FDED subroutine sends the output to active peripherals and when combined with a BRUN, in this program, causes the computer to "hang."

Often, it is necessary to transfer a Binary file from one diskette to another. There are many programs that can accomplish such a task, but it is handy to know how to find the address and length of a Binary file if you do not have access to one of those Binary file transfer programs. (FID on DOS 3.3 system master diskettes is one such program.)

Type carefully:

BLOAD APPLE7

If you have a 48k Apple type the following:

PRINT PEEK (43634) + PEEK (43635) * 256

The computer should respond with:

768

This is the decimal starting address; i.e., the number that goes after the "A".

Then type:

PRINT PEEK (43616) + PEEK (43617) * 256

The computer should respond with:

67

This is the length in decimal, the number that goes after the "L". Finally, insert the diskette that is to contain the copy of the binary file and type:

BSAVE APPLE7, A768, L67

You have just transferred a copy of the Binary file APPLE7 to another diskette. If you have a 32K Apple, the numbers you must use are:

27250 and 27251 for the first set.

and

27232 and 27233 for the second set.

CHAPTER 14   BINARY FILES                                           265

For a 16K system, use:

   10866 and 10867 for the first set.
                    and
   10848 and 10849 for the second set.

If your file is small (under 255 bytes), you can have the computer figure the length of the file in hexadecimal. When you are in the monitor (an asterisk prompt), type in the larger number, a minus sign, the smaller number, and hit return. The computer will respond with the correct answer as long as the answer is positive and less than 255 ($FF). There are small machine code programs that can be used to figure the length when the length grows past 255 bytes. Of course, there is always the stand-by of subtracting the hex numbers on paper or counting the total number of bytes of information either by decimal or hex.

You have now had some experience with Binary files, BSAVE, BLOAD, and BRUN. You have also been exposed to entering hexadecimal information directly and listing and changing that information. It is somewhat complicated to understand, but if you review this chapter, things should become clear eventually. The Appendix contains a brief tutorial on RWTS (Read/Write Track/Sector) for those interested in more information.

A very good introduction to machine language and assembly language can be found in the Inman and Inman book, *Apple Machine Language* from Reston Publishing Company and *Assembly Lines: The Book (A Beginner's Guide to 6502 Programing on the Apple II)* by Robert Wagner from Softalk Publishing. For the advanced assembly language programmer, *Beneath Apple DOS,* by Don Worth and Pieter Lechner (Quality Software) provides excellent information. This book includes example programs indicating how to create text files or Binary files from within an assembly language program.

## QUESTIONS

1. In a CATALOG listing, what letter indicates Binary files?
2. Which of the three main Binary DOS commands require additional parameters?
3. What are these parameters?
4. How are Binary files displayed?
5. TRUE or FALSE: A Binary file containing only graphic information can be used with a BRUN command.
6. The computer operates according to ---------------.
7. Give the progression of computer languages.
8. What must you type to get to the monitor?
9. What must you type to begin the operation of a program located at $300 while in the MONITOR?
10. Give the two ways of listing a program beginning at $300.
11. What word, when used with PRINT, shows the contents of memory as a decimal value?

## ANSWERS

1. B
2. BLOAD
3. A for starting address and L for length of file
4. Binary files are displayed in hexadecimal.
5. FALSE
6. on/off switches
7. a) on/off switches
   b) ones and zeros (base 2)
   c) hexadecimal (base 16) or machine language
   d) assembly language (mnemonics)
   e) higher level languages such as BASIC
8. CALL -151
9. 300G---The G following an address instructs the computer to begin operation at that address.
10. a) 300.400---The period between two addresses instructs the computer to provide a hex dump from the first address to the last address.
    b) 300L---The L following an address instructs the computer to provide an assembly listing.
11. PEEK

# Appendices

# APPENDIX A.

## READ/WRITE TRACK/SECTOR (RWTS)

There are several things that you should know before we get into the process of looking directly at the contents of a diskette. The information stored on the diskette, when viewed in its most direct way, appears as hexadecimal numbers. The letter "A" appears as "C1", "B" as "C2", etc.

The process we are going through is more completely handled by some utility programs available for purchase. I am going to show how you can look at and read a diskette. The process is not easy or quick, but it does work. If your needs do not go beyond this explanation, you might find that you do not not need an expensive utility program.

We are going to look at the CATALOG listing on the diskette first. We will not see the file names as they are normally displayed, but as they look to the computer when it reads the diskette. The CATALOG's first part of information is on track $11, sector $0F. Make certain that the diskette in the drive contains the binary file APPLE7 as the second file. If APPLE7 is not the second file on the diskette, count the number of files preceding APPLE7 in such a way that you know which group of seven the APPLE7 file is in. The following program is set up for computers with a single disk drive in slot 6. Type very carefully.

    CALL -151

You should get the *, and then:

    300: A9 03 A0 0A 20 D9 03 60
    308: 00 FF 01 60 01 00 11 0F
    310: 20 0C 00 60 00 00 01 00
    318: 00 60 01 00 00 00 00 00
    320: 00 01 EF D8

If the file APPLE7 is in the second set of seven files on the diskette, change the contents of $30F to 0E instead of the 0F given above, or 0D if it is in the third set of seven files on the diskette. Check your work carefully. This program is a modification of the one given in the

DOS manuals. Location $30B is the slot number of the disk drive (6) multiplied by 16 decimal with the result (96) expressed as a hex number ($60). $30C is the drive number, either a $01 or $02. $30E contains the value of the track number we are accessing (in hex). $30F is the sector number, and $311 and $312 contain the locations of the starting address of the data or program in memory in reverse order. $312 contains the first part of the address, called the high-order byte, and $311 contains the last part of the address or the low-order byte. The location of $6000 will receive information from the disk. By changing location $316 to a $02, we would transfer any information starting at location $6000 from the computer's memory to the diskette. In other words, $316 is the memory location whose value controls the read/write decision. A value of $01 in $316 reads information on the disk, while a value of $02 writes information to the diskette. For our examples, I will only READ information on the diskette. You should be very certain that you understand what is happening before attempting to write information to a disk with this routine.

Now type:

3D0G

You should be back in BASIC.

BSAVE RWTS, A$300, L$24

CALL-151

Make sure your diskette is in the drive.

300G

Remember, this runs the program located at $300. The disk comes on briefly and then the cursor returns. Now the real work begins. Type:

6000.6050

This command lists the contents of the computer's memory between these two numbers. We have brought the first seven file names into the computer's memory along with information about the actual diskette location of those files. You must examine this information closely. The file name we are looking for is APPLE7, but the letters are listed as hexadecimal numbers, so instead we are looking for this sequence:

C1 D0 D0 CC C5 B7

which means:

A   P   P   L   E   7

If you do not find this sequence, type:

6050.6100

Look carefully through this information for the above sequence. If you still do not find it, then either you have missed the sequence or APPLE7 is not one of the first seven files on this diskette. If it is not in the first seven, make the following change:

30F: 0E

That is all.

300G again!

6000.6050

We now have the next seven file names.
Examine the contents for the above sequence.
If it is not there, type:

6050.6100

If it is still not there, repeat the process placing "0D" in memory location 30F. That will bring in the third set of seven file names on this diskette. Finding the file name turns out to be the hardest part of what we are doing, especially if you have not worked with hexadecimal numbers before. If you have gone through the third set of seven file names, and still have not found the APPLE7 sequence, you should go back to the CATALOG and make certain that the binary file APPLE7 is really on the diskette. At some point you should come across this sequence.

A0 A0 02 00 13 0F 04 C1 D0 D0 CC C5 B7

If APPLE7 is not the second file on the diskette, the numbers 02 and 13 0F may be different. First, the 04 preceding the C1 or the "A" (of APPLE7) indicates the file type.

04 is a Binary file

00 is a Text file

02 is an Applesoft program file

01 is an Integer program file

In other words, the hexadecimal number immediately preceding the first letter of the file name will always give the type of the file. There are four other file types possible with codes of:

08—for an S file

10—for an R file

20—for a new A file

40—for a new B file

# READ/WRITE TRACK/SECTOR (RWTS)

Immediately preceding the file type indicator is the address of the "track/sector list". As I indicated, these numbers may be different on your diskette. In the example, the 13 indicates the track number and the 0F the sector number. These are the numbers we are really interested in. Type the following:

30E:13 0F (or whatever your numbers are!)

Then type:

300G
6000.6050

You should see:

00 00 00 00 13 0E 00 00

Your numbers may be different, but whatever they are, you will need the two numbers between the zeros. The first number (13) is the track number for the start of the file data. The second number (0E) is the sector number for the start of the file data. Once again, type:

30E:13 0E (or whatever your numbers are!)

Then

300G
6000.6050

Now you should see exactly the following:

```
6000- 00 03 43 00 20 58 FC A2
6008- 00 BD 11 03 20 F0 FD E8
6010- E0 32 D0 F5 60 C8 C5 CC
6018- CC CF 8D CD D9 A0 CE C1
6020- CD C5 A0 C9 D3 A0 C1 D0
6028- D0 CC C5 A0 C9 C9 8D C9
6030- A0 C1 CD A0 C1 A0 D3 CD
6038- C1 D2 D4 A0 C3 CF CD D0
6040- D5 D4 C5 D2 8D 8D 8D 00
6048- 00 00 00 00 00 00 00 00
6050- 00
```

The first two numbers are the starting address of the APPLE7 file in reverse order—00 03 equals $0300. The next two numbers are the length of the APPLE7 file in reverse order—43 00 equals $0043 or simply $43. The rest of the listing is the data in the file or the program instructions. With a little practice, you can quickly examine the CATALOG listing on the diskette, determine where that file information is on the diskette, and finally examine the data itself.

# APPENDIX B.

## TAPE FILES

I have included this section on tape files for those without disk drives. With the necessary modifications, most of the programs in the sequential section of the book should also work with tape. The biggest problem is the lack of string arrays with the Applesoft STORE and RECALL commands. This means that any character to be saved must be translated into its ASC equivalent. When the file is retrieved from tape, every ASC value must then be translated back into its alphanumeric equivalent. Apart from these necessary translations, and the fact that you are saving to tape rather than disk, the sequential access programs should work much the same. Since the introduction of the disks, tape has been largely ignored. Although much slower than disk, tape can be a very powerful computer tool.

I chose the DRILL & PRACTICE programs for the examples since the largest number of Apples without disk drives probably exist in the schools.

## TAPE CREATE QA

```
10   REM  **--INPUT Q & A--**
11 :
12 :
40   DIM Q$(50),A$(50)
60   I = 1
70 :
75 :
100  REM  **--INPUT ROUTINE--**
105  HOME : VTAB 10
120  PRINT : PRINT
140  PRINT "QUESTION # ";I: INPUT Q$(I)
160  IF Q$(I) = "END" THEN 300
180  INPUT "ANSWER ";A$(I)
```

# TAPE CREATE QA

```
200 PRINT : PRINT : PRINT Q$(I)
220 PRINT : PRINT A$(I)
230 PRINT
240 INPUT "IS THIS CORRECT? ";Y$
250 PRINT
260 IF Y$ = "N" THEN 140
280 I = I + 1: GOTO 140
290 :
295 :
300 REM **--TRANSLATION ROUTINE--**
305 Z = 1:X = 1
310 FOR J = 1 TO I - 1
320 L1 = L1 + LEN (Q$(J)): REM QUEST. LEN.
330 L2 = L2 + LEN (A$(J)): REM ANS. LEN.
340 NEXT J
345 L1 = L1 + 50:L2 = L2 + 50: REM 50 Q&A
350 DIM Q(L1),A(L2)
351 :
352 :
355 REM **--QUESTION TRANSLATION--**
360 FOR J = 1 TO I - 1
410 FOR K = 1 TO LEN (Q$(J))
425 Q(Z) = ASC ( MID$ (Q$(J),K,1))
427 Z = Z + 1
430 NEXT K
431 :
432 :
435 REM **--ANSWER TRANSLATION--**
440 FOR K = 1 TO LEN (A$(J))
455 A(X) = ASC ( MID$ (A$(J),K,1))
457 X = X + 1
460 NEXT K
465 Q(Z) = 13:A(X) = 13
467 Z = Z + 1:X = X + 1
470 NEXT J
471 :
472 :
475 REM **--TAPE ROUTINE--**
480 INPUT "TURN ON THE TAPE RECORDER ";L$
500 QP(1) = L1
505 AP(1) = L2
508 PRINT "SAVING QUESTION POINTER"
510 STORE QP
513 PRINT "SAVING ANSWER POINTER"
```

```
515 STORE AP
518 PRINT "SAVING QUESTIONS"
520 STORE Q
530 PRINT "SAVING ANSWERS"
580 STORE A
600 PRINT "ALL FINISHED. TURN OFF TAPE RECORDER."
```

## DRILL QA TAPE

```
10  REM ***--DRILL & PRACTICE--***
11 :
12 :
40  DIM Q$(50),A$(50)
50  DIM QP(11),AP(11)
90 :
95 :
100 REM **--TAPE FILE POINTER--**
103 PRINT "GETTING QUESTIONS POINTER"
105 RECALL QP: REM QUESTION POINTER
108 PRINT "GETTING ANSWERS POINTER"
110 RECALL AP: REM ANSWER POINTER
115 :
116 :
120 REM **--TAPE QUEST. & ANS.---**
125 DIM Q(QP(1)),A(AP(1))
128 PRINT "GETTING QUESTIONS"
130 RECALL Q
135 PRINT "GETTING ANSWERS"
140 RECALL A
150 Y = 1:X = 1
151 :
152 :
155 REM **--QUESTION TRANSLATION--**
160 FOR I = 1 TO QP(1)
170 Q$(Y) = Q$(Y) + CHR$ (Q(I))
180 IF Q(I) = 13 THEN Y = Y + 1
190 NEXT I
191 :
192 :
195 REM **--ANSWER TRANSLATION--**
200 FOR I = 1 TO AP(1)
205 IF A(I) = 13 THEN X = X + 1: GOTO 220
```

```
210 A$(X) = A$(X) + CHR$ (A(I))
220 NEXT I
230 J = Y - 1
245 :
246 :
250 REM **--GET Q & A--**
260 I = RND (1) * 10: I = INT (I)
280 IF I > J OR I < 1 THEN 260
300 PRINT Q$(I)
320 PRINT : PRINT
340 INPUT "YOUR ANSWER IS ";S$
360 IF S$ = "END" THEN 600
380 IF S$ = A$(I) THEN PRINT "CORRECT": A = A + 1: GOTO
    540
400 IF Z > 0 THEN 500
420 PRINT "NO, TRY ONCE MORE"
440 Z = 1
460 A2 = A2 + 1
480 GOTO 340
500 PRINT "NO, THE ANSWER IS ";A$(I)
520 M = M + 1
540 Z = 0
560 PRINT : PRINT
580 GOTO 260
590 :
595 :
600 REM **--DISPLAY SCORE--**
610 A2 = A2 - M
620 A = A - A2
640 HOME : VTAB 10
660 PRINT "YOU GOT ";A;" RIGHT ON THE FIRST TRY"
680 PRINT : PRINT
700 PRINT "YOU GOT ";A2;" RIGHT ON THE SECOND TRY"
720 PRINT : PRINT
740 PRINT "YOU MISSED ";M;" ANSWERS"
```

# APPENDIX C.

## MAILING LIST SYSTEM PROGRAMS

### MAILING LIST MENU

```
10  REM ***--MAILING LIST PROGRAM MENU--***
11  :
12  :
20  D$ = CHR$ (4): REM CONTROL D
25  :
26  :
30  REM **--MENU ROUTINE--**
40  HOME : VTAB 5
60  HTAB 17: PRINT "PROGRAM MENU"
80  PRINT : PRINT
100 HTAB 8: PRINT "1. FILE CREATION PROGRAM"
120 PRINT
140 HTAB 8: PRINT "2. FILE ADDITION PROGRAM"
160 PRINT
180 HTAB 8: PRINT "3. FILE DISPLAY PROGRAM"
200 PRINT
220 HTAB 8: PRINT "4. FILE CORRECTION PROGRAM"
240 PRINT
300 HTAB 8: PRINT "5. CATALOG"
320 PRINT
340 HTAB 8: PRINT "6. END"
360 PRINT : PRINT
380 HTAB 8: INPUT "WHICH PROGRAM NUMBER? ";NUMBER
400 IF NUMBER < 1 OR NUMBER > 6 THEN 380
420 IF NUMBER = 1 THEN 1000
440 IF NUMBER = 2 THEN 2000
460 IF NUMBER = 3 THEN PRINT D$; "RUN MAILING LIST READER"
```

```
 470 IF NUMBER = 4 THEN PRINT D$;"RUN MAILING
     LIST CORRECTOR"
 480 IF NUMBER = 5 THEN PRINT D$;"CATALOG": INPUT
     "HIT RETURN TO GO TO MENU ";L$: GOTO 40
 500 IF NUMBER = 6 THEN END
 600 :
 700 :
1000 REM **--FILE CREATOR PROGRAM--**
1020 PRINT : PRINT "IF THE ADDRESS FILE ALREADY EXISTS"
1040 PRINT : PRINT "DO NOT RUN THIS PROGRAM!!"
1060 PRINT : PRINT "DO YOU WANT THE FILE CREATION
     PROGRAM?"
1070 PRINT
1080 INPUT "TYPE 'YES' IF YOU DO: ";YES$
1100 IF YES$ = "YES" THEN PRINT D$;"RUN MAILING
     LIST CREATOR"
1120 GOTO 40
1140 :
1160 :
2000 REM **--FILE ADDITION PROGRAM--**
2020 PRINT : PRINT "YOU WANT TO ADD TO THE EXISTING"
2040 PRINT : PRINT "ADDRESS FILE. IS THIS CORRECT?"
2060 PRINT : INPUT "TYPE 'YES' IF IT IS. ";YES$
2080 IF YES$ = "YES" THEN PRINT D$;"RUN MAILING LIST
     ADDER2"
2100 GOTO 40
```

## MAILING LIST CREATOR

```
10  REM  **--MAILING LIST CREATOR--**
11  :
12  :
20  D$ = CHR$ (4) : REM CONTROL D
40  DIM NAME$(20)
60  K = 1: REM LINE COUNTER
65  :
66  :
70  REM  **--INPUT ROUTINE--**
80  HOME : VTAB 5
100 PRINT "TYPE NAME AND ADDRESS AS IF ADDRESSING
      AN ENVELOPE. ";
120 PRINT "DO NOT USE A COMMA OR COLON. "
140 PRINT : PRINT "TYPE 'END' WHEN FINISHED"
160 PRINT : PRINT "TYPE IN LINE  ";K
180 INPUT NAME$(K)
200 IF NAME$(K) = "END" THEN 300
220 K = K + 1
240 GOTO 160: REM GO BACK FOR ANOTHER LINE
300 NAME$(K) = "*": REM SEPARATOR FOR PHONE NUMBER
320 K = K + 1
340 PRINT "PHONE: ";: PRINT "TYPE 'RETURN' IF NONE. "
360 INPUT NAME$(K)
380 K = K + 1
400 NAME$(K) = "!": REM SEPARATOR BETWEEN
      SETS OF INFORMATION
405 :
406 :
410 REM  **--CORRECTION ROUTINE--**
420 HOME : VTAB 5
440 PRINT "DO NOT CHANGE THE LINE WITH THE '*'"
460 PRINT "THIS SYMBOL IS USED AS A SEPARATOR. "
480 PRINT
500 FOR I = 1 TO K - 1
520 PRINT I;" ";NAME$(I)
540 NEXT I
560 PRINT
580 INPUT "CHANGE ANY LINE? TYPE 'Y' OR 'N' ";YES$
600 IF YES$ = "Y" THEN 640
620 GOTO 740: REM GO TO FILE CREATION ROUTINE
640 INPUT "CHANGE WHICH LINE ";LINE
```

```
660 IF LINE > K - 1 THEN PRINT "NUMBER TOO LARGE": GOTO
    640
680 PRINT "OLD LINE = ";NAME$(LINE)
700 INPUT "CORRECT LINE = ";NAME$(LINE)
720 GOTO 420
725 :
726 :
730 REM **--FILE CREATION ROUTINE--**
740 PRINT D$;"OPEN ADDRESS FILE"
760 PRINT D$;"DELETE ADDRESS FILE"
780 PRINT D$;"OPEN ADDRESS FILE"
800 PRINT D$;"WRITE ADDRESS FILE"
820 PRINT K: REM NUMBER OF LINES
840 PRINT "      ": REM 5 SPACES FOR INCREASING COUNTER
860 FOR I = 1 TO K
880 PRINT NAME$(I)
900 NEXT I
920 PRINT D$;"CLOSE ADDRESS FILE"
940 END
```

## MAILING LIST ADDER1

```
10  REM **--MAILING LIST ADDER1--**
11 :
12 :
20  D$ = CHR$ (4): REM CONTROL D
40  DIM NAME$(20)
60  K = 1: REM LINE COUNTER
65 :
66 :
70  REM **--INPUT ROUTINE--**
80  HOME : VTAB 5
100 PRINT "TYPE NAME AND ADDRESS AS IF ADDRESSING
    AN ENVELOPE. ";
120 PRINT "DO NOT USE A COMMA OR COLON. "
140 PRINT : PRINT "TYPE 'END' WHEN FINISHED"
160 PRINT : PRINT "TYPE IN LINE ";K
180 INPUT NAME$(K)
200 IF NAME$(K) = "END" THEN 300
220 K = K + 1
240 GOTO 160: REM GO BACK FOR ANOTHER LINE
300 NAME$(K) = "*": REM SEPARATOR FOR PHONE NUMBER
```

```
320 K = K + 1
340 PRINT "PHONE: " ; : PRINT "TYPE 'RETURN' IF NONE. "
360  INPUT NAME$(K)
380 K = K + 1
400  NAME$(K) = "!": REM  SEPARATOR BETWEEN SETS OF
     INFORMATION
405 :
406 :
410  REM **--CORRECTION ROUTINE--**
420  HOME : VTAB 5
440  PRINT "DO NOT CHANGE THE LINE WITH THE '*'"
460  PRINT "THIS SYMBOL IS USED AS A SEPARATOR. "
480  PRINT
500  FOR I = 1 TO K - 1
520  PRINT I; " "; NAME$(I)
540  NEXT I
560  PRINT
580  INPUT "CHANGE ANY LINE? TYPE 'Y' OR 'N' "; YES$
600  IF YES$ = "Y" THEN 640
620  GOTO 2000: REM GO TO FILE CREATION ROUTINE
640  INPUT "CHANGE WHICH LINE "; LINE
660  IF LINE > K - 1 THEN PRINT "NUMBER TOO LARGE": GOTO
     640
680  PRINT "OLD LINE = "; NAME$(LINE)
700  INPUT "CORRECT LINE = "; NAME$(LINE)
720  GOTO 410
740 :
760 :
1000  REM **--REPEAT ROUTINE--**
1020  HOME : VTAB 5
1040  PRINT "DO YOU WANT TO ADD MORE INFO?"
1060  INPUT "TYPE 'Y' OR 'N' "; YES$
1080  IF YES$ = "Y" THEN RUN
1100  PRINT D$; "RUN MENU"
1120 :
1140 :
2000  REM **--FILE ADDITION ROUTINE--**
2020  PRINT D$; "OPEN ADDRESS FILE"
2040  PRINT D$; "READ ADDRESS FILE"
2060  INPUT REC
2080  PRINT D$; "CLOSE ADDRESS FILE"
2100 REC = REC + K
2120  PRINT D$; "OPEN ADDRESS FILE"
2140  PRINT D$; "WRITE ADDRESS FILE"
```

```
2160  PRINT REC
2180  PRINT D$; "CLOSE ADDRESS FILE"
2200  PRINT D$; "APPEND ADDRESS FILE"
2220  PRINT D$; "WRITE ADDRESS FILE"
2240  FOR I = 1 TO K
2260  PRINT NAME$(I)
2280  NEXT I
2300  PRINT D$; "CLOSE ADDRESS FILE"
2320  GOTO 1000: REM REPEAT ROUTINE
```

## MAILING LIST ADDER2

```
10   REM ***--MAILING LIST ADDER2--***
11 :
12 :
20   D$ = CHR$ (4): REM CONTROL D
40   DIM NAME$(20), LINE$(100)
60   K = 1: REM LINE COUNTER
65 :
66 :
70   REM **--INPUT ROUTINE--**
80   HOME : VTAB 5
100  PRINT "TYPE NAME AND ADDRESS AS IF ADDRESSING  AN
     ENVELOPE.";
120  PRINT "DO NOT USE A COMMA OR COLON."
140  PRINT : PRINT "TYPE 'END' WHEN FINISHED"
160  PRINT : PRINT "TYPE IN LINE ";K
180  INPUT NAME$(K)
200  IF NAME$(K) = "END" THEN 300
220  K = K + 1
240  GOTO 160: REM GO BACK FOR ANOTHER LINE
300  NAME$(K) = "*": REM SEPARATOR FOR PHONE NUMBER
320  K = K + 1
340  PRINT "PHONE: ";: PRINT "TYPE 'RETURN' IF NONE."
360  INPUT NAME$(K)
380  K = K + 1
400  NAME$(K) = "!": REM SEPARATOR BETWEEN SETS OF
     INFORMATION
405 :
406 :
410  REM **--CORRECTION ROUTINE--**
420  HOME : VTAB 5
440  PRINT "DO NOT CHANGE THE LINE WITH THE '*'"
```

```
460 PRINT "THIS SYMBOL IS USED AS A SEPARATOR. "
480 PRINT
500 FOR I = 1 TO K - 1
520 PRINT I; " "; NAME$(I)
540 NEXT I
560 PRINT
580 INPUT "CHANGE ANY LINE? TYPE 'Y' OR 'N' "; YES$
600 IF YES$ = "Y" THEN 640
620 GOTO 800: REM  PRINT LABEL ROUTINE
640 INPUT "CHANGE WHICH LINE "; LINE
660 IF LINE > K - 1 THEN PRINT "NUMBER TOO LARGE": GOTO 640
680 PRINT "OLD LINE = "; NAME$(LINE)
700 INPUT "CORRECT LINE = "; NAME$(LINE)
720 GOTO 410
740 :
760 :
800 REM **--PRINT LABEL ROUTINE--**
810 PRINT "DO YOU WANT TO PRINT A LABEL NOW"
820 INPUT "TYPE 'Y' OR 'N' "; YES$
840 IF YES$ = "Y" THEN 880
860 GOTO 1000: REM  REPEAT ROUTINE
880 PR# 1
900 FOR I = 1 TO K
920 IF NAME$(I) = "*" THEN I = I + 1: GOTO 980
940 IF NAME$(I) = "!" THEN 980
960 PRINT NAME$(I)
980 NEXT I
990 PR# 0: GOTO 800
995 :
996 :
1000 REM **--REPEAT ROUTINE--**
1020 HOME : VTAB 5
1023 FOR I = 1 TO K
1026 LINE$(TK + I) = NAME$(I)
1033 NEXT I
1036 TK = TK + K
1040 PRINT "DO YOU WANT TO ADD MORE INFO? "
1060 INPUT "TYPE 'Y' OR 'N' "; YES$
1080 IF YES$ = "Y" THEN GOTO 60
1100 :
1111 :
2000 REM **--FILE ADDITION ROUTINE--**
2020 PRINT D$; "OPEN ADDRESS FILE"
```

```
2040  PRINT D$; "READ ADDRESS FILE"
2060  INPUT REC
2080  PRINT D$; "CLOSE ADDRESS FILE"
2100 REC = REC + TK
2120  PRINT D$; "OPEN ADDRESS FILE"
2140  PRINT D$; "WRITE ADDRESS FILE"
2160  PRINT REC
2180  PRINT D$; "CLOSE ADDRESS FILE"
2200  PRINT D$; "APPEND ADDRESS FILE"
2220  PRINT D$; "WRITE ADDRESS FILE"
2240  FOR I = 1 TO TK
2260  PRINT LINE$(I)
2280  NEXT I
2300  PRINT D$; "CLOSE ADDRESS FILE"
3000  PRINT D$; "RUN MENU"
```

## MAILING LIST READER 2

```
10  REM  ***--MAILING LIST READER--***
11 :
12 :
20 D$ = CHR$(4) : REM  CONTROL D
25 :
26 :
30  REM  **--INPUT ROUTINE--**
40  PRINT D$; "OPEN ADDRESS FILE"
60  PRINT D$; "READ ADDRESS FILE"
80  INPUT K
100  DIM NAME$(K), AD$(K), ND$(K), L(K), R(K)
120  FOR I = 1 TO K
140  INPUT NAME$(I)
160  NEXT I
180  PRINT D$; "CLOSE ADDRESS FILE"
190 :
191 :
200  REM  **--MENU ROUTINE--**
210  HOME : VTAB 2
220  HTAB 17: PRINT "MENU"
230  PRINT : PRINT
240  HTAB 3: PRINT "1. DISPLAY INFO--ORIG. ORDER"
250  PRINT
260  HTAB 3: PRINT "2. DISPLAY NAMES ONLY"
```

```
270 PRINT
280 HTAB 3: PRINT "3. DISPLAY INFO- -NO PHONE"
290 PRINT
300 HTAB 3: PRINT "4. DISPLAY SPECIFIC NAME"
310 PRINT
320 HTAB 3: PRINT "5. DISPLAY SPECIFIC NAME- -NO PHONE"
330 PRINT
340 HTAB 3: PRINT "6. DISPLAY INFO- -RANGE"
350 PRINT
360 HTAB 3: PRINT "7. DISPLAY INFO- -ALPHABETICAL"
365 PRINT
370 HTAB 3: PRINT "8. RETURN TO PROGRAM MENU"
375 PRINT
380 HTAB 3: INPUT "WHICH NUMBER? ";NUMBER
385 IF NUMBER < 1 OR NUMBER > 8 THEN PRINT
    "INCORRECT NUMBER": GOTO 380
410 IF NUMBER = 1 THEN 1000
420 IF NUMBER = 2 THEN 2000
430 IF NUMBER = 3 THEN 3000
440 IF NUMBER = 4 THEN 4000
450 IF NUMBER = 5 THEN 5000
460 IF NUMBER = 6 THEN 6000
470 IF NUMBER = 7 THEN 7000
480 IF NUMBER = 8 THEN PRINT D$; "RUN MENU"
600 :
700 :
1000 REM **- -ORIGINAL ORDER ROUTINE- -**
1020 GOSUB 10000: REM PRINTER ROUTINE
1040 HOME : VTAB 5
1060 FOR I = 1 TO K
1080 IF NAME$(I) = "*" THEN 1140
1100 IF NAME$(I) = "!" THEN PRINT : GOTO 1140
1120 PRINT NAME$(I)
1140 NEXT I
1160 GOTO 20000: REM RETURN TO MENU ROUTINE
1500 :
1600 :
2000 REM **- -NAME ONLY ROUTINE- -**
2020 GOSUB 10000: REM PRINTER ROUTINE
2040 HOME : VTAB 5
2060 FOR I = 1 TO K - 1
2080 IF NAME$(I) = NAME$(2) THEN PRINT I; " ";NAME$(I)
2100 IF NAME$(I) = "!" THEN PRINT I; " "; NAME$(I + 1)
2120 NEXT I
```

```
2140  GOTO 20000: REM RETURN TO MENU ROUTINE
2500 :
2600 :
3000  REM **--NO PHONE ROUTINE--**
3020  GOSUB 10000: REM PRINTER ROUTINE
3040  HOME : VTAB 5
3060  FOR I = 1 TO K
3080  IF NAME$(I) = "*" THEN I = I + 1: GOTO 3140
3100  IF NAME$(I) = "!" THEN PRINT : GOTO 3140
3120  PRINT NAME$(I)
3140  NEXT I
3160  GOTO 20000: REM  RETURN TO MENU ROUTINE
3500 :
3600 :
4000  REM **--SEARCH ROUTINE--**
4020  HOME : VTAB 5
4040  INPUT "NAME TO FIND? ";FIND$
4060  IF FIND$ = "END" THEN 4400
4070  GOSUB 10000: REM  PRINTER ROUTINE
4080  PRINT
4100  FOR I = 1 TO K
4120  IF NAME$(I) = FIND$ THEN 4160
4140  GOTO 4340
4160  IF NAME$(I) = "*" THEN 4340
4180  IF NAME$(I) = "!" THEN PRINT : GOTO 4340
4200  PRINT NAME$(I)
4220  PRINT NAME$(I + 1)
4240  PRINT NAME$(I + 2)
4260  IF NAME$(I + 3) < > "*" THEN PRINT NAME$(I + 3)
4280  IF NAME$(I + 4) = "*" THEN 4320
4300  PRINT NAME$(I + 4): GOTO 4340
4320  PRINT NAME$(I + 5)
4340  NEXT I
4360  PRINT
4380  GOTO 4040
4400  GOTO 20000: REM RETURN TO MENU ROUTINE
4500 :
4600 :
5000  REM **--SEARCH ROUTINE NO PHONE--**
5020  HOME : VTAB 5
5040  PRINT "TYPE 'END' WHEN FINISHED"
5050  INPUT "NAME TO FIND? ";FIND$
5060  IF FIND$ = "END" THEN 5400
5070  GOSUB 10000: REM  PRINTER ROUTINE
```

```
5080 PRINT
5100 FOR I = 1 TO K
5120 IF NAME$(I) = FIND$ THEN 5160
5140 GOTO 5340
5160 IF NAME$(I) = "*" THEN I = I + 1: GOTO 5340
5180 IF NAME$(I) = "!" THEN PRINT : GOTO 5340
5200 PRINT NAME$(I)
5220 PRINT NAME$(I + 1)
5240 PRINT NAME$(I + 2)
5260 IF NAME$(I + 3) < > "*" THEN PRINT NAME$(I + 3)
5270 IF NAME$(I + 3) = "*" THEN I = I + 1: GOTO 5340
5280 IF NAME$(I + 4) = "*" THEN I = I + 1 : GOTO 5340
5300 PRINT NAME$(I + 4): GOTO 5340
5320 PRINT NAME$(I + 5)
5340 NEXT I
5360 PRINT
5380 GOTO 5040
5400 GOTO 20000: REM  RETURN TO MENU ROUTINE
5500 :
5600 :
6000 REM  **--RANGE ROUTINE--**
6020 HOME : VTAB 5
6040 INPUT "TYPE BEGINNING LINE NUMBER ";BL
6060 PRINT
6080 IF BL < 2 THEN PRINT "NUMBER TOO SMALL": GOTO 6040
6100 INPUT "TYPE ENDING LINE NUMBER ";EL
6120 PRINT
6140 IF EL > K THEN PRINT "NUMBER TOO LARGE": GOTO 6100
6160 GOSUB 10000: REM  PRINTER ROUTINE
6180 FOR I = BL TO EL
6200 IF NAME$(I) = "*" THEN I = I + 1: GOTO 6260
6220 IF NAME$(I) = "!" THEN PRINT : GOTO 6260
6240 PRINT NAME$(I)
6260 NEXT I
6280 GOTO 20000: REM  RETURN TO MENU ROUTINE
6500 :
6600 :
6700 :
6800 :
6900 :
7000 REM  **--ALPHABETICAL ORDER ROUTINE--**
7040 HOME : VTAB 5
7060 PRINT "WORKING--PLEASE DONT TOUCH!!"
7065 :
```

## MAILING LIST READER 2

```
7066 :
7070 REM  GET FIRST INFO-LINE
7080 FOR I = 2 TO K - 1
7100 IF NAME$(I) = NAME$(2) THEN 7160
7120 IF NAME$(I) = "!" THEN I = I + 1: GOTO 7160
7140 GOTO 7340
7145 :
7146 :
7150 REM  REVERSE ORDER
7160 LN = LEN (NAME$(I))
7180 FOR J1 = 1 TO LN: IF MID$ (NAME$(I), J1, 1) = " " THEN
     J2 = J1
7200 NEXT J1
7210 IF J2 = 0 OR J2 > LN THEN AD$(I) = NAME$(I) :
     GOTO 7240
7220 AD$(I) = MID$ (NAME$(I), J2 + 1, LN - J2) + " " +
     LEFT$ (NAME$(I), J2)
7240 AD$(I) = AD$(I) + "**" + NAME$(I + 1) + "**" +
     NAME$(I + 2)
7260 IF NAME$(I + 3) < > "*" THEN AD$(I) = AD$(I) + "**"
     + NAME$(I + 3)
7280 IF NAME$(I + 4) = "*" THEN 7320
7300 AD$(I) = AD$(I) + "**" + NAME$(I + 4) : GOTO 7340
7320 AD$(I) = AD$(I) + "**" + NAME$(I + 5)
7340 NEXT I
7345 :
7346 :
7350 REM  RENUMBER FOR SORT
7360 J = 1
7380 FOR I = 1 TO K
7400 IF LEN (AD$(I)) > 0 THEN ND$(J) = AD$(I) : J = J + 1
7420 NEXT I
7440 N = J - 1
7445 :
7446 :
7460 REM  ***--QUICKSORT--***
7480 S1 = 1
7500 PRINT "WORKING--PLEASE DONT TOUCH!!"
7520 L(1) = 1
7540 R(1) = N
7560 L1 = L(S1)
7580 R1 = R(S1)
7600 S1 = S1 - 1
```

```
7620 L2 = L1
7640 R2 = R1
7660 X$ = ND$( INT ((L1 + R1) / 2))
7680 C = C + 1
7700  IF ND$(L2) = X$ OR ND$(L2) > X$ THEN 7760
7720 L2 = L2 + 1
7740  GOTO 7680
7760 C = C1
7780  IF X$ = ND$(R2) OR X$ > ND$(R2) THEN 7840
7800 R2 = R2 - 1
7820  GOTO 7760
7840  IF L2 > R2 THEN 7980
7860 S = S + 1
7880 T$ = ND$(L2)
7900 ND$(L2) = ND$(R2)
7920 ND$(R2) = T$
7940 L2 = L2 + 1
7960 R2 = R2 - 1
7980  IF L2 = R2 OR L2 < R2 THEN 7680
8000  IF L2 = R1 OR L2 > R1 THEN 8080
8020 S1 = S1 + 1
8040 L(S1) = L2
8060 R(S1) = R1
8080 R1 = R2
8100  IF L1 < R1 THEN 7620
8120  IF S1 > 0 THEN 7560
8140 REM  SORT COMPLETED
8142 :
8143 :
8145 REM **--DISPLAY--**
8150 GOSUB 10000: REM  PRINTER ROUTINE
8160 FOR I = 1 TO N
8180 PRINT ND$(I)
8200 PRINT
8220 NEXT I
8240 GOTO 20000: REM  RETURN TO MENU ROUTINE
8500 :
8600 :
8700 :
8800 :
8900 :
10000  REM **--PRINT ROUTINE--**
10020  PRINT "DO YOU WANT A PAPER PRINT OUT?"
10040  INPUT "TYPE 'Y' OR 'N' ";YES$
```

```
10060 IF YES$ = "Y" THEN 10120
10080 SPEED = 150
10100 RETURN
10120 PR# 1
10140 RETURN
15000 :
16000 :
20000 REM **--RETURN TO MENU ROUTINE--**
20020 PR# 0: SPEED = 255
20040 INPUT "HIT RETURN TO GO TO MENU ";L$
20060 GOTO 200: REM  MENU
```

## MAILING LIST CORRECTOR

```
10 REM **--MAILING LIST CORRECTOR--**
11 :
12 :
20 D$ = CHR$ (4): REM  CONTROL D
25 :
26 :
30 REM **--INPUT ROUTINE--**
40 PRINT D$;"OPEN ADDRESS FILE"
60 PRINT D$;"READ ADDRESS FILE"
80 INPUT K
100 DIM NAME$(K),LINE$(K)
120 FOR I = 1 TO K
140 INPUT NAME$(I)
160 NEXT I
180 PRINT D$;"CLOSE ADDRESS FILE"
190 :
191 :
200 REM **--MENU ROUTINE--**
220 HOME : VTAB 5
240 HTAB 19
260 PRINT "MENU"
280 PRINT : PRINT
300 HTAB 8: PRINT "1. CHANGE OR CORRECT INFO"
320 PRINT
340 HTAB 8: PRINT "2. DELETE INFO"
360 PRINT
380 HTAB 8: PRINT "3. WRITE REVISED FILE"
400 PRINT
```

```
420  HTAB 8: PRINT "4. RETURN TO PROGRAM MENU"
430  PRINT : PRINT
440  HTAB 8: INPUT "WHICH NUMBER ";NB
460  IF NB < 0 OR NB > 4 THEN 440
510  IF NB = 1 THEN 1000
520  IF NB = 2 THEN 2000
530  IF NB = 3 THEN 3000
540  IF NB = 4 THEN PRINT D$;"RUN MENU"
600 :
700 :
1000 REM **--CORRECTION ROUTINE--**
1020 HOME : VTAB 5
1040 PRINT "TYPE '0' WHEN FINISHED"
1060 INPUT "DISPLAY WHICH LINE ";NUMBER
1080 IF NUMBER = 0 THEN 200
1100 PRINT
1120 PRINT NUMBER;"  ";NAME$(NUMBER)
1140 PRINT
1160 PRINT "IS THIS CORRECT? ";
1180 INPUT "TYPE 'Y' OR 'N' ";YES$
1200 IF YES$ = "Y" THEN 1020
1220 PRINT
1240 PRINT "TYPE IN THE CORRECT INFORMATION"
1260 PRINT
1280 PRINT NUMBER;"  ";: INPUT CN$
1300 PRINT : NAME$(NUMBER) = CN$
1320 PRINT NUMBER;"  ";NAME$(NUMBER)
1340 PRINT
1360 GOTO 1160
1500 :
1600 :
2000 REM **--DELETE ROUTINE--**
2020 HOME : VTAB 5
2040 PRINT "TYPE '0' WHEN FINISHED"
2060 INPUT "DELETE WHICH LINE ";LINE
2080 IF LINE = 0 THEN 200
2100 PRINT
2120 PRINT LINE;"  ";NAME$(LINE)
2140 PRINT
2220 PRINT "ARE YOU SURE? TYPE 'YES' IF SURE";
2240 INPUT YES$
2260 IF YES$ = "YES" THEN 2300
2280 GOTO 2000
2300 J = LINE
```

```
2320  IF NAME$(J) = "!" THEN 2360
2340  J = J + 1: GOTO 2320
2360  FOR I = LINE TO J
2380  PRINT I;" ";NAME$(I)
2400  NAME$(I) = "DELETED":D = D + 1
2420  NEXT I
2440  PRINT
2460  PRINT "DELETING THIS INFORMATION"
2480  Q = 2
2500  FOR I = 2 TO K
2520  IF NAME$(I) = "DELETED" THEN 2580
2540  NAME$(Q) = NAME$(I)
2560  Q = Q + 1
2580  NEXT I
2600  K = K - D
2620  D = 0:J = 0
2700  GOTO 200
2800 :
2900 :
3000  REM **--FILE ROUTINE--**
3020  PRINT D$;"OPEN ADDRESS FILE BACKUP"
3040  PRINT D$;"DELETE ADDRESS FILE BACKUP"
3060  PRINT D$;"RENAME ADDRESS FILE, ADDRESS FILE BACKUP"
3080  PRINT D$;"OPEN ADDRESS FILE"
3100  PRINT D$;"WRITE ADDRESS FILE"
3120  PRINT K
3140  FOR I = 1 TO K
3160  PRINT NAME$(I)
3180  NEXT I
3200  PRINT D$;"CLOSE ADDRESS FILE"
3220  PRINT D$;"RUN MENU"
```

## MAILING LIST READER1

```
10  REM ***--MAILING LIST READER--***
11 :
12 :
20  D$ = CHR$(4): REM CONTROL D
25 :
26 :
30  REM **--INPUT ROUTINE--**
40  PRINT D$;"OPEN ADDRESS FILE"
```

```
60  PRINT D$;"READ ADDRESS FILE"
80  INPUT K
100  DIM NAME$(K)
120  FOR I = 1 TO K
140  INPUT NAME$(I)
160  NEXT I
180  PRINT D$;"CLOSE ADDRESS FILE"
190 :
191 :
200  REM **--DISPLAY ROUTINE--**
220  HOME : VTAB 5
240  FOR I = 1 TO K
260  PRINT NAME$(I)
280  NEXT I
```

## MENU CHAIN

```
10  REM  MAILING LIST PROGRAM MENU
20  D$ = CHR$ (4): REM  CONTROL D
40  HOME : VTAB 5
60  HTAB 17: PRINT "PROGRAM MENU"
80  PRINT : PRINT
100  HTAB 8: PRINT "1. FILE CREATION PROGRAM"
120  PRINT
140  HTAB 8: PRINT "2. FILE ADDITION PROGRAM"
160  PRINT
180  HTAB 8: PRINT "3. FILE DISPLAY PROGRAM"
200  PRINT
220  HTAB 8: PRINT "4. FILE CORRECTION PROGRAM"
240  PRINT
300  HTAB 8: PRINT "5. CATALOG"
320  PRINT
340  HTAB 8: PRINT "6. END"
360  PRINT : PRINT
380  HTAB 8: INPUT "WHICH PROGRAM NUMBER? ";NUMBER
400  IF NUMBER < 1 OR NUMBER > 6 THEN 380
420  IF NUMBER = 1 THEN 1000
440  IF NUMBER = 2 THEN 2000
460  IF NUMBER = 3 THEN PRINT D$;"RUN MAILING
     LIST READER"
470  IF NUMBER = 4 THEN 4000
480  IF NUMBER = 5 THEN PRINT D$;"CATALOG": INPUT
     "HIT RETURN TO GO TO MENU ";L$: GOTO 40
```

```
500  IF NUMBER = 6 THEN END
1000 REM  FILE CREATOR PROGRAM
1020 PRINT : PRINT "IF THE ADDRESS FILE ALREADY EXISTS"
1040 PRINT : PRINT "DO NOT RUN THIS PROGRAM! ! "
1060 PRINT :  PRINT "DO YOU WANT THE FILE CREATION
     PROGRAM? "
1070 PRINT
1080 INPUT "TYPE 'YES' IF YOU DO: "; YES$
1100 IF YES$ = "YES" THEN PRINT D$; "RUN MAILING LIST
     CREATOR"
1120 GOTO 40
2000 REM  FILE ADDITION PROGRAM
2020 PRINT : PRINT "YOU WANT TO ADD TO THE EXISTING"
2040 PRINT : PRINT "ADDRESS FILE. IS THIS CORRECT? "
2060 PRINT : INPUT "TYPE 'YES' IF IT IS. " ; YES$
2080 IF YES$ = "YES" THEN PRINT D$; "RUN MAILING
     LIST ADDER2"
2100 GOTO 40
4000 PRINT D$; "BLOAD CHAIN, A520"
4020 CALL 520 "MAILING LIST CORRECTOR"
```

## INTEGER MAILING LIST READER

```
10 REM  INTEGER MAILING LIST READER
20 D$ = "": REM  CONTROL D
40 DIM NAME$(50), A$(50), BLANK$(10)
50 REM  INPUT ROUTINE
60 Q=1
80 PRINT D$; "OPEN ADDRESS FILE"
100 PRINT D$; "READ ADDRESS FILE"
120 INPUT K
140 INPUT BLANK$
160 DIM A(K*15)
180 FOR I=1 TO K-1
200 INPUT NAME$
220 IF NAME$ = "" THEN 380
240 LN = LEN (NAME$)
260 FOR J=1 TO LN
280 A(Q) = ASC (NAME$(J)) : REM  CONVERT TO NUMBER
300 Q=Q+1
320 NEXT J
```

```
340 A(Q) = 161
360 Q = Q + 1
380 NEXT I
400 PRINT D$; "CLOSE ADDRESS FILE"
500 REM  DISPLAY ROUTINE
520 T = Q-1
540 FOR Q = 1 TO T
560 GOSUB 5000: REM  CONVERT TO CHARACTER
580 IF A$ = "!" THEN PRINT
600 IF A$ = "!" THEN 660
620 IF A$ = "*" THEN 660
640 PRINT A$;
660 NEXT Q
1000 END
```

## CHARACTER ROUTINE1

```
5000 REM  CHARACTER ROUTINE
5160 IF A(Q) = 160 THEN 6160
5161 IF A(Q) = 161 THEN 6161
5162 IF A(Q) = 162 THEN 6162
5163 IF A(Q) = 163 THEN 6163
5164 IF A(Q) = 164 THEN 6164
5165 IF A(Q) = 165 THEN 6165
5166 IF A(Q) = 166 THEN 6166
5167 IF A(Q) = 167 THEN 6167
5168 IF A(Q) = 168 THEN 6168
5169 IF A(Q) = 169 THEN 6169
5170 IF A(Q) = 170 THEN 6170
5171 IF A(Q) = 171 THEN 6171
5172 IF A(Q) = 172 THEN 6172
5173 IF A(Q) = 173 THEN 6173
5174 IF A(Q) = 174 THEN 6174
5175 IF A(Q) = 175 THEN 6175
5176 IF A(Q) = 176 THEN 6176
5177 IF A(Q) = 177 THEN 6177
5178 IF A(Q) = 178 THEN 6178
5179 IF A(Q) = 179 THEN 6179
5180 IF A(Q) = 180 THEN 6180
5181 IF A(Q) = 181 THEN 6181
5182 IF A(Q) = 182 THEN 6182
5183 IF A(Q) = 183 THEN 6183
```

```
5184 IF A(Q) = 184 THEN 6184
5185 IF A(Q) = 185 THEN 6185
5186 IF A(Q) = 186 THEN 6186
5187 IF A(Q) = 187 THEN 6187
5188 IF A(Q) = 188 THEN 6188
5189 IF A(Q) = 189 THEN 6189
5190 IF A(Q) = 190 THEN 6190
5191 IF A(Q) = 191 THEN 6191
5192 IF A(Q) = 192 THEN 6192
5193 IF A(Q) = 193 THEN 6193
5194 IF A(Q) = 194 THEN 6194
5195 IF A(Q) = 195 THEN 6195
5196 IF A(Q) = 196 THEN 6196
5197 IF A(Q) = 197 THEN 6197
5198 IF A(Q) = 198 THEN 6198
5199 IF A(Q) = 199 THEN 6199
5200 IF A(Q) = 200 THEN 6200
5201 IF A(Q) = 201 THEN 6201
5202 IF A(Q) = 202 THEN 6202
5203 IF A(Q) = 203 THEN 6203
5204 IF A(Q) = 204 THEN 6204
5205 IF A(Q) = 205 THEN 6205
5206 IF A(Q) = 206 THEN 6206
5207 IF A(Q) = 207 THEN 6207
5208 IF A(Q) = 208 THEN 6208
5209 IF A(Q) = 209 THEN 6209
5210 IF A(Q) = 210 THEN 6210
5211 IF A(Q) = 211 THEN 6211
5212 IF A(Q) = 212 THEN 6212
5213 IF A(Q) = 213 THEN 6213
5214 IF A(Q) = 214 THEN 6214
5215 IF A(Q) = 215 THEN 6215
5216 IF A(Q) = 216 THEN 6216
5217 IF A(Q) = 217 THEN 6217
5218 IF A(Q) = 218 THEN 6218
6160 A$ = " " : RETURN
6161 A$ = "!" : RETURN
6162 A$ = "'" : RETURN
6163 A$ = "#" : RETURN
6164 A$ = "$" : RETURN
6165 A$ = "%" : RETURN
6166 A$ = "&" : RETURN
6167 A$ = "'" : RETURN
6168 A$ = "(" : RETURN
```

```
6169 A$ = ")" : RETURN
6170 A$ = "*" : RETURN
6171 A$ = "+" : RETURN
6172 A$ = "," : RETURN
6173 A$ = "-" : RETURN
6174 A$ = "." : RETURN
6175 A$ = "/" : RETURN
6176 A$ = "0" : RETURN
6177 A$ = "1" : RETURN
6178 A$ = "2" : RETURN
6179 A$ = "3" : RETURN
6180 A$ = "4" : RETURN
6181 A$ = "5" : RETURN
6182 A$ = "6" : RETURN
6183 A$ = "7" : RETURN
6184 A$ = "8" : RETURN
6185 A$ = "9" : RETURN
6186 A$ = ":" : RETURN
6187 A$ = ";" : RETURN
6188 A$ = "<" : RETURN
6189 A$ = "=" : RETURN
6190 A$ = ">" : RETURN
6191 A$ = "?" : RETURN
6192 A$ = "@" : RETURN
6193 A$ = "A" : RETURN
6194 A$ = "B" : RETURN
6195 A$ = "C" : RETURN
6196 A$ = "D" : RETURN
6197 A$ = "E" : RETURN
6198 A$ = "F" : RETURN
6199 A$ = "G" : RETURN
6200 A$ = "H" : RETURN
6201 A$ = "I" : RETURN
6202 A$ = "J" : RETURN
6203 A$ = "K" : RETURN
6204 A$ = "L" : RETURN
6205 A$ = "M" : RETURN
6206 A$ = "N" : RETURN
6207 A$ = "O" : RETURN
6208 A$ = "P" : RETURN
6209 A$ = "Q" : RETURN
6210 A$ = "R" : RETURN
6211 A$ = "S" : RETURN
6212 A$ = "T" : RETURN
```

```
6213 A$ = "U" : RETURN
6214 A$ = "V" : RETURN
6215 A$ = "W" : RETURN
6216 A$ = "X" : RETURN
6217 A$ = "Y" : RETURN
6218 A$ = "Z" : RETURN
```

## CHARACTER ROUTINE2

```
5000 REM  CHARACTER ROUTINE
5160 IF A(Q) = 160 THEN A$ = " "
5161 IF A(Q) = 161 THEN A$ = "!"
5162 IF A(Q) = 162 THEN A$ = " ' "
5163 IF A(Q) = 163 THEN A$ = "#"
5164 IF A(Q) = 164 THEN A$ = "$"
5165 IF A(Q) = 165 THEN A$ = "%"
5166 IF A(Q) = 166 THEN A$ = "&"
5167 IF A(Q) = 167 THEN A$ = " ' "
5168 IF A(Q) = 168 THEN A$ = "("
5169 IF A(Q) = 169 THEN A$ = ")"
5170 IF A(Q) = 170 THEN A$ = "*"
5171 IF A(Q) = 171 THEN A$ = "+"
5172 IF A(Q) = 172 THEN A$ = ","
5173 IF A(Q) = 173 THEN A$ = "-"
5174 IF A(Q) = 174 THEN A$ = "."
5175 IF A(Q) = 175 THEN A$ = "/"
5176 IF A(Q) = 176 THEN A$ = "0"
5177 IF A(Q) = 177 THEN A$ = "1"
5178 IF A(Q) = 178 THEN A$ = "2"
5179 IF A(Q) = 179 THEN A$ = "3"
5180 IF A(Q) = 180 THEN A$ = "4"
5181 IF A(Q) = 181 THEN A$ = "5"
5182 IF A(Q) = 182 THEN A$ = "6"
5183 IF A(Q) = 183 THEN A$ = "7"
5184 IF A(Q) = 184 THEN A$ = "8"
5185 IF A(Q) = 185 THEN A$ = "9"
5186 IF A(Q) = 186 THEN A$ = ":"
5187 IF A(Q) = 187 THEN A$ = ";"
5188 IF A(Q) = 188 THEN A$ = "<"
5189 IF A(Q) = 189 THEN A$ = "="
5190 IF A(Q) = 190 THEN A$ = ">"
5191 IF A(Q) = 191 THEN A$ = "?"
```

```
5192 IF A(Q) = 192 THEN A$ = "@"
5193 IF A(Q) = 193 THEN A$ = "A"
5194 IF A(Q) = 194 THEN A$ = "B"
5195 IF A(Q) = 195 THEN A$ = "C"
5196 IF A(Q) = 196 THEN A$ = "D"
5197 IF A(Q) = 197 THEN A$ = "E"
5198 IF A(Q) = 198 THEN A$ = "F"
5199 IF A(Q) = 199 THEN A$ = "G"
5200 IF A(Q) = 200 THEN A$ = "H"
5201 IF A(Q) = 201 THEN A$ = "I"
5202 IF A(Q) = 202 THEN A$ = "J"
5203 IF A(Q) = 203 THEN A$ = "K"
5204 IF A(Q) = 204 THEN A$ = "L"
5205 IF A(Q) = 205 THEN A$ = "M"
5206 IF A(Q) = 206 THEN A$ = "N"
5207 IF A(Q) = 207 THEN A$ = "O"
5208 IF A(Q) = 208 THEN A$ = "P"
5209 IF A(Q) = 209 THEN A$ = "Q"
5210 IF A(Q) = 210 THEN A$ = "R"
5211 IF A(Q) = 211 THEN A$ = "S"
5212 IF A(Q) = 212 THEN A$ = "T"
5213 IF A(Q) = 213 THEN A$ = "U"
5214 IF A(Q) = 214 THEN A$ = "V"
5215 IF A(Q) = 215 THEN A$ = "W"
5216 IF A(Q) = 216 THEN A$ = "X"
5217 IF A(Q) = 217 THEN A$ = "Y"
5218 IF A(Q) = 218 THEN A$ = "Z"
6000 RETURN
```

# ALPHABETICAL FILE CREATOR

```
10  REM  ALPHABETICAL FILE CREATOR
20  D$ = CHR$ (4): REM  CONTROL D
40  PRINT D$; "OPEN ADDRESS FILE"
60  PRINT D$; "READ ADDRESS FILE"
80  INPUT K
100  DIM NAME$ (K), AD$ (K), ND$ (K), L (K), R (K)
120  FOR I = 1 TO K
140  INPUT NAME$ (I)
160  NEXT I
180  PRINT D$; "CLOSE ADDRESS FILE"
7000  REM  ALPHABETICAL ORDER ROUTINE
7040  HOME : VTAB 5
7060  PRINT "WORKING- -PLEASE DONT TOUCH!! "
7080  FOR I = 2 TO K - 1
7100  IF NAME$ (I) = NAME$ (2) THEN 7160
7120  IF NAME$ (I) = "!" THEN I = I + 1: GOTO 7160
7140  GOTO 7340
7160  LN = LEN (NAME$ (I))
7180  FOR J1 = 1 TO LN: IF MID$ (NAME$ (I), J1, 1) = " " THEN
      J2 = J1
7200  NEXT J1
7210  IF J2 = 0 OR J2 > LN THEN AD$ (I) = NAME$ (I):
      GOTO 7240
7220  AD$ (I) = MID$ (NAME$ (I), J2 + 1, LN - J2) + " " +
      LEFT$ (NAME$ (I), J2)
7240  AD$ (I) = AD$ (I) + "**" + NAME$ (I + 1) + "**" +
      NAME$ (I + 2)
7260  IF NAME$ (I + 3) < > "*" THEN AD$ (I) = AD$ (I) + "**"
      + NAME$ (I + 3)
7280  IF NAME$ (I + 4) = "*" THEN 7320
7300  AD$ (I) = AD$ (I) + "**" + NAME$ (I + 4): GOTO 7340
7320  AD$ (I) = AD$ (I) + "**" + NAME$ (I + 5)
7340  NEXT I
7360  J = 1
7380  FOR I = 1 TO K
7400  IF LEN (AD$ (I)) > 0 THEN ND$ (J) = AD$ (I): J = J + 1
7420  NEXT I
7440  N = J - 1
7460  REM QUICKSORT 2
7480  S1 = 1
7500  PRINT "WORKING- -PLEASE DONT TOUCH!! "
```

```
7520 L(1) = 1
7540 R(1) = N
7560 L1 = L(S1)
7580 R1 = R(S1)
7600 S1 = S1 - 1
7620 L2 = L1
7640 R2 = R1
7660 X$ = ND$( INT ((L1 + R1) / 2))
7680 C = C + 1
7700  IF ND$(L2) = X$ OR ND$(L2) > X$ THEN 7760
7720 L2 = L2 + 1
7740  GOTO 7680
7760 C = C1
7780  IF X$ = ND$(R2) OR X$ > ND$(R2) THEN 7840
7800 R2 = R2 - 1
7820  GOTO 7760
7840  IF L2 > R2 THEN 7980
7860 S = S + 1
7880 T$ = ND$(L2)
7900 ND$(L2) = ND$(R2)
7920 ND$(R2) = T$
7940 L2 = L2 + 1
7960 R2 = R2 - 1
7980  IF L2 = R2 OR L2 < R2 THEN 7680
8000  IF L2 = R1 OR L2 > R1 THEN 8080
8020 S1 = S1 + 1
8040 L(S1) = L2
8060 R(S1) = R1
8080 R1 = R2
8100  IF L1 < R1 THEN 7620
8120  IF S1 > 0 THEN 7560
8140  REM  SORT COMPLETED
9000  REM  FILE CREATION ROUTINE
9020  HOME : VTAB 5
9040  INPUT "NAME FOR ALPHABETIZED FILE ";FILE$
9060  PRINT D$;"OPEN";FILE$
9080  PRINT D$;"WRITE";FILE$
9100  FOR I = 1 TO N
9120  PRINT ND$(I)
9140  NEXT I
9160  PRINT D$;"CLOSE";FILE$
9200  END
```

## R&B PROG1

```
 20 D$ = CHR$ (4): REM CONTROL D
 40 HOME : VTAB 5
 60 INPUT "READ WHICH RECORD ";R
 80 PRINT D$;"OPEN ADDRESS FILE"
100 PRINT D$;"POSITION ADDRESS FILE,R";R
120 PRINT D$;"READ ADDRESS FILE"
140 INPUT LINE$
160 PRINT D$;"CLOSE ADDRESS FILE"
180 PRINT
200 PRINT LINE$
220 PRINT
240 GOTO 60
```

## R&B PROG2

```
 20 D$ = CHR$ (4): REM CONTROL D
 40 HOME : VTAB 5
 60 INPUT "READ WHICH RECORD ",R
 80 PRINT D$;"OPEN ADDRESS FILE"
100 PRINT D$;"POSITION ADDRESS FILE,R";R
120 PRINT D$;"READ ADDRESS FILE"
140 INPUT LINE$
145 PRINT D$;"POSITION ADDRESS FILE,R5"
150 PRINT D$;"READ ADDRESS FILE"
155 INPUT LINE2$
160 PRINT D$;"CLOSE ADDRESS FILE"
180 PRINT
200 PRINT LINE$
210 PRINT LINE2$
220 PRINT
240 GOTO 60
```

## R&B PROG3

```
20 D$ = CHR$ (4) : REM  CONTROL D
40 HOME : VTAB 5
60 INPUT "READ WHICH BYTE "; B
80 PRINT D$; "OPEN ADDRESS FILE"
120 PRINT D$; "READ ADDRESS FILE, B"; B
140 INPUT LINE$
160 PRINT D$; "CLOSE ADDRESS FILE"
180 PRINT
200 PRINT LINE$
220 PRINT
240 GOTO 60
```

## EXEC FILE CREATOR

```
10 REM ***--EXEC FILE CREATOR--***
11 :
12 :
20 D$ = CHR$ (4) : REM  CONTROL D
30 Q$ = CHR$ (34) : REM  QUOTATION MARKS
31 :
32 :
35 REM **--FILE CREATION--**
40 PRINT D$; "OPEN DEMO"
60 PRINT D$; "DELETE DEMO"
80 PRINT D$; "OPEN DEMO"
100 PRINT D$; "WRITE DEMO"
101 :
102 :
104 REM **--SET UP--**
105 PRINT "SPEED = 50"
107 PRINT "MON I, O, C"
108 :
109 REM **--MENU--**
110 PRINT "RUN MENU"
120 PRINT "1"
140 PRINT "N"
160 PRINT "2"
180 PRINT "N"
200 PRINT "3"
205 :
```

# EXEC FILE CREATOR

```
206 :
210 REM **--READER--**
220 PRINT "1"
240 PRINT "N"
260 PRINT "N"
280 PRINT "2"
300 PRINT "N"
320 PRINT "N"
340 PRINT "4"
360 PRINT "RON WISE"
380 PRINT "N"
400 PRINT "END"
410 PRINT "N"
420 PRINT "5"
440 PRINT "RON WISE"
460 PRINT "N"
480 PRINT "END"
490 PRINT "N"
500 PRINT "6"
520 PRINT "50"
540 PRINT "75"
560 PRINT "N"
580 PRINT "N"
600 PRINT "7"
620 PRINT "N"
640 PRINT "N"
660 PRINT "8"
665 :
666 :
670 REM **--MENU--**
680 PRINT "4"
685 :
686 :
690 REM **--CORRECTOR--**
700 PRINT "1"
720 PRINT "100"
725 PRINT "N"
730 PRINT "TEST OF CORRECTION PROGRAM"
740 PRINT "Y"
750 PRINT "0"
760 PRINT "2"
780 PRINT "135"
800 PRINT "YES"
820 PRINT "4"
```

```
825 :
826 :
830 REM **--MENU--**
840 PRINT "6"
845 :
846 :
850 REM **--RESTORE--**
860 PRINT "SPEED=255"
880 PRINT "NOMON I,O,C"
900 PRINT "PRINT"Q$"THE ADDRESS FILE WAS NOT CHANGED"Q$
920 PRINT "PRINT"Q$"BY THIS DEMO. "Q$
1000 PRINT D$;"CLOSE DEMO"
```

## PROGRAM CAPTURE

```
10  REM ***--PROGRAM CAPTURE--***
15  D$ = CHR$ (4): REM CONTROL D
16  PRINT D$;"OPEN PROGRAM CAPTURE"
17  PRINT D$;"WRITE PROGRAM CAPTURE"
18  POKE 33,33
19  LIST 1,2100: PRINT D$;"CLOSE PROGRAM CAPTURE" : TEXT :
    END
20  D$ = CHR$ (4): REM CONTROL D
40  HOME : VTAB 5
60  HTAB 17: PRINT "PROGRAM MENU"
80  PRINT : PRINT
100 HTAB 8: PRINT "1. FILE CREATION PROGRAM"
120 PRINT
140 HTAB 8: PRINT "2. FILE ADDITION PROGRAM"
160 PRINT
180 HTAB 8: PRINT "3. FILE DISPLAY PROGRAM"
200 PRINT
220 HTAB 8: PRINT "4. FILE CORRECTION PROGRAM"
240 PRINT
300 HTAB 8: PRINT "5. CATAlOG"
320 PRINT
340 HTAB 8: PRINT "6. END"
360 PRINT : PRINT
380 HTAB 8: INPUT "WHICH PROGRAM NUMBER? ";NUMBER
400 IF NUMBER < 1 OR NUMBER > 6 THEN 380
420 IF NUMBER = 1 THEN 1000
440 IF NUMBER = 2 THEN 2000
```

```
460 IF NUMBER = 3 THEN PRINT D$; "RUN MAILING LIST
    READER"
470 IF NUMBER = 4 THEN PRINT D$; "RUN MAILING LIST
    CORRECTOR"
480 IF NUMBER = 5 THEN PRINT D$; "CATALOG": INPUT "HIT
    RETURN TO GO TO MENU "; L$: GOTO 40
500 IF NUMBER = 6 THEN END
1000 REM FILE CREATOR PROGRAM
1020 PRINT : PRINT "IF THE ADDRESS FILE ALREADY EXISTS"
1040 PRINT : PRINT "DO NOT RUN THIS PROGRAM!!"
1060 PRINT : PRINT "DO YOU WANT THE FILE CREATION
     PROGRAM?"
1070 PRINT
1080 INPUT "TYPE 'YES' IF YOU DO: "; YES$
1100 IF YES$ = "YES" THEN PRINT D$; "RUN MAILING LIST
     CREATOR"
1120 GOTO 40
2000 REM FILE ADDITION PROGRAM
2020 PRINT : PRINT "YOU WANT TO ADD TO THE EXISTING"
2040 PRINT : PRINT "ADDRESS FILE. IS THIS CORRECT?"
2060 PRINT : INPUT "TYPE 'YES' IF IT IS. "; YES$
2080 IF YES$ = "YES" THEN PRINT D$; "RUN MAILING LIST
     ADDER2"
2100 GOTO 40
```

# APPENDIX D.

## MATH SYSTEM PROGRAMS

### MATH MENU

```
10  REM   ***--MATH MENU--***
20  D$ = CHR$ (4): REM CONTROL D
40  HOME : VTAB 2
60  HTAB 14: PRINT "MATH MENU"
80  PRINT : PRINT
100 HTAB 8: PRINT "1. ADDITION"
120 PRINT
140 HTAB 8: PRINT "2. SUBTRACTION"
160 PRINT
180 HTAB 8: PRINT "3. MULTIPLICATION"
200 PRINT
220 HTAB 8: PRINT "4. DIVISION"
240 PRINT
300 HTAB 8: PRINT "5. SCORES"
320 PRINT
340 HTAB 8: PRINT "6. INFORMATION"
345 PRINT
350 HTAB 8: PRINT "7. END"
360 PRINT : PRINT
380 HTAB 8: INPUT "WHICH PROGRAM NUMBER? "; NUMBER
400 IF NUMBER < 1 OR NUMBER > 7 THEN 380
420 IF NUMBER = 1 THEN PRINT D$; "RUN ADD"
440 IF NUMBER = 2 THEN PRINT D$; "RUN SUBTRACT"
460 IF NUMBER = 3 THEN PRINT D$; "RUN MULTIPLY"
470 IF NUMBER = 4 THEN PRINT D$; "RUN DIVIDE"
480 IF NUMBER = 5 THEN PRINT D$; "RUN SCORES"
500 IF NUMBER = 6 THEN 1000
520 IF NUMBER = 7 THEN END
```

```
1000 REM **--INFORMATION--**
1020 HOME
1040 PRINT "THIS IS A SERIES OF MATH DRILL AND"
1050 PRINT
1060 PRINT "PRACTICE PROGRAMS. IT IS DESIGNED TO"
1070 PRINT
1080 PRINT "ALLOW FOR AS MUCH FLEXIBILITY AS"
1090 PRINT
1100 PRINT "POSSIBLE. THE QUESTION ABOUT THE "
1110 PRINT
1120 PRINT "NUMBER OF DIGITS MIGHT, AT FIRST, "
1130 PRINT
1140 PRINT "SEEM CONFUSING. THE QUESTION SIMPLY"
1150 PRINT
1160 PRINT "ASKS FOR THE GREATEST NUMBER OF "
1170 PRINT
1180 PRINT "DIGITS POSSIBLE IN EITHER FIGURE. "
1190 PRINT
1200 PRINT "THE NEXT TWO QUESTIONS FURTHER ALLOW"
1210 PRINT
1220 PRINT "YOU TO LIMIT THE POSSIBLE PROBLEMS. "
1230 GOSUB 5000
1240 PRINT "FOR EXAMPLE, IF YOU WANTED TO PRACTICE"
1250 PRINT
1260 PRINT "MULTIPLYING BY '5', YOU COULD CHOOSE"
1270 PRINT
1280 PRINT "THREE DIGIT NUMBERS AND THEN ANSWER"
1290 PRINT
1300 PRINT "WITH A '5' FOR EACH OF THE NEXT TWO"
1310 PRINT
1320 PRINT "QUESTIONS. YOU WOULD THEN BE GIVEN"
1330 PRINT
1340 PRINT "PROBLEMS LIKE: 345 X 5 OR 823 X 5. "
1350 GOSUB 5000
1360 PRINT "ANOTHER EXAMPLE WOULD BE TO ADD TWO"
1370 PRINT
1380 PRINT "DIGIT NUMBERS BY ANSWERING THE"
1390 PRINT
1400 PRINT "QUESTIONS IN THIS WAY: "
1410 PRINT
1420 PRINT "HOW MANY DIGITS--2"
1430 PRINT
1440 PRINT "LARGEST NUMBER--99"
1450 PRINT
```

```
1460 PRINT "SMALLEST NUMBER- -1"
1470 PRINT
1480 PRINT "YOU COULD THEN GET PROBLEMS LIKE: "
1490 PRINT
1500 PRINT "58 + 34 OR 87 + 9. "
1510 GOSUB 5000
1520 PRINT "TRYING THE DIFFERENT POSSIBILITIES"
1530 PRINT
1540 PRINT "WILL SOON INDICATE THE FLEXIBILITY. "
1550 PRINT
1560 PRINT "THE DIVISION SECTION WILL ONLY GIVE"
1570 PRINT
1580 PRINT "PROBLEMS THAT COME OUT EVEN. "
1590 PRINT
1600 PRINT "YOU MAY HAVE TO WAIT A SHORT TIME. "
1610 PRINT
1620 PRINT "FOR THE NEXT PROBLEM. THIS"
1630 PRINT
1640 PRINT "IS BECAUSE THE NUMBERS GENERATED"
1650 PRINT
1660 PRINT "MUST MEET CERTAIN SPECIFICATIONS. "
1670 GOSUB 5000
1680 PRINT "THIS IS NOT A PROFESSIONAL PROGRAM"
1690 PRINT
1700 PRINT "AND THEREFORE DOES NOT DO A LOT OF"
1710 PRINT
1720 PRINT "ERROR CHECKING. YOU CAN CRASH THE "
1730 PRINT
1740 PRINT "PROGRAMS WITH CONFUSING ANSWERS"
1750 PRINT
1760 PRINT "OR MISTAKES IN TYPING. TYPING A "
1770 PRINT
1780 PRINT "'CTRL' 'C' WILL END ANY PROGRAM. "
1790 PRINT
1800 PRINT "YOU MUST THEN START OVER. "
1810 PRINT
1820 PRINT "THIS SERIES OF PROGRAMS WAS DONE"
1830 PRINT
1840 PRINT "MAINLY TO DEMONSTRATE, IN A USEFUL"
1850 PRINT
1860 PRINT "MANNER, CERTAIN FILE HANDLING"
1870 PRINT
1880 PRINT "CAPABILITIES. "
2000 GOSUB 5000
```

```
2020  GOTO 40
5000  PRINT
5020  INPUT "HIT RETURN TO CONTINUE ";L$
5040  HOME
5060  RETURN
```

## ADD

```
10  REM ***--ADDITION--***
11 :
12 :
20  REM **--VARIABLE LIST--**
21  REM  A = BOTTOM NUMBER
22  REM  B = TOP NUMBER
23  REM  C = CORRECT ANSWER
24  REM  D = STUDENT'S ANSWER
25  REM  Q = COUNTER
26  REM  W = PREVIOUS ANSWER
27  REM  Z = NUMBER OF TRIES
28  REM  CR = CORRECT ANSWERS
29  REM  WR = WRONG ANSWERS
30  REM  DT = # OF DIGITS
31  REM  LA = # OF DIGITS IN A
32  REM  LB = # OF DIGITS IN B
33  REM  LN = # OF DIGITS IN C
34  REM  OTHER VARIABLES ARE
35  REM  DESCRIPTIVE
36 :
37 :
40  HOME : VTAB 5
60  INPUT "HOW MANY DIGITS ";DIGIT
80  PRINT
100  PRINT "WHAT IS THE LARGEST FIGURE FOR THE"
120  PRINT
140  INPUT "NUMBER YOU ARE ADDING BY? ";MAX
160  PRINT
180  PRINT "WHAT IS THE SMALLEST FIGURE FOR THE"
200  PRINT
220  INPUT "NUMBER YOU ARE ADDING BY? ";MN
240  DT = DIGIT: DIGIT = 10 ^ DIGIT
260  PRINT
280  INPUT "WHAT IS YOUR NAME? ";NAME$
```

```
290 :
295 :
300  REM  **--CREATE PROBLEM--**
320  HOME
340  HTAB 10: VTAB 2
360  PRINT "TYPE 'END' WHEN FINISHED"
380  VTAB 10
382 MAX$ = STR$ (MAX)
384 LM = LEN (MAX$)
386  IF DT = LM + 1 OR DT < LM + 1 THEN 400
388 LM = 10 ^ LM
390 A = INT ( RND (1) * LM)
392  IF A < MN THEN 390
394  IF A > MAX THEN 390
396  GOTO 480
400 Z = 1
420 A = INT ( RND (1) * DIGIT)
440  IF A < MN THEN 420
460  IF A > MAX THEN 420
480 B = INT ( RND (1) * DIGIT)
500 C = B + A
520 S$ = "+"
540  IF C < 0 THEN 480
560  IF C = W THEN 480
580 W = C
600 A$ = STR$ (A)
620 LA = LEN (A$)
640 B$ = STR$ (B)
660 LB = LEN (B$)
680  HTAB 22 - LB: PRINT B
700  HTAB 22 - (LA + 1) : PRINT S$; A
720 C$ = STR$ (C)
740 LN = LEN (C$)
760 Q = 1
780  IF LA < LB THEN Q = 0
800  HTAB 22 - (DT + Q) : FOR I = 1 TO DT + Q : PRINT CHR$
     (95); : NEXT I
810 :
815 :
820  REM **--GET ANSWER--**
840  PRINT : HTAB 22 - (LN + 1) : INPUT ANSWER$
860  IF ANSWER$ = "END" THEN 1060
880 D = VAL (ANSWER$)
```

```
900  IF D = C THEN PRINT : PRINT : PRINT : HTAB 19: PRINT
     "GOOD ": FOR I = 1 TO 1000: NEXT I: CR = CR + 1: GOTO 320
920  IF Z < 3 THEN PRINT : HTAB 10: PRINT "NO, PLEASE TRY
     AGAIN. ": Z = Z + 1: PRINT : WR = WR + 1: GOTO 660
940  PRINT
960  PRINT "NO, THE ANSWER IS "; C
980  PRINT : PRINT B; " "; S$; " "; A; " = "; C
1000 PRINT : Z = 1: WR = WR + 1
1020 INPUT "HIT RETURN WHEN READY TO GO ON "; L$
1040 GOTO 320
1050 :
1055 :
1060 REM **--TOTAL ROUTINE--**
1080 HOME : VTAB 5
1100 PRINT "YOU GOT "; CR; " RIGHT!"
1120 PRINT
1140 PRINT "YOU MISSED "; WR
1160 :
1180 :
2000 REM **--FILE ROUTINE--**
2020 D$ = CHR$ (4)
2040 ONERR GOTO 2180
2060 PRINT D$; "APPEND"; NAME$
2080 PRINT D$; "WRITE"; NAME$
2085 PRINT S$
2090 PRINT DT
2100 PRINT CR
2120 PRINT WR
2140 PRINT D$; "CLOSE"; NAME$
2160 PRINT D$; "RUN MATH MENU"
2180 POKE 216, 0
2200 PRINT D$; "OPEN"; NAME$
2220 GOTO 2080
```

## SUBTRACT

```
10  REM  ***--SUBTRACTION--***
11  :
12  :
20  REM  **--VARIABLE LIST--**
21  REM  A = BOTTOM NUMBER
22  REM  B = TOP NUMBER
23  REM  C = CORRECT ANSWER
24  REM  D = STUDENT'S ANSWER
25  REM  Q = COUNTER
26  REM  W = PREVIOUS ANSWER
27  REM  Z = NUMBER OF TRIES
28  REM  CR = CORRECT ANSWERS
29  REM  WR = WRONG ANSWERS
30  REM  DT = # OF DIGITS
31  REM  LA = # OF DIGITS IN A
32  REM  LB = # OF DIGITS IN B
33  REM  LN = # OF DIGITS IN C
34  REM  OTHER VARIABLES ARE
35  REM  DESCRIPTIVE
36  :
37  :
40  HOME : VTAB 5
60  INPUT "HOW MANY DIGITS ";DIGIT
80  PRINT
100  PRINT "WHAT IS THE LARGEST FIGURE FOR THE"
120  PRINT
140  INPUT "NUMBER YOU ARE SUBTRACTING BY? ";MAX
160  PRINT
180  PRINT "WHAT IS THE SMALLEST FIGURE FOR THE"
200  PRINT
220  INPUT "NUMBER YOU ARE SUBTRACTING BY? ";MN
240  DT = DIGIT: DIGIT = 10 ^ DIGIT
260  PRINT
280  INPUT "WHAT IS YOUR NAME? ";NAME$
290  :
295  :
300  REM  **--CREATE PROBLEM--**
320  HOME
340  HTAB 10: VTAB 2
360  PRINT "TYPE 'END' WHEN FINISHED"
380  VTAB 10
```

## SUBTRACT

```
382 MAX$ = STR$ (MAX)
384 LM = LEN (MAX$)
386 IF DT = LM + 1 OR DT < LM + 1 THEN 400
388 LM = 10 ^ LM
390 A = INT ( RND (1) * LM)
392  IF A < MN THEN 390
394  IF A > MAX THEN 390
396  GOTO 480
400 Z = 1
420 A = INT ( RND (1) * DIGIT)
440  IF A < MN THEN 420
460  IF A > MAX THEN 420
480 B = INT ( RND (1) * DIGIT)
500 C = B - A
520 S$ = "-"
540  IF C < 0 THEN 480
560  IF C = W THEN 480
580 W = C
600 A$ = STR$ (A)
620 LA = LEN (A$)
640 B$ = STR$ (B)
660 LB = LEN (B$)
680  HTAB 22 - LB: PRINT B
700  HTAB 22 - (LA + 1): PRINT S$;A
720 C$ = STR$ (C)
740 LN = LEN (C$)
760 Q = 1
780  IF LA < LB THEN Q = 0
800  HTAB 22 - (DT + Q): FOR I = 1 TO DT + Q :
     PRINT CHR$ (95);: NEXT I
810 :
815 :
820  REM **--GET ANSWER--**
840  PRINT : HTAB 22 - (LN + 1): INPUT ANSWER$
860  IF ANSWER$ = "END" THEN 1060
880 D = VAL (ANSWER$)
900  IF D = C THEN PRINT : PRINT : PRINT : HTAB 19: PRINT
     "GOOD ": FOR I = 1 TO 1000: NEXT I: CR = CR + 1: GOTO 32(
920  IF Z < 3 THEN PRINT : HTAB 10: PRINT "NO, PLEASE TRY
     AGAIN. " :Z = Z + 1: PRINT :WR = WR + 1: GOTO 660
940  PRINT
960  PRINT "NO, THE ANSWER IS ";C
980  PRINT : PRINT B;" ";S$;" ";A;" = ";C
1000  PRINT :Z = 1:WR = WR + 1
```

```
1020  INPUT "HIT RETURN WHEN READY TO GO ON "; L$
1040  GOTO 320
1050  :
1055  :
1060  REM **--TOTAL ROUTINE--**
1080  HOME : VTAB 5
1100  PRINT "YOU GOT "; CR; " RIGHT!"
1120  PRINT
1140  PRINT "YOU MISSED "; WR
1160  :
1180  :
2000  REM **--FILE ROUTINE--**
2020  D$ = CHR$ (4)
2040  ONERR GOTO 2180
2060  PRINT D$; "APPEND"; NAME$
2080  PRINT D$; "WRITE"; NAME$
2085  PRINT S$
2090  PRINT DT
2100  PRINT CR
2120  PRINT WR
2140  PRINT D$; "CLOSE"; NAME$
2160  PRINT D$; "RUN MATH MENU"
2180  POKE 216,0
2200  PRINT D$; "OPEN"; NAME$
2220  GOTO 2080
```

# MULITIPLY

```
10 REM ***--MULTIPLICATION--***
11 :
12 :
20 REM **--VARIABLE LIST--**
21 REM  A = BOTTOM NUMBER
22 REM  B = TOP NUMBER
23 REM  C = CORRECT ANSWER
24 REM  D = STUDENT'S ANSWER
25 REM  Q = COUNTER
26 REM  W = PREVIOUS ANSWER
27 REM  Z = NUMBER OF TRIES
28 REM  CR = CORRECT ANSWERS
29 REM  WR = WRONG ANSWERS
30 REM  DT = # OF DIGITS
31 REM  LA = # OF DIGITS IN A
32 REM  LB = # OF DIGITS IN B
33 REM  LN = # OF DIGITS IN C
34 REM  OTHER VARIABLES ARE
35 REM  DESCRIPTIVE
36 :
37 :
40 HOME : VTAB 5
60 INPUT "HOW MANY DIGITS ";DIGIT
80 PRINT
100 PRINT "WHAT IS THE LARGEST FIGURE FOR THE"
120 PRINT
140 INPUT "NUMBER YOU ARE MULTIPLYING BY? ";MAX
160 PRINT
180 PRINT "WHAT IS THE SMALLEST FIGURE FOR THE"
200 PRINT
220 INPUT "NUMBER YOU ARE MULTIPLYING BY? ";MN
240 DT = DIGIT:DIGIT = 10 ^ DIGIT
260 PRINT
280 INPUT "WHAT IS YOUR NAME? ";NAME$
290 :
295 :
300 REM **--CREATE PROBLEM--**
320 HOME
340 HTAB 10: VTAB 2
360 PRINT "TYPE 'END' WHEN FINISHED"
380 VTAB 10
```

```
382 MAX$ = STR$ (MAX)
384 LM = LEN (MAX$)
386  IF DT = LM + 1 OR DT < LM + 1 THEN 400
388 LM = 10 ^ LM
390 A = INT ( RND (1) * LM)
392  IF A < MN THEN 390
394  IF A > MAX THEN 390
396  GOTO 480
400 Z = 1
420 A = INT ( RND (1) * DIGIT)
440  IF A < MN THEN 420
460  IF A > MAX THEN 420
480 B = INT ( RND (1) * DIGIT)
500 C = B * A
520 S$ = "X"
540  IF C < 0 THEN 480
560  IF C = W THEN 480
580 W = C
600 A$ = STR$ (A)
620 LA = LEN (A$)
640 B$ = STR$ (B)
660 LB = LEN (B$)
680  HTAB 22 - LB: PRINT B
700  HTAB 22 - (LA + 1) : PRINT S$; A
720 C$ = STR$ (C)
740 LN = LEN (C$)
760 Q = 1
780  IF LA < LB THEN Q = 0
800  HTAB 22 - (DT + Q): FOR I = 1 TO DT + Q: PRINT
     CHR$ (95); : NEXT I
810 :
815 :
820 REM **--GET ANSWER--**
840 PRINT : HTAB 22 - (LN + 1) : INPUT ANSWER$
860  IF ANSWER$ = "END" THEN 1060
880 D = VAL (ANSWER$)
900  IF D = C THEN PRINT : PRINT : PRINT : HTAB 19 : PRINT
     "GOOD ": FOR I = 1 TO 1000: NEXT I: CR = CR + 1: GOTO 320
920  IF Z < 3 THEN PRINT : HTAB 10: PRINT "NO, PLEASE
     TRY AGAIN. ": Z = Z + 1: PRINT : WR = WR + 1: GOTO 660
940 PRINT
960 PRINT "NO, THE ANSWER IS "; C
980 PRINT : PRINT B; " "; S$; " "; A; " = "; C
1000  PRINT : Z = 1: WR = WR + 1
```

```
1020  INPUT "HIT RETURN WHEN READY TO GO ON "; L$
1040  GOTO 320
1050 :
1055 :
1060  REM **--TOTAL ROUTINE--**
1080  HOME : VTAB 5
1100  PRINT "YOU GOT "; CR; " RIGHT! "
1120  PRINT
1140  PRINT "YOU MISSED "; WR
1160 :
1180 :
2000  REM **--FILE ROUTINE--**
2020  D$ = CHR$ (4)
2040  ONERR GOTO 2180
2060  PRINT D$; "APPEND"; NAME$
2080  PRINT D$; "WRITE"; NAME$
2085  PRINT S$
2090  PRINT DT
2100  PRINT CR
2120  PRINT WR
2140  PRINT D$; "CLOSE"; NAME$
2160  PRINT D$; "RUN MATH MENU"
2180  POKE 216, 0
2200  PRINT D$; "OPEN"; NAME$
2220  GOTO 2080
```

## DIVIDE

```
10  REM  ***--DIVISION--***
11 :
12 :
20  REM  **--VARIABLE LIST--**
21  REM  A = DIVISOR
22  REM  B = DIVIDEND
23  REM  C = CORRECT ANSWER
24  REM  D = STUDENT'S ANSWER
25  REM  Q = COUNTER
26  REM  W = PREVIOUS ANSWER
27  REM  Z = NUMBER OF TRIES
28  REM  CR = CORRECT ANSWERS
29  REM  WR = WRONG ANSWERS
30  REM  DT = # OF DIGITS
31  REM  LA = # OF DIGITS IN A
32  REM  LB = # OF DIGITS IN B
33  REM  LN = # OF DIGITS IN C
34  REM  OTHER VARIABLES ARE
35  REM  DESCRIPTIVE
36 :
37 :
40  HOME : VTAB 5
60  INPUT "HOW MANY DIGITS ";DIGIT
80  PRINT
100 PRINT "WHAT IS THE LARGEST FIGURE FOR THE NO. "
120 PRINT
140 INPUT "YOU ARE DIVIDING BY (DIVISOR)? ";MAX
160 PRINT
180 PRINT "WHAT IS THE SMALLEST FIGURE FOR THE NO. "
200 PRINT
220 INPUT "YOU ARE DIVIDING BY (DIVISOR)? ";MN
240 DT = DIGIT: DIGIT = 10 ^ DIGIT
260 PRINT
280 INPUT "WHAT IS YOUR NAME? ";NAME$
290 :
295 :
300 REM **--CREATE PROBLEM--**
310 MAX$ = STR$ (MAX)
320 LM = LEN (MAX$)
330 IF DT = LM + 1 OR DT < LM + 1 THEN 400
340 LM = 10 ^ LM
```

```
350 A = INT ( RND (1) * LM)
360  IF A < MN THEN 350
370  IF A > MAX THEN 350
380  GOTO 480
400 Z = 1
420 A = INT ( RND (1) * DIGIT)
440  IF A < MN THEN 420
460  IF A > MAX THEN 420
480 B = INT ( RND (1) * DIGIT)
485  IF B = 0 OR B < (A) THEN 480
490  DEF  FN MOD (C) = INT ( (B / A - INT (B / A) )
     * A + .05) * SGN (B / A)
500 C = INT (B) / (A)
510 C = INT (C)
520 S$ = "/"
540  IF C < 0 THEN 420
560  IF C = W THEN 420
570  IF FN MOD (RM) < > 0 THEN 480
580 W = C
600 A$ = STR$ (A)
620 LA = LEN (A$)
640 B$ = STR$ (B)
660 LB = LEN (B$)
662  HOME
664  HTAB 10: VTAB 2
666  PRINT "TYPE 'END' WHEN FINISHED"
668  VTAB 10
670  HTAB 22: FOR I = 1 TO DT + 1 : PRINT CHR$ (95) ; : NEXT I
675  PRINT
680  HTAB 22 - LA: PRINT A; ")"; B
720 C$ = STR$ (C)
740 LN = LEN (C$)
760 Q = 1
780  IF LB < DT THEN LN = LN + (DT - LB)
810 :
815 :
820  REM **--GET ANSWER--**
830  VTAB 9
840  PRINT : HTAB (22 + DT) - (LN - 1) : INPUT ""; ANSWER$
860  IF ANSWER$ = "END" THEN 1060
880 D = VAL (ANSWER$)
900  IF D = C THEN PRINT : PRINT : PRINT : HTAB 19: PRINT
     "GOOD": FOR I = 1 TO 500: NEXT I : CR = CR + 1: GOTO 320
```

```
 920  IF Z < 3 THEN PRINT : PRINT : PRINT : HTAB 10: PRINT
      "NO, PLEASE TRY AGAIN. " : Z = Z + 1: PRINT : WR = WR + 1:
      FOR WT = 1 TO 1000: NEXT WT: VTAB 10: GOTO 660
 940  PRINT : PRINT : PRINT
 960  PRINT "NO, THE ANSWER IS "; C
 980  PRINT : PRINT B; " "; S$; " "; A; " = "; C
1000  PRINT : Z = 1: WR = WR + 1
1020  INPUT "HIT RETURN WHEN READY TO GO ON "; L$
1040  GOTO 320
1050  :
1055  :
1060  REM **--TOTAL ROUTINE--**
1080  HOME : VTAB 5
1100  PRINT "YOU GOT "; CR; " RIGHT! "
1120  PRINT
1140  PRINT "YOU MISSED "; WR
1160  :
1180  :
2000  REM **--FILE ROUTINE--**
2020 D$ = CHR$ (4)
2040  ONERR GOTO 2180
2060  PRINT D$; "APPEND"; NAME$
2080  PRINT D$; "WRITE"; NAME$
2085  PRINT S$
2090  PRINT DT
2100  PRINT CR
2120  PRINT WR
2140  PRINT D$; "CLOSE"; NAME$
2160  PRINT D$; "RUN MATH MENU"
2180  POKE 216, 0
2200  PRINT D$; "OPEN"; NAME$
2220  GOTO 2080
```

# SCORES

```
10  REM ***--SCORES--***
11 :
12 :
20  D$ = CHR$ (4)
40  ONERR GOTO 380
60  DIM S$(100),DT(100)
80  DIM CR(100),WR(100)
100 I = 1
120 HOME : VTAB 5
140 INPUT "STUDENT'S NAME PLEASE? ";NAME$
160 :
180 :
200 REM **--FILE ROUTINE--**
220 PRINT D$;"OPEN";NAME$
240 PRINT D$;"READ";NAME$
260 INPUT S$(I)
280 INPUT DT(I)
300 INPUT CR(I)
320 INPUT WR(I)
340 I = I + 1
360 GOTO 260
380 POKE 216,0
400 PRINT D$;"CLOSE";NAME$
420 :
440 :
460 REM **--DISPLAY ROUTINE--**
480 HOME : VTAB 1: HTAB 19: PRINT NAME$: PRINT : PRINT
500 PRINT "SESS. ";
520 HTAB 7: PRINT "OPERATION";
540 HTAB 18: PRINT "DIGITS";
560 HTAB 26: PRINT "CORRECT";
580 HTAB 35: PRINT "WRONG"
590 POKE 34,4: REM SET TOP WINDOW
600 FOR K = 1 TO I - 1
620 IF S$(K) = "+" THEN S$(K) = "ADD"
640 IF S$(K) = "-" THEN S$(K) = "SUB"
660 IF S$(K) = "X" THEN S$(K) = "MLT"
680 IF S$(K) = "/" THEN S$(K) = "DIV"
700 HTAB 3: PRINT K;
720 HTAB 10: PRINT S$(K);
740 HTAB 20: PRINT DT(K);
```

```
760 IF CR(K) > 9 THEN L = - 1
780 HTAB 29 + L: PRINT CR(K);
800 L = 0
820 IF WR(K) > 9 THEN L = - 1
840 HTAB 37 + L: PRINT WR(K)
860 L = 0
880 NEXT K
900 PRINT : INPUT "HIT RETURN WHEN FINISHED ";L$
910 TEXT
920 PRINT D$; "RUN MATH MENU"
```

## SCORES--DIF

```
10  REM  ***--SCORES.DIF--***
11 :
12 :
20  D$ = CHR$ (4): REM  CONTROL D
30  Q$ = CHR$ (34): REM  QUOTATION MARK
40  ONERR GOTO 380
60  DIM S$(100),DT(100)
80  DIM CR(100),WR(100)
100 I = 1
120 HOME : VTAB 5
140 INPUT "STUDENT'S NAME PLEASE? ";NAME$
160 :
180 :
200 REM  **--FILE ROUTINE--**
220 PRINT D$; "OPEN";NAME$
240 PRINT D$; "READ";NAME$
260 INPUT S$(I): REM  SIGN
280 INPUT DT(I): REM  DIGITS
300 INPUT CR(I): REM  # RIGHT
320 INPUT WR(I): REM  # WRONG
330 IF S$(I) = "+" THEN S$(I) = "ADD"
331 IF S$(I) = "-" THEN S$(I) = "SUB"
332 IF S$(I) = "X" THEN S$(I) = "MLT"
333 IF S$(I) = "/" THEN S$(I) = "DIV"
340 I = I + 1
360 GOTO 260: REM  GET ANOTHER
380 POKE 216,0: REM  RESET ERR FLAG
400 PRINT D$; "CLOSE";NAME$
420 :
```

```
440 :
450 REM **--DIF ROUTINE--**
460 J = I - 1: NV = 5: NT = I - 1
470 FILE$ = NAME$
480 NAME$ = NAME$ + ".DIF"
500 PRINT D$; "OPEN"; NAME$
510 PRINT D$; "WRITE"; NAME$
511 :
515 REM **--HEADER SECTION--**
516 :
520 PRINT "TABLE"
530 PRINT "0,1"
540 PRINT Q$FILE$Q$
545 :
550 PRINT "VECTORS"
560 PRINT "0,"; NV
570 PRINT Q$Q$
575 :
580 PRINT "TUPLES"
590 PRINT "0,"; NT
600 PRINT Q$Q$
605 :
610 PRINT "LABEL"
620 PRINT "1,0"
630 PRINT Q$"SESSION #"Q$
635 :
640 PRINT "LABEL"
650 PRINT "2,0"
660 PRINT Q$"OPERATION"Q$
665 :
670 PRINT "LABEL"
680 PRINT "3,0"
690 PRINT Q$"DIGITS"Q$
695 :
700 PRINT "LABEL"
710 PRINT "4,0"
720 PRINT Q$"CORRECT"Q$
725 :
730 PRINT "LABEL"
740 PRINT "5,0"
750 PRINT Q$"WRONG"Q$
755 :
760 PRINT "DATA"
770 PRINT "0,0"
```

```
780 PRINT Q$Q$
785 :
800 REM **--DATA SECTION--**
805 :
810 PRINT "-1,0"
820 PRINT "BOT"
825 :
830 PRINT "1,0"
840 PRINT Q$"SESSION #"Q$
845 :
850 PRINT "1,0"
860 PRINT Q$"OPERATION"Q$
865 :
870 PRINT "1,0"
880 PRINT Q$"DIGITS"Q$
885 :
890 PRINT "1,0"
900 PRINT Q$"CORRECT"Q$
905 :
910 PRINT "1,0"
920 PRINT Q$"WRONG"Q$
925 :
930 FOR I = 1 TO J
935 :
940 PRINT "-1,0"
950 PRINT "BOT"
955 :
960 PRINT "0,";I
970 PRINT "V"
975 :
980 PRINT "1,0"
990 PRINT S$(I)
995 :
1000 PRINT "0,";DT(I)
1010 PRINT "V"
1015 :
1020 PRINT "0,";CR(I)
1030 PRINT "V"
1035 :
1040 PRINT "0,";WR(I)
1050 PRINT "V"
1055 :
1060 NEXT I
1065 :
```

```
1070 PRINT "-1,0"
1080 PRINT "EOD"
1085 :
1100 PRINT D$; "CLOSE"; NAME$
```

# APPENDIX E.

## RECIPE AND DRILL & PRACTICE PROGRAMS

**RECIPES**

```
2  REM  ***--RECIPES--***
3 :
4 :
5  REM  **--VARIABLES LIST--**
6  REM  RECNBR=NUMBER OF RECORDS
7  REM  INGNBR=TOTAL # OF INGRED.
8  REM  ING$ & ID$ = INGRED.
9  REM  REC$ & RC$ = RECIPES
10 REM  IG$ = CURRENT SESS. INGRED.
11 REM  RZ$ = RECIPE NAMES ONLY
12 :
13 :
14 REM  **--INITIALIZATION--**
15 DIM REC(100),ING$(50),IG$(100,50),RC$(100),RZ$
   (100),ID$(50)
20 D$ = CHR$ (4): REM  CONTROL D
25 TB = 8: REM  HTAB VALUE
27  ONERR GOTO 10000
30  PRINT D$;"OPEN RECIPE NAMES"
35  PRINT D$;"READ RECIPE NAMES"
40  INPUT NUMBERS$
45  PRINT D$;"CLOSE RECIPE NAMES"
47  POKE 216,0: REM  RESET ERROR FLAG
50 LR = LEN (NUMBERS$)
55 T = 1
60  IF MID$ (NUMBERS$,T,1) = "*" THEN 70
65 T = T + 1: GOTO 60
70 RECNBR = VAL ( LEFT$ (NUMBERS$,T - 1))
75 INGNBR = VAL ( MID$ (NUMBERS$,T + 1,LR - T))
```

# RECIPES

```
 97 :
 98 :
 99 :
100  REM  **--RECIPE MENU--**
120  HOME : VTAB 5: HTAB 15
140  PRINT "RECIPE MENU"
160  PRINT : PRINT
180  HTAB TB
200  PRINT "1. ADD RECIPE TO LIST"
220  PRINT : HTAB TB
240  PRINT "2. SELECT RECIPE FROM LIST"
260  PRINT : HTAB TB
280  PRINT "3. END PROGRAM"
380  PRINT : HTAB TB
400  INPUT "WHICH NUMBER? ";NB
420  IF NB < 1 OR NB > 5 THEN PRINT "INCORRECT
     NUMBER!": GOTO 380
510  IF NB = 1 THEN 1000
520  IF NB = 2 THEN 2000
530  IF NB = 3 THEN END
970 :
980 :
990 :
1000  REM  **--ADD TO RECIPE LIST--**
1002 R = 1
1005  HOME : VTAB 5
1010  INPUT "NAME OF RECIPE  ";REC$(R)
1015  I = 1
1020  PRINT : PRINT "TYPE 'END' WHEN FINISHED.": VTAB 10
1025  PRINT "TYPE IN INGREDIENT #";I;" BELOW THIS LINE."
1030  INPUT ING$(I)
1035  IF ING$(I) = "END" THEN 1050
1040  I = I + 1
1045  HOME : VTAB 6: GOTO 1020
1050  HOME : VTAB 5: PRINT REC$(R): PRINT : PRINT
1055  FOR J = 1 TO I - 1
1060  PRINT J;"  ";ING$(J)
1065  NEXT J
1070  PRINT
1075  INPUT "IS THIS CORRECT? ";YES$
1080  IF YES$ = "Y" THEN 1110
1085  PRINT
1090  INPUT "WHICH NUMBER IS WRONG? ";WR
1095  PRINT "TYPE IN CORRECT INFO. FOR INGREDIENT #";WR
```

```
1100  INPUT ING$(WR)
1105  GOTO 1050
1110  REC$(R) = REC$(R) + "!" + STR$ (INGNBR) + "*" +
      STR$ (I - 1)
1111  FOR J = 1 TO I - 1
1112  IG$(R,J) = ING$(J)
1113  NEXT J
1115  INGNBR = INGNBR + I - 1
1120  R = R + 1
1125  PRINT
1130  INPUT "ADD MORE RECIPES? ";YES$
1135  IF YES$ = "Y" THEN 1005
1136  RECNBR = RECNBR + R - 1
1137  NUMBERS$ = STR$ (RECNBR) + "*" + STR$ (INGNBR)
1140  PRINT D$;"APPEND RECIPE NAMES"
1145  PRINT D$;"APPEND INGRED"
1146  FOR K = 1 TO R - 1
1150  PRINT D$;"WRITE RECIPE NAMES"
1155  PRINT REC$(K)
1160  LN = LEN (REC$(K))
1165  T = 1
1170  IF MID$ (REC$(K),T,1) = "*" THEN 1180
1175  T = T + 1: GOTO 1170
1180  Q = VAL ( MID$ (REC$(K),T + 1,LN - T))
1185  PRINT D$;"WRITE INGRED"
1190  FOR H = 1 TO Q
1195  PRINT IG$(K,H)
1200  NEXT H
1220  NEXT K
1225  PRINT D$;"CLOSE"
1265  PRINT D$;"OPEN RECIPE NAMES"
1270  PRINT D$;"WRITE RECIPE NAMES"
1275  PRINT NUMBERS$
1280  PRINT D$;"CLOSE"
1285  GOTO 100: REM  MENU
1970 :
1980 :
1990 :
2000  REM **--SELECT RECIPE--**
2002  HOME : VTAB 5
2005  PRINT D$;"OPEN RECIPE NAMES"
2007  PRINT D$;"POSITION RECIPE NAMES,R2"
2010  PRINT D$;"READ RECIPE NAMES"
2015  FOR I = 1 TO RECNBR
```

```
2020  INPUT RC$(I)
2021 T = 1
2022  IF MID$ (RC$(I),T,1) = "!" THEN 2024
2023 T = T + 1: GOTO 2022
2024 RZ$(I) = LEFT$ (RC$(I),T - 1)
2025  NEXT I
2030 PRINT D$;"CLOSE RECIPE NAMES"
2032  FOR I = 1 TO RECNBR
2033  PRINT I;"  ";RZ$(I)
2034  NEXT I
2035  PRINT : PRINT
2040  INPUT "WHICH RECIPE? ";RC
2045 LN = LEN (RC$(RC))
2050 T = 1
2055  IF MID$ (RC$(RC),T,1) = "!" THEN 2065
2060 T = T + 1: GOTO 2055
2065 T1 = T
2070  IF MID$ (RC$(RC),T1,1) = "*" THEN 2080
2075 T1 = T1 + 1: GOTO 2070
2080 IGNB = VAL ( MID$ (RC$(RC),T + 1,T1 - 1)) + 1
2085 LIGNB = VAL ( MID$ (RC$(RC),T1 + 1,LN - T1))
2087  HOME : PRINT RZ$(RC)
2090  PRINT D$;"OPEN INGRED"
2095  PRINT D$;"POSITION INGRED,R";IGNB
2100  PRINT D$;"READ INGRED"
2105  FOR K = 1 TO LIGNB
2110  INPUT ID$(K)
2115  PRINT ID$(K)
2120  NEXT K
2125  PRINT D$;"CLOSE INGRED"
2130  PRINT : PRINT
2135  INPUT "HIT RETURN WHEN READY  ";L$
2140  INPUT "SELECT ANOTHER RECIPE? ";YES$
2145  IF YES$ = "Y" THEN 2155
2150  GOTO 100: REM  MENU
2155  HOME
2160  GOTO 2032: REM SELECT ANOTHER
2970 :
2980 :
2990 :
10000  REM **--FIRST TIME--**
10002  POKE 216,0: REM  RESET ERROR FLAG
10005  PRINT D$;"WRITE RECIPE NAMES"
10010  PRINT "0*0-----------"
```

```
10015  PRINT D$; "CLOSE RECIPE NAMES"
10020  PRINT D$; "OPEN INGRED"
10025  PRINT D$; "WRITE INGRED"
10030  PRINT "RECIPE INGREDENTS"
10040  PRINT D$; "CLOSE INGRED"
10045  GOTO 30
```

## CREATE Q & A

```
10   REM  **--INPUT Q & A--**
11   :
12   :
20   D$ = CHR$ (4): REM  CONTROL D
40   DIM Q$(50),A$(50)
60   I = 1
70   :
75   :
100  REM  **--INPUT ROUTINE--**
105  HOME : VTAB 10
110  INPUT "SUBJECT NAME ";SUB$
120  PRINT : PRINT
140  PRINT "QUESTION # ";I: INPUT Q$(I)
160  IF Q$(I) = "END" THEN 300
180  INPUT "ANSWER ";A$(I)
200  PRINT : PRINT : PRINT Q$(I)
220  PRINT : PRINT A$(I)
230  PRINT
240  INPUT "IS THIS CORRECT? ";Y$
250  PRINT
260  IF Y$ = "N" THEN 140
280  I = I + 1: GOTO 140
290  :
295  :
300  REM  **--FILE ROUTINE--**
310  PRINT D$; "OPEN" + SUB$
320  PRINT D$; "WRITE" + SUB$
340  PRINT I - 1
360  FOR J = 1 TO I - 1
380  PRINT Q$(J)
400  PRINT A$(J)
420  NEXT J
440  PRINT D$; "CLOSE"
```

# DRILL Q & A

```
10  REM ***--DRILL & PRACTICE--***
11 :
12 :
20  D$ = CHR$ (4): REM  CONTROL D
40  DIM Q$(50),A$(50)
60  D$ = CHR$ (4)
90 :
95 :
100 REM **--FILE ROUTINE--**
105 HOME : VTAB 10
110 INPUT "SUBJECT NAME  ";SUB$
115 PRINT : PRINT
120 PRINT D$;"OPEN" + SUB$
140 PRINT D$;"READ" + SUB$
160 INPUT J
180 FOR I = 1 TO J
200 INPUT Q$(I),A$(I)
220 NEXT I
240 PRINT D$;"CLOSE"
245 :
246 :
250 REM **--GET Q & A--**
260 I = RND (1) * 10: I = INT (I)
280 IF I > J OR I < 1 THEN 260
300 PRINT Q$(I)
320 PRINT : PRINT
340 INPUT "YOUR ANSWER IS ";S$
360 IF S$ = "END" THEN 600
380 IF S$ = A$(I) THEN PRINT "CORRECT": A = A + 1
    : GOTO 540
400 IF Z > 0 THEN 500
420 PRINT "NO, TRY ONCE MORE"
440 Z = 1
460 A2 = A2 + 1
480 GOTO 340
500 PRINT "NO, THE ANSWER IS ";A$(I)
520 M = M + 1
540 Z = 0
560 PRINT : PRINT
580 GOTO 260
590 :
```

```
595 :
600 REM **--DISPLAY SCORE--**
610 A2 = A2 - M
620 A = A - A2
640 HOME : VTAB 10
660 PRINT "YOU GOT ";A;" RIGHT ON THE FIRST TRY"
680 PRINT : PRINT
700 PRINT "YOU GOT ";A2;" RIGHT ON THE SECOND TRY"
720 PRINT : PRINT
740 PRINT "YOU MISSED ";M;" ANSWERS"
```

# APPENDIX F.

## DIF PROGRAMS

**SCORES--DIF**

```
10  REM ***--SCORES.DIF--***
11 :
12 :
20  D$ = CHR$ (4): REM  CONTROL D
30  Q$ = CHR$ (34): REM  QUOTATION MARK
40  ONERR GOTO 380
60  DIM S$(100),DT(100)
80  DIM CR(100),WR(100)
100 I = 1
120 HOME : VTAB 5
140 INPUT "STUDENT'S NAME PLEASE? ";NAME$
160 :
180 :
200 REM **--FILE ROUTINE--**
220 PRINT D$;"OPEN";NAME$
240 PRINT D$;"READ";NAME$
260 INPUT S$(I): REM SIGN
280 INPUT DT(I): REM DIGITS
300 INPUT CR(I): REM # RIGHT
320 INPUT WR(I): REM # WRONG
330 IF S$(I) = "+" THEN S$(I) = "ADD"
331 IF S$(I) = "-" THEN S$(I) = "SUB"
332 IF S$(I) = "X" THEN S$(I) = "MLT"
333 IF S$(I) = "/" THEN S$(I) = "DIV"
340 I = I + 1
360 GOTO 260: REM  GET ANOTHER
380 POKE 216,0: REM  RESET ERR FLAG
400 PRINT D$;"CLOSE";NAME$
```

```
420 :
440 :
450 REM **--DIF ROUTINE--**
460 J = I - 1:NV = 5:NT = I - 1
470 FILE$ = NAME$
480 NAME$ = NAME$ + ".DIF"
500 PRINT D$;"OPEN";NAME$
510 PRINT D$;"WRITE";NAME$
511 :
515 REM **--HEADER SECTION--**
516 :
520 PRINT "TABLE"
530 PRINT "0,1"
540 PRINT Q$FILE$Q$
545 :
550 PRINT "VECTORS"
560 PRINT "0,";NV
570 PRINT Q$Q$
575 :
580 PRINT "TUPLES"
590 PRINT "0,";NT
600 PRINT Q$Q$
605 :
610 PRINT "LABEL"
620 PRINT "1,0"
630 PRINT Q$"SESSION #"Q$
635 :
640 PRINT "LABEL"
650 PRINT "2,0"
660 PRINT Q$"OPERATION"Q$
665 :
670 PRINT "LABEL"
680 PRINT "3,0"
690 PRINT Q$"DIGITS"Q$
695 :
700 PRINT "LABEL"
710 PRINT "4,0"
720 PRINT Q$"CORRECT"Q$
725 :
730 PRINT "LABEL"
740 PRINT "5,0"
750 PRINT Q$"WRONG"Q$
755 :
760 PRINT "DATA"
```

```
770 PRINT "0,0"
780 PRINT Q$Q$
785 :
800 REM **--DATA SECTION--**
805 :
810 PRINT "-1,0"
820 PRINT "BOT"
825 :
830 PRINT "1,0"
840 PRINT Q$"SESSION #"Q$
845 :
850 PRINT "1,0"
860 PRINT Q$"OPERATION"Q$
865 :
870 PRINT "1,0"
880 PRINT Q$"DIGITS"Q$
885 :
890 PRINT "1,0"
900 PRINT Q$"CORRECT"Q$
905 :
910 PRINT "1,0"
920 PRINT Q$"WRONG"Q$
925 :
930 FOR I = 1 TO J
935 :
940 PRINT "-1,0"
950 PRINT "BOT"
955 :
960 PRINT "0,";I
970 PRINT "V"
975 :
980 PRINT "1,0"
990 PRINT S$(I)
995 :
1000 PRINT "0,";DT(I)
1010 PRINT "V"
1015 :
1020 PRINT "0,";CR(I)
1030 PRINT "V"
1035 :
1040 PRINT "0,";WR(I)
1050 PRINT "V"
1055 :
1060 NEXT I
```

```
1065 :
1070 PRINT "-1,0"
1080 PRINT "EOD"
1085 :
1100 PRINT D$;"CLOSE";NAME$
```

## DIF READER

```
10  REM ***--DIF READER--***
11 :
12 :
20  D$ = CHR$ (4) : REM  CONTROL D
40  DIM A$(200),S(200),N(200)
60  I = 1
80  HOME : VTAB 5
100 INPUT "FILE NAME PLEASE? ";FILE$
120 :
140 :
160 REM **--INPUT ROUTINE--**
180 PRINT D$;"OPEN";FILE$
200 PRINT D$;"READ";FILE$
220 INPUT T$: REM  READ THE TOPIC NAME
240 INPUT S,N: REM  READ THE VECTOR #,VALUE
260 INPUT S$: REM  THE STRING VALUE
280 IF T$ = "VECTORS" THEN NV = N
300 IF T$ = "TUPLES" THEN NT = N
320 IF T$ < > "DATA" THEN 220
340 K = 1
360 INPUT S(K),N(K)
380 INPUT A$(K)
400 IF A$(K) = "EOD" THEN 440
420 K = K + 1: GOTO 360
440 PRINT D$;"CLOSE";FILE$
460 :
480 :
500 REM **-DISPLAY ROUTINE-**
520 FOR J = 1 TO K
540 PRINT S(J);",";N(J)
560 PRINT A$(J)
580 NEXT J
600 END
```

# DIF TRANSLATOR 1

```
10  REM  ***--DIF TRANSLATOR--***
11  :
12  :
13  REM  **--VARIABLES--**
14  REM  L1$=HEAD. SECT. LINE 1
16  REM  L3$=HEAD. SECT. LINE 3
18  REM  A()=HEAD. SECT. 2ND LINE
19  REM       FIRST VALUE
20  REM  B()=HEAD. SECT. 2ND LINE
21  REM       SECOND VALUE
22  REM  I=SET COUNTER
24  REM  RL=RECORD LENGTH
26  REM  LL=LABEL LENGTH
28  REM  DA$()=ACTUAL DATA VALUE
30  REM  J=ARRAY COUNTER
32  REM  Q=ARRAY COUNTER
34  REM  A$=DATA SECT. 1ST LINE
36  REM       VARIABLE TYPE
38  REM  B$=DATA SECT. 1ST LINE
40  REM       SECOND VALUE
42  REM  C$=DATA SECTION
44  REM       STRING VALUE
46  :
48  :
50  D$ = CHR$ (4): REM  CTRL D
60  DIM L1$(99),A(99),B(99),L3$(99)
80  HOME : VTAB 5
100  INPUT "DIF FILE NAME PLEASE! ";FILE$
110  INPUT "DO YOU WANT A PAPER PRINTOUT? ";YES$
115  IF YES$ = "Y" THEN PRINT D$;"PR#1"
120  IF RIGHT$ (FILE$,4) < > ".DIF" THEN FILE$ = FILE$+
     ".DIF"
121  :
122  :
130  REM  **--INPUT HEAD. SECT--**
140  PRINT D$;"OPEN";FILE$
160  PRINT D$;"READ";FILE$
180  I = 1
200  INPUT L1$(I)
220  INPUT A(I),B(I)
240  INPUT L3$(I)
```

```
260  IF L1$(I) = "DATA" THEN 305
280  I = I + 1
300  GOTO 200
301 :
302 :
305  REM **-DISPLAY HEAD. SECT-**
310  HOME : PRINT "    LABEL";: HTAB 23: PRINT "FIELD"
312  PRINT "    NAME";: HTAB 24: PRINT "SIZE"
315  POKE 34,3: PRINT : REM SET WINDOW
320  FOR K = 2 TO I
340  IF LEN (L3$(K)) < > 0 THEN PRINT A(K);" ";L3$(K);:
     HTAB 25: PRINT B(K + 1)
360  LL = LL + LEN (L3$(K))
420  IF L1$(K) = "SIZE" THEN RL = RL + B(K)
440  NEXT K
445  RL = RL + B(K - 2)
450  PRINT : PRINT
460  PRINT "RECORD LENGTH = ";RL
480  PRINT "LABEL LENGTH = ";LL
490  PRINT L1$(2);" = ";B(2)
495  PRINT L1$(3);" OR NUMBER OF RECORDS = ";B(3)
496  PRINT : POKE 34,22: REM SET WINDOW
498 :
499 :
500  REM **-INPUT DATA SECT-**
510  VTAB 24: FLASH : PRINT "READING DIF FILE--DON'T
     TOUCH!!!": NORMAL
520  DIM DA$(B(3),B(2))
540  J = 0:Q = 0
560  INPUT A$,B$
580  INPUT C$
600  IF C$ = "BOT" THEN 740
620  IF C$ = "EOD" THEN 770
625 :
630  REM  IF ALPHABETICAL THEN
635  REM  SAVE C$
640  IF A$ = "1" AND B$ = "0" THEN DA$(Q,J) = C$:
     GOTO 700
645 :
650  REM  IF NUMERICAL THEN
655  REM  SAVE B$
660  IF A$ = "0" AND C$= "V" THEN DA$(Q,J) = B$:
     GOTO 700
665 :
```

```
670  REM  IF NOT "V" THEN
675  REM  SAVE BOTH
680  DA$(Q,J) = B$ + C$
685 :
700  J = J + 1
720  GOTO 560
740  J = 1:Q = Q + 1
760  GOTO 560
768 :
769 :
770  REM **--WRITE NEW FILE--**
775  VTAB 24: FLASH : PRINT "WRITING NEW FILE": NORMAL
780  PRINT D$;"OPEN";FILE$ + ".UP";",L";RL
800  FOR K = 1 TO Q
840  PRINT D$;"WRITE";FILE$ + ".UP";",R";K
850  FOR W = 1 TO J - 1
860  PRINT DA$(K,W)
880  NEXT W
900  NEXT K
920  PRINT D$;"CLOSE";FILE$ + ".UP"
1000 PRINT D$;"CLOSE"
1010 PRINT "ALL FINISHED"
1020 TEXT
1030 NORMAL
1040 PRINT D$;"PR#0"
1060 END
```

## VARIABLE CREATOR

```
 10 REM **-CREATE VARIABLES-**
 11 :
 12 :
 13 REM **- -VARIABLES- -**
 14 REM  L1$=HEAD. SECT. LINE 1
 16 REM  L3$=HEAD. SECT. LINE 3
 18 REM  A() =HEAD. SECT. 2ND LINE
 19 REM        FIRST VALUE
 20 REM  B() =HEAD. SECT. 2ND LINE
 21 REM        SECOND VALUE
 22 REM  I=SET COUNTER
 24 REM  RL=RECORD LENGTH
 26 REM  LL=LABEL LENGTH
 46 :
 48 :
 50 D$ = CHR$ (4): REM  CTRL D
 60 DIM L1$(99),A(99),B(99),L3$(99)
 80 HOME : VTAB 5
100  INPUT "DIF FILE NAME PLEASE! ";FILE$
110  INPUT "DO YOU WANT A PAPER PRINTOUT? ";YES$
115  IF YES$ = "Y" THEN PRINT D$;"PR#1"
120  IF RIGHT$ (FILE$,4) < > ".DIF" THEN FILE$ = FILE$
     + ".DIF"
121 :
122 :
130  REM **- -INPUT HEAD. SECT- -**
140  PRINT D$;"OPEN";FILE$
160  PRINT D$;"READ";FILE$
180  I = 1
200  INPUT L1$(I)
220  INPUT A(I),B(I)
240  INPUT L3$(I)
260  IF L1$(I) = "DATA" THEN 305
280  I = I + 1
300  GOTO 200
301 :
302 :
305  REM **-DISPLAY HEAD. SECT-**
310  HOME : PRINT "   LABEL";: HTAB 23: PRINT "FIELD"
312  PRINT "   NAME";: HTAB 24: PRINT "SIZE"
315  POKE 34,3: PRINT : REM SET WINDOW
```

# VARIABLE CREATOR

```
320  FOR K = 2 TO I
340  IF LEN (L3$(K)) < > 0 THEN PRINT A(K);"   ";L3$(K);:
     HTAB 25: PRINT B(K + 1)
360  LL = LL + LEN (L3$(K))
420  IF L1$(K) = "SIZE" THEN RL = RL + B(K)
440  NEXT K
445  RL = RL + B(K - 2)
450  PRINT : PRINT
460  PRINT "RECORD LENGTH = ";RL
480  PRINT "LABEL LENGTH = ";LL
490  PRINT L1$(2);" = ";B(2)
495  PRINT L1$(3);" OR NUMBER OF RECORDS = ";B(3)
496  PRINT : POKE 34,22: REM  SET WINDOW
498  :
499  :
500  REM **-WRITE LABELS INFO-**
510  NF$ = FILE$ + ".UP"
512  TN = B(3)
520  PRINT D$;"OPEN VARIABLES"
540  PRINT D$;"WRITE VARIABLES"
560  PRINT NF$
580  PRINT TN
600  PRINT RL
640  PRINT B(2)
700  FOR K = 2 TO I
720  IF LEN (L3$(K)) < > 0 THEN PRINT L3$(K):
     PRINT B(K + 1)
740  NEXT K
1000 PRINT D$;"CLOSE"
1010 PRINT "ALL FINISHED"
1020 TEXT
1030 NORMAL
1040 PRINT D$;"PR#0"
1060 END
```

## READ NEW FILE

```
20 D$ = CHR$ (4)
25 PRINT D$;"OPEN VARIABLES"
27 PRINT D$;"READ VARIABLES"
29 INPUT NF$
31 INPUT TN
33 INPUT RL
35 INPUT LABEL
36 DIM LABEL$(LABEL),FIELDSIZE(LABEL),DA$(TN,LABEL)
37 FOR K = 1 TO LABEL
39 INPUT LABEL$(K)
41 INPUT FIELDSIZE(K)
43 NEXT K
45 PRINT D$;"CLOSE VARIABLES"
50 FILE$ = NF$
60 PRINT D$;"OPEN";FILE$;",L";RL
80 FOR I = 1 TO TN
100 PRINT D$;"READ";FILE$;",R";I
120 FOR K = 1 TO LABEL
140 INPUT DA$(I,K)
160 NEXT K
180 NEXT I
200 PRINT D$;"CLOSE"
220 FOR I = 1 TO TN
240 FOR K = 1 TO LABEL
250 PRINT LABEL$(K);: HTAB 20: PRINT FIELDSIZE(K)
260 PRINT DA$(I,K)
280 NEXT K
300 NEXT I
320 END
```

# APPENDIX G.

## MEDICAL RECORDS SYSTEM PROGRAMS

**MEDICAL RECORDS**

```
10  REM **--MEDICAL RECORDS--**
11  :
12  :
20  D$ = CHR$ (4): REM CONTROL D
60  TB = 15: REM  HTAB VALUE
80  :
90  :
100 REM **--MENU ROUTINE--**
110 HOME : VTAB 5
120 HTAB TB
140 PRINT "MEDICAL RECORDS"
160 PRINT : PRINT : PRINT
180 HTAB TB
200 PRINT "1. WRITE RECORD"
210 PRINT : HTAB TB
220 PRINT "2. READ RECORD"
230 PRINT : HTAB TB
240 PRINT "3. SEARCH RECORDS"
250 PRINT : HTAB TB
260 PRINT "4. END"
280 PRINT : HTAB TB
300 INPUT "WHICH NUMBER ";NUMBER
320 IF NUMBER < 1 OR NUMBER > 4 THEN 280
410 IF NUMBER = 1 THEN 1000
420 IF NUMBER = 2 THEN 2000
430 IF NUMBER = 3 THEN 3000
440 IF NUMBER = 4 THEN END
980 :
```

```
990 :
1000 REM **--WRITE ROUTINE--**
1020 HOME : VTAB 10
1030 HTAB TB
1040 INPUT "NAME ";NAME$
1050 IF LEN (NAME$) > 14 THEN NAME$ = LEFT$ (NAME$,14)
1060 PRINT
1070 HTAB TB
1080 INPUT "DATE ";DT$
1100 PRINT
1120 HOME : VTAB 5
1140 HTAB TB
1160 PRINT "TYPE OF RECORD"
1180 PRINT : PRINT
1200 HTAB TB
1220 PRINT "D--DR. VISIT"
1230 PRINT : HTAB TB
1240 PRINT "M--MEDICATION"
1250 PRINT : HTAB TB
1260 PRINT "I--ILLNESS"
1270 PRINT : HTAB TB
1280 PRINT "A--ACCIDENT/INJURY"
1290 PRINT : HTAB TB
1300 PRINT "S--SHOT/IMMUNIZATION"
1310 PRINT : HTAB TB
1320 PRINT "X--X-RAY"
1330 PRINT : HTAB TB
1340 INPUT "WHICH TYPE OF RECORD ";TYPE$
1360 HOME : VTAB 5
1365 HTAB 8: PRINT "TYPE IN ANY MISC. INFO. "
1368 VTAB 10
1370 HTAB 8
1380 FOR I = 1 TO 22
1400 PRINT CHR$ (95);: REM UNDERLINE
1420 NEXT I
1430 HTAB 8
1440 VTAB 10
1460 INPUT " ";MISC$: REM INPUT OVER UNDERLINE
1480 IF LEN (MISC$) > 22 THEN PRINT "TOO LONG": PRINT : PRINT "DO NOT GO BEYOND THE END OF THE DASHES": FOR I = TO 3000: NEXT I: GOTO 1360
1500 HOME : VTAB 5
1520 PRINT NAME$
1530 PRINT
```

```
1540 PRINT DT$
1550 PRINT
1555 TP$ = TYPE$: GOSUB 10000: REM TYPE SUBROUTINE
1560 PRINT TP$
1570 PRINT
1580 PRINT MISC$
1590 PRINT
1600 INPUT "IS THIS CORRECT? ";YES$
1620 IF YES$ < > "Y" THEN 1000: REM START OVER
1640 :
1660 :
1700 REM **--OUTPUT ROUTINE--**
1710 ONERR GOTO 1950: REM FOR FIRST USE
1720 PRINT D$;"OPEN MEDICAL FILE, L50"
1740 PRINT D$;"READ MEDICAL FILE, R0"
1760 INPUT PTR
1780 PTR = PTR + 1
1800 PRINT D$;"WRITE MEDICAL FILE, R";PTR;",B";1
1810 PRINT NAME$
1820 PRINT D$;"WRITE MEDICAL FILE, R";PTR;",B";15
1830 PRINT DT$
1840 PRINT D$;"WRITE MEDICAL FILE, R";PTR;",B";25
1850 PRINT TYPE$
1860 PRINT D$;"WRITE MEDICAL FILE, R";PTR;",B";27
1870 PRINT MISC$
1880 PRINT D$;"WRITE MEDICAL FILE, R0"
1900 PRINT PTR
1920 PRINT D$;"CLOSE MEDICAL FILE"
1930 POKE 216,0: REM RESET ERR FLAG
1940 GOTO 100: REM MENU
1945 :
1946 :
1950 REM CREATE PTR FOR FIRST TIME
1955 POKE 216,0: REM RESET ERR FLAG
1960 PRINT D$;"WRITE MEDICAL FILE, R0"
1970 PRINT "0"
1975 GOTO 1700: REM BEGIN AGAIN
1980 :
1990 :
2000 REM **--READ ROUTINE--**
2020 PRINT D$;"OPEN MEDICAL FILE, L50"
2040 PRINT D$;"READ MEDICAL FILE, R0"
2060 INPUT PTR
2080 FOR I = 1 TO PTR
```

```
2100  PRINT D$; "READ MEDICAL FILE, R"; I; ",B"; 1
2120  INPUT NAME$
2140  PRINT D$; "READ MEDICAL FILE, R"; I; ",B"; 15
2160  INPUT DT$
2180  PRINT D$; "READ MEDICAL FILE, R"; I; ",B"; 25
2200  INPUT TYPE$
2220  PRINT D$; "READ MEDICAL FILE, R"; I; ",B"; 27
2240  INPUT MISC$
2260  TP$ = TYPE$: GOSUB 10000: TYPE$ = TP$
2280  PRINT NAME$;
2300  HTAB 10
2320  PRINT DT$;
2340  HTAB 20
2360  PRINT TYPE$
2380  HTAB 20
2400  PRINT MISC$
2420  PRINT
2440  NEXT I
2460  PRINT D$; "CLOSE MEDICAL FILE"
2480  INPUT "HIT RETURN TO GO TO MENU "; L$
2500  GOTO 100: REM  MENU
2980 :
2990 :
3000  REM **--SEARCH ROUTINE--**
3020  HOME : VTAB 5: HTAB TB
3030  PRINT "SEARCH FOR..."
3035  PRINT
3040  HTAB TB
3060  PRINT "1. NAME"
3070  PRINT : HTAB TB
3080  PRINT "2. DATE"
3090  PRINT : HTAB TB
3100  PRINT "3. TYPE"
3110  PRINT : HTAB TB
3120  PRINT "4. MISC"
3130  PRINT : HTAB TB
3135  PRINT "5. END SEARCH"
3137  PRINT : HTAB TB
3140  INPUT "WHICH NUMBER? "; NB
3160  IF NB < 1 OR NB > 5 THEN 3137
3180  IF NB = 1 THEN BYTE = 1: B$ = "NAME"
3200  IF NB = 2 THEN BYTE = 15: B$ = "DATE"
3220  IF NB = 3 THEN BYTE = 25: B$ = "TYPE"
3240  IF NB = 4 THEN BYTE = 27: B$ = "MISC"
```

```
3260  IF NB = 5 THEN 100: REM  MENU
3265  PRINT : HTAB TB
3270  PRINT "WHICH ";B$;: INPUT "?";SRCH$
3271  HOME : VTAB 2
3272 :
3273 :
3275  REM **--INPUT ROUTINE--**
3280  PRINT D$;"OPEN MEDICAL FILE,L50"
3300  PRINT D$;"READ MEDICAL FILE,R0"
3320  INPUT PTR
3340  FOR I = 1 TO PTR
3360  PRINT D$;"READ MEDICAL FILE,R";I;",B";BYTE
3380  INPUT FIND$
3400  IF SRCH$ < > FIND$ THEN 3640: REM  NEXT RECORD
3420  FOR K = 1 TO 4
3440  IF K = 1 THEN BT = 1
3460  IF K = 2 THEN BT = 15
3480  IF K = 3 THEN BT = 25
3500  IF K = 4 THEN BT = 27
3520  PRINT D$;"READ MEDICAL FILE,R";I;",B";BT
3540  INPUT A$(K)
3560  IF K = 3 THEN TP$ = A$(3): GOSUB 10000:A$(3) = TP$
3580  PRINT A$(K)
3600  NEXT K
3620  PRINT
3640  NEXT I
3660  PRINT D$;"CLOSE MEDICAL FILE"
3680  INPUT "HIT RETURN WHEN READY ";L$
3700  GOTO 3000: REM  SEARCH AGAIN
9998 :
9999 :
10000 REM **--SUBROUTINES--**
10100 REM    *--TYPE SUBROUTINE--*
10120 IF TP$ = "D" THEN TP$ = "DR. VISIT"
10140 IF TP$ = "M" THEN TP$ = "MEDICATION"
10160 IF TP$ = "I" THEN TP$ = "ILLNESS"
10180 IF TP$ = "A" THEN TP$ = "ACCIDENT/INJURY"
10200 IF TP$ = "S" THEN TP$ = "SHOT/IMMUNIZATION"
10220 IF TP$ = "X" THEN TP$ = "X-RAYS"
10240 RETURN
```

## MEDICAL RECORDS W/ARRAYS

```
10  REM  **--MEDICAL RECORDS--**
11 :
12 :
20  D$ = CHR$ (4): REM  CONTROL D
40  DIM NAME$(50),DT$(50),TYPE$(50),MISC$(50)
60  TB = 15: REM  HTAB VALUE
80 :
90 :
100 REM  **--MENU ROUTINE--**
110 HOME : VTAB 5
120 HTAB TB
140 PRINT "MEDICAL RECORDS"
160 PRINT : PRINT : PRINT
180 HTAB TB
200 PRINT "1. WRITE RECORD"
210 PRINT : HTAB TB
220 PRINT "2. READ RECORD"
230 PRINT : HTAB TB
240 PRINT "3. SEARCH RECORDS"
250 PRINT : HTAB TB
260 PRINT "4. END"
280 PRINT : HTAB TB
300 INPUT "WHICH NUMBER ";NUMBER
320 IF NUMBER < 1 OR NUMBER > 4 THEN 280
410 IF NUMBER = 1 THEN 1000
420 IF NUMBER = 2 THEN 2000
430 IF NUMBER = 3 THEN 3000
440 IF NUMBER = 4 THEN END
980 :
990 :
1000 REM  **--WRITE ROUTINE--**
1020 HOME : VTAB 10
1030 HTAB TB
1040 INPUT "NAME ";NAME$
1050 IF LEN (NAME$) > 14 THEN NAME$ = LEFT$ (NAME$,14)
1060 PRINT
1070 HTAB TB
1080 INPUT "DATE ";DT$
1100 PRINT
1120 HOME : VTAB 5
1140 HTAB TB
```

# MEDICAL RECORDS W/ARRAYS

```
1160  PRINT "TYPE OF RECORD"
1180  PRINT : PRINT
1200  HTAB TB
1220  PRINT "D- -DR. VISIT"
1230  PRINT : HTAB TB
1240  PRINT "M- -MEDICATION"
1250  PRINT : HTAB TB
1260  PRINT "I- -ILLNESS"
1270  PRINT : HTAB TB
1280  PRINT "A- -ACCIDENT/INJURY"
1290  PRINT : HTAB TB
1300  PRINT "S- -SHOT/IMMUNIZATION"
1310  PRINT : HTAB TB
1320  PRINT "X- -X-RAY"
1330  PRINT : HTAB TB
1340  INPUT "WHICH TYPE OF RECORD ";TYPE$
1360  HOME : VTAB 5
1365  HTAB 8: PRINT "TYPE IN ANY MISC. INFO. "
1368  VTAB 10
1370  HTAB 8
1380  FOR I = 1 TO 22
1400  PRINT CHR$ (95);: REM UNDERLINE
1420  NEXT I
1430  HTAB 8
1440  VTAB 10
1460  INPUT " ";MISC$: REM INPUT OVER UNDERLINE
1480  IF LEN (MISC$) > 22 THEN PRINT "TOO LONG": PRINT :
      PRINT "DO NOT GO BEYOND THE END OF THE DASHES": FOR I =
      TO 3000: NEXT I: GOTO 1360
1500  HOME : VTAB 5
1520  PRINT NAME$
1530  PRINT
1540  PRINT DT$
1550  PRINT
1555  TP$ = TYPE$: GOSUB 10000: REM  TYPE SUBROUTINE
1560  PRINT TP$
1570  PRINT
1580  PRINT MISC$
1590  PRINT
1600  INPUT "IS THIS CORRECT? ";YES$
1620  IF YES$ < > "Y" THEN 1000: REM  START OVER
1640 :
1660 :
1700  REM **- -OUTPUT ROUTINE- -**
```

```
1710  ONERR GOTO 1950: REM  FOR FIRST USE
1720  PRINT D$; "OPEN MEDICAL FILE, L50"
1740  PRINT D$; "READ MEDICAL FILE, R0"
1760  INPUT PTR
1780 PTR = PTR + 1
1800  PRINT D$; "WRITE MEDICAL FILE, R"; PTR; ", B"; 1
1810  PRINT NAME$
1820  PRINT D$; "WRITE MEDICAL FILE, R"; PTR; ", B"; 15
1830  PRINT DT$
1840  PRINT D$; "WRITE MEDICAL FILE, R"; PTR; ", B"; 25
1850  PRINT TYPE$
1860  PRINT D$; "WRITE MEDICAL FILE, R"; PTR; ", B"; 27
1870  PRINT MISC$
1880  PRINT D$; "WRITE MEDICAL FILE, R0"
1900  PRINT PTR
1920  PRINT D$; "CLOSE MEDICAL FILE"
1930  POKE 216, 0: REM  RESET ERR FLAG
1940  GOTO 100: REM  MENU
1945 :
1946 :
1950  REM  CREATE PTR FOR FIRST TIME
1955  POKE 216, 0: REM  RESET ERR FLAG
1960  PRINT D$; "WRITE MEDICAL FILE, R0"
1970  PRINT "0"
1975  GOTO 1700: REM  BEGIN AGAIN
1980 :
1990 :
2000  REM  **--READ ROUTINE--**
2001 :
2002 :
2010  REM  **--INPUT ROUTINE--**
2020  PRINT D$; "OPEN MEDICAL FILE, L50"
2040  PRINT D$; "READ MEDICAL FILE, R0"
2060  INPUT PTR
2080  FOR I = 1 TO PTR
2100  PRINT D$; "READ MEDICAL FILE, R"; I; ", B"; 1
2120  INPUT NAME$
2140  PRINT D$; "READ MEDICAL FILE, R"; I; ", B"; 15
2160  INPUT DT$
2180  PRINT D$; "READ MEDICAL FILE, R"; I; ", B"; 25
2200  INPUT TYPE$
2220  PRINT D$; "READ MEDICAL FILE, R"; I; ", B"; 27
2240  INPUT MISC$
2260 TP$ = TYPE$: GOSUB 10000: TYPE$ = TP$
```

```
2280  PRINT NAME$;
2300  HTAB 10
2320  PRINT DT$;
2340  HTAB 20
2360  PRINT TYPE$
2380  HTAB 20
2400  PRINT MISC$
2420  PRINT
2440  NEXT I
2460  PRINT D$; "CLOSE MEDICAL FILE"
2480  INPUT "HIT RETURN TO GO TO MENU "; L$
2500  GOTO 100: REM  MENU
2980 :
2990 :
3000  REM **--SEARCH ROUTINE--**
3010  Q = 0
3020  HOME : VTAB 5: HTAB TB
3030  PRINT "SEARCH FOR..."
3035  PRINT
3040  HTAB TB
3060  PRINT "1. NAME"
3070  PRINT : HTAB TB
3080  PRINT "2. DATE"
3090  PRINT : HTAB TB
3100  PRINT "3. TYPE"
3110  PRINT : HTAB TB
3120  PRINT "4. MISC"
3130  PRINT : HTAB TB
3135  PRINT "5. END SEARCH"
3137  PRINT : HTAB TB
3140  INPUT "WHICH NUMBER? "; NB
3160  IF NB < 1 OR NB > 5 THEN 3137
3180  IF NB = 1 THEN BYTE = 1: B$ = "NAME"
3200  IF NB = 2 THEN BYTE = 15: B$ = "DATE"
3220  IF NB = 3 THEN BYTE = 25: B$ = "TYPE"
3240  IF NB = 4 THEN BYTE = 27: B$ = "MISC"
3260  IF NB = 5 THEN 100: REM  MENU
3265  PRINT : HTAB TB
3270  PRINT "WHICH "; B$; : INPUT "?"; SRCH$
3271 :
3272 :
3275  REM **--INPUT ROUTINE--**
3280  PRINT D$; "OPEN MEDICAL FILE, L50"
3300  PRINT D$; "READ MEDICAL FILE, R0"
```

```
3320  INPUT PTR
3340  FOR I = 1 TO PTR
3360  PRINT D$; "READ MEDICAL FILE, R"; I; ", B"; BYTE
3380  INPUT FIND$
3400  IF SRCH$ < > FIND$ THEN 3600: REM  NEXT RECORD
3420 K = 1
3430  FOR K = 1 TO 4
3440  IF K = 1 THEN BT = 1
3450  IF K = 2 THEN BT = 15
3460  IF K = 3 THEN BT = 25
3470  IF K = 4 THEN BT = 27
3480  PRINT D$; "READ MEDICAL FILE, R"; I; ", B"; BT
3500  INPUT A$(K)
3520  NEXT K
3540 Q = Q + 1
3560 NAME$(Q) = A$(1)
3570 DT$(Q) = A$(2)
3580 TYPE$(Q) = A$(3) : TP$ = TYPE$(Q) :
     GOSUB 10000: TYPE$(Q) = TP$
3590 MISC$(Q) = A$(4)
3600  NEXT I
3620  PRINT D$; "CLOSE MEDICAL FILE"
3621 :
3622 :
3630  REM  **--DISPLAY ROUTINE--**
3640  HOME : VTAB 10
3700  FOR I = 1 TO Q
3720  PRINT NAME$(I);
3730  HTAB 10
3740  PRINT DT$(I);
3750  HTAB 20
3760  PRINT TYPE$(I)
3770  HTAB 20
3780  PRINT MISC$(I)
3790  PRINT
3800  NEXT I
3820  INPUT "HIT RETURN WHEN READY"; L$
3840  GOTO 3000: REM  SEARCH AGAIN
9998 :
9999 :
10000  REM  **--SUBROUTINES--**
10100  REM   *--TYPE SUBROUTINE--*
10120  IF TP$ = "D" THEN TP$ = "DR. VIST"
10140  IF TP$ = "M" THEN TP$ = "MEDICATION"
```

```
10160 IF TP$ = "I" THEN TP$ = "ILLNESS"
10180 IF TP$ = "A" THEN TP$ = "ACCIDENT/INJURY"
10200 IF TP$ = "S" THEN TP$ = "SHOT/IMMUNIZATION"
10220 IF TP$ = "X" THEN TP$ = "X-RAYS"
10240 RETURN
```

# APPENDIX H.

## HOME INVENTORY AND BACK ORDER SYSTEM PROGRAMS

### HOME MENU

```
10  REM  **--HOME INVENTORY SYSTEM--**
11 :
12 :
20  D$ = CHR$ (4): REM  CONTROL D
40  TB = 8: REM  HTAB VALUE
60 :
80 :
100 REM  **--MENU ROUTINE--**
120 HOME : VTAB 5
140 HTAB TB
160 PRINT "HOME INVENTORY SYSTEM"
180 PRINT : PRINT : PRINT
200 HTAB TB
220 PRINT "1. WRITE RECORD"
240 PRINT : HTAB TB
260 PRINT "2. READ RECORD"
280 PRINT : HTAB TB
300 PRINT "3. SEARCH RECORDS"
320 PRINT : HTAB TB
340 PRINT "4. CORRECT RECORD"
360 PRINT : HTAB TB
380 PRINT "5. END"
400 PRINT : HTAB TB
420 INPUT "WHICH NUMBER ";NUMBER
440 IF NUMBER < 1 OR NUMBER > 5 THEN 400
460 IF NUMBER = 1 THEN 1000
480 IF NUMBER = 2 THEN 2000
500 IF NUMBER = 3 THEN 3000
```

```
520  IF NUMBER = 4 THEN 4000
540  IF NUMBER = 5 THEN END
560 :
580 :
1000 REM  **--WRITE RECORD--**
1020 PRINT D$; "RUN CREATE HOME INVENTORY"
1998 :
1999 :
2000 REM  **--READ RECORD--**
2020 PRINT D$; "RUN DISPLAY HOME INVENTORY"
2998 :
2999 :
3000 REM  **--SEARCH RECORDS--**
3020 PRINT D$; "RUN SEARCH HOME INVENTORY"
3998 :
3999 :
4000 REM  **--CORRECT RECORDS--**
4020 PRINT D$; "RUN CORRECT HOME INVENTORY"
```

## CREATE HOME INVENTORY

```
10  REM  **--CREATE HOME INVENTORY--**
11 :
12 :
20  D$ = CHR$ (4): REM  CONTROL D
40  TB = 8: REM  HTAB VALUE
60 :
80 :
100 REM  **--INPUT ROUTINE--**
120 HOME : VTAB 5
140 HTAB TB
160 PRINT "CREATE HOME INVENTORY"
180 PRINT : PRINT : PRINT
200 POKE 34, 7: POKE 32, 7: REM  SET WINDOW
220 :
240 :
260 HOME
280 PRINT "ITEM NAME PLEASE. "
300 PRINT : PRINT
320 SP = 25
340 GOSUB 5000: REM  INPUT SUBROUTINE
360 INPUT " "; ITEM$
```

```
380 IF LEN (ITEM$) > SP THEN PRINT " " : GOTO 260:
    REM 5 CTRL G'S
400 :
420 :
440 HOME
460 PRINT "ITEM SERIAL NUMBER PLEASE. "
480 PRINT : PRINT
500 SP = 15
520 GOSUB 5000: REM  INPUT SUBROUTINE
540 INPUT " ";SERIAL$
560 IF LEN (SERIAL$) > SP THEN PRINT "": GOTO 440:
    REM  5 CTRL G'S
580 :
600 :
620 HOME
640 PRINT "ITEM COST PLEASE"
660 PRINT : PRINT
680 SP = 10
700 GOSUB 5000: REM  INPUT SUBROUTINE
720 INPUT " ";CST$
740 IF LEN (CST$) > SP THEN PRINT " ": GOTO 620:
    REM  5 CTRL G'S
760 :
780 :
800 HOME
820 PRINT "ROOM OF ITEM"
840 PRINT : PRINT
860 SP = 20
880 GOSUB 5000: REM  INPUT SUBROUTINE
900 INPUT " ";ROOM$
920 IF LEN (ROOM$) > SP THEN PRINT " ": GOTO 800:
    REM  5 CTRL G'S
940 :
960 :
980 HOME
1000 PRINT "ITEM DESCRIPTION"
1020 PRINT : PRINT
1040 SP = 30
1060 GOSUB 5000: REM  INPUT SUBROUTINE
1080 INPUT " ";DESC$
1100 IF LEN (DESC$) > SP THEN PRINT " ": GOTO 980:
     REM  5 CTRL G'S
1120 :
1140 :
```

## CREATE HOME INVENTORY

```
1400  REM **--DISPLAY FOR CORRECTION--**
1420  HOME
1440  PRINT "1. "; ITEM$
1460  PRINT "2. "; SERIAL$
1480  PRINT "3. "; CST$
1500  PRINT "4. "; ROOM$
1520  PRINT "5. "; DESC$
1540  PRINT : PRINT
1560  INPUT "IS THIS CORRECT ('Y' OR 'N') "; YES$
1580  IF YES$ = "Y" THEN 2000: REM  FILE ROUTINE
1600  INPUT "WHICH NUMBER IS WRONG "; NB
1610  IF NB < 1 OR NB > 5 THEN PRINT "INCORRECT
      CHOICE": GOTO 1600
1620  IF NB = 1 THEN SP = 25
1640  IF NB = 2 THEN SP = 15
1660  IF NB = 3 THEN SP = 10
1680  IF NB = 4 THEN SP = 20
1700  IF NB = 5 THEN SP = 30
1720  PRINT
1740  PRINT "TYPE IN CORRECT INFO "
1760  INPUT CT$(NB)
1780  IF LEN (CT$(NB)) > SP THEN PRINT "TOO LONG--TRY
      AGAIN PLEASE": GOTO 1740
1800  IF NB = 1 THEN ITEM$ = CT$(NB)
1820  IF NB = 2 THEN SERIAL$ = CT$(NB)
1840  IF NB = 3 THEN CST$ = CT$(NB)
1860  IF NB = 4 THEN ROOM$ = CT$(NB)
1880  IF NB = 5 THEN DESC$ = CT$(NB)
1900  GOTO 1400: REM  CHECK AGAIN
1998  :
1999  :
2000  REM **--FILE ROUTINE--**
2020  TEXT
2040  ONERR GOTO 2580: REM  FIRST USE ONLY
2060  PRINT D$; "OPEN INVENTORY, L100"
2080  PRINT D$; "READ INVENTORY, R0"
2100  INPUT PTR
2120  PTR = PTR + 1: POKE 216,0: REM  RESET ERROR FLAG
2140  PRINT D$; "WRITE INVENTORY, R"; PTR; ",B"; 0
2160  PRINT ITEM$
2180  PRINT D$; "WRITE INVENTORY, R"; PTR; ",B"; 25
2200  PRINT SERIAL$
2220  PRINT D$; "WRITE INVENTORY, R"; PTR; ",B"; 40
2240  PRINT CST$
```

```
2260 PRINT D$;"WRITE INVENTORY,R";PTR;",B";50
2280 PRINT ROOM$
2300 PRINT D$;"WRITE INVENTORY,R";PTR;",B";70
2320 PRINT DESC$
2340 PRINT D$;"WRITE INVENTORY,R0"
2360 PRINT PTR
2380 PRINT D$;"CLOSE INVENTORY"
2400 TEXT : HOME
2420 VTAB 5
2440 PRINT "DO YOU WANT TO ADD MORE ITEMS?"
2460 PRINT
2480 INPUT "TYPE 'NO' TO STOP ";NO$
2500 IF NO$ = "NO" THEN PRINT D$;"RUN HOME MENU"
2520 GOTO 100: REM BEGIN AGAIN
2540 :
2560 :
2580 REM **--FIRST USE ONLY--**
2600 POKE 216,0: REM RESET ERROR FLAG
2620 PRINT D$;"WRITE INVENTORY,R0"
2640 PRINT "0"
2660 PRINT D$;"CLOSE INVENTORY"
2680 GOTO 2000: REM BEGIN FILE ROUTINE AGAIN
2700 :
2720 :
5000 REM **--SUBROUTINE--**
5040 HTAB 1
5060 FOR I = 1 TO SP
5080 PRINT CHR$ (95);: REM UNDERLINE
5100 NEXT I
5120 HTAB 1
5160 RETURN
```

## DISPLAY HOME INVENTORY

```
10  REM **--DISPLAY HOME INVENTORY--**
11 :
12 :
20  D$ = CHR$ (4): REM CONTROL D
40  PRINT D$;"OPEN INVENTORY, L100"
60  PRINT D$;"READ INVENTORY, R0"
80  INPUT PTR
90  HOME
100 FOR I = 1 TO PTR
120 PRINT D$;"READ INVENTORY, R";I;",B";0
140 INPUT ITEM$
160 PRINT D$;"READ INVENTORY, R";I;",B";25
180 INPUT SERIAL$
200 PRINT D$;"READ INVENTORY, R";I;",B";40
220 INPUT CST$
240 PRINT D$;"READ INVENTORY, R";I;",B";50
260 INPUT ROOM$
280 PRINT D$;"READ INVENTORY, R";I;",B";70
300 INPUT DESC$
320 PRINT D$
340 PRINT I;"  ";ITEM$;
360 HTAB 25: PRINT SERIAL$
380 PRINT "$";CST$;
400 HTAB 15: PRINT ROOM$
420 PRINT DESC$
440 PRINT : PRINT
450 TTLCST = TTLCST + VAL (CST$)
460 NEXT I
480 PRINT D$;"CLOSE INVENTORY"
500 PRINT : PRINT : PRINT "TOTAL VALUE OF ITEMS = $";TTLCST
510 PRINT
520 INPUT "HIT RETURN TO RETURN TO MENU ";L$
540 PRINT D$;"RUN HOME MENU"
```

## SEARCH HOME INVENTORY

```
10  REM  ***--SEARCH/SORT RECORDS--***
11 :
12 :
13 :
20  D$ = CHR$ (4): REM  CONTROL D
22  PRINT D$;"OPEN INVENTORY, L100"
24  PRINT D$;"READ INVENTORY, R0"
26  INPUT PTR
28  PRINT D$: DIM C$(PTR)
30 :
32 :
40  REM  **--MENU ROUTINE--**
50  HOME : VTAB 3
60  TB = 8: HTAB 12
80  PRINT "SEARCH/SORT MENU"
100  PRINT : PRINT
120  HTAB TB
140  PRINT "1. SEARCH FOR ITEM"
160  PRINT : HTAB TB
180  PRINT "2. SEARCH FOR SERIAL #"
200  PRINT : HTAB TB
220  PRINT "3. SEARCH FOR COST"
240  PRINT : HTAB TB
260  PRINT "4. SEARCH FOR ROOM ITEMS"
280  PRINT : HTAB TB
300  PRINT "5. SORT ITEMS ALPHABETICALLY"
320  PRINT : HTAB TB
340  PRINT "6. SORT ITEMS BY SERIAL #"
360  PRINT : HTAB TB
380  PRINT "7. RETURN TO MAIN MENU"
400  PRINT : HTAB TB
420  INPUT "WHICH NUMBER ";NUMBER
440  IF NUMBER < 1 OR NUMBER > 7 THEN PRINT "INCORRECT
     NUMBER!": GOTO 400

510  IF NUMBER = 1 THEN 1000
520  IF NUMBER = 2 THEN 2000
530  IF NUMBER = 3 THEN 3000
540  IF NUMBER = 4 THEN 4000
550  IF NUMBER = 5 THEN 5000
560  IF NUMBER = 6 THEN 6000
```

## SEARCH HOME INVENTORY

```
570  IF NUMBER = 7 THEN 7000
970 :
980 :
990 :
1000 REM **--SEARCH FOR ITEM--**
1020 HOME : VTAB 5
1040 HTAB TB
1060 INPUT "WHICH ITEM? ";SRCH$
1080 I = 1:BYTE = 0
1100 GOSUB 10000: REM  SEARCH ROUTINE
1120 PRINT ITEM$;: HTAB 25: PRINT SERIAL$
1140 PRINT CST$;: HTAB 15: PRINT ROOM$
1160 PRINT DESC$
1180 PRINT : ITEM$ = "":SERIAL$ = "":CST$ = ""
     :ROOM$ = "":DESC $ = ""
1200 IF I = PTR OR I > PTR THEN 1260
1220 INPUT "SEARCH FOR MORE? ";YES$
1240 IF YES$ = "Y" THEN GOTO 1100
1260 PRINT
1280 GOSUB 9000: REM  HOUSKEEPING
1300 GOTO 40: REM  MENU
1970 :
1980 :
1990 :
2000 REM **--SEARCH FOR SERIAL #--**
2020 HOME : VTAB 5
2040 HTAB TB
2060 INPUT "WHICH SERIAL # ";SRCH$
2080 I = 1:BYTE = 25
2100 GOSUB 10000: REM  SEARCH ROUTINE
2120 PRINT SERIAL$;: HTAB 15: PRINT ITEM$
2140 PRINT
2160 GOSUB 9000: REM  HOUSEKEEPING
2180 GOTO 40: REM  MENU
2970 :
2980 :
2990 :
3000 REM **--SEARCH FOR COST--**
3020 HOME : VTAB 5:BYTE = 40:TTLAMT = 0:FIND$ = ""
3040 HTAB 14
3060 PRINT "SEARCH FOR ITEMS..."
3080 PRINT : HTAB TB
3100 PRINT "A...ABOVE A CERTAIN AMOUNT"
3120 PRINT : HTAB TB
```

```
3140 PRINT "B...BELOW A CERTAIN AMOUNT"
3160 PRINT : HTAB TB
3180 INPUT "WHICH LETTER 'A' OR 'B' ";LT$
3190 IF LT$ = "A" THEN 3220
3200 IF LT$ = "B" THEN 3500
3210 PRINT "INCORRECT CHOICE ": GOTO 3160
3211 :
3212 :
3220 REM **--ITEMS ABOVE $ AMOUNT--**
3230 HTAB TB
3240 INPUT "ABOVE WHICH AMOUNT? ";AMT
3250 HOME : VTAB 2: HTAB 14
3260 PRINT "ITEMS ABOVE $";AMT
3270 FOR I = 1 TO PTR
3280 PRINT D$; "READ INVENTORY,R"; I; ",B";BYTE
3290 INPUT FIND$
3300 IF FIND$ = "D" THEN 3360
3310 IF AMT > VAL (FIND$) THEN 3360
3320 PRINT D$; "READ INVENTORY,R"; I; ",B"; 0
3330 INPUT ITEM$
3340 TTLAMT = TTLAMT + VAL (FIND$)
3350 PRINT ITEM$;: HTAB 30: PRINT FIND$
3360 NEXT I
3370 PRINT
3380 PRINT "TOTAL VALUE = $";TTLAMT
3390 PRINT : GOSUB 9000: REM HOUSEKEEPING
3400 GOTO 40: REM MENU
3496 :
3497 :
3500 REM **--ITEMS BELOW $ AMOUNT--**
3510 HTAB TB
3520 INPUT "BELOW WHICH AMOUNT ";AMT
3530 HOME : VTAB 2: HTAB 14
3540 PRINT "ITEMS BELOW $";AMT
3550 FOR I = 1 TO PTR
3560 PRINT D$; "READ INVENTORY,R"; I; ",B";BYTE
3570 INPUT FIND$
3580 IF FIND$ = "D" THEN 3640
3590 IF AMT < VAL (FIND$) THEN 3640
3600 PRINT D$; "READ INVENTORY,R"; I; ",B"; 0
3610 INPUT ITEM$
3620 TTLAMT = TTLAMT + VAL (FIND$)
3630 PRINT ITEM$;: HTAB 30: PRINT FIND$
3640 NEXT I
```

## SEARCH HOME INVENTORY

```
3650  PRINT
3660  PRINT "TOTAL VALUE = $"; TTLAMT
3670  PRINT
3680  GOSUB 9000: REM  HOUSEKEEPING
3690  GOTO 40: REM  MENU
3970 :
3980 :
3990 :
4000  REM **--SEARCH FOR ROOM ITEMS--**
4020  HOME : VTAB 5: TLROOM = 0
4040  HTAB TB
4060  INPUT "WHICH ROOM "; SRCH$
4080  I = 1: BYTE = 50: HOME : VTAB 5
4100  HTAB 14: PRINT SRCH$: PRINT : PRINT
4120  GOSUB 10000: REM  SEARCH ROUTINE
4140  PRINT ITEM$; : HTAB 25: PRINT SERIAL$
4160  PRINT CST$;: HTAB 11: PRINT DESC$
4180  TLROOM = TLROOM + VAL (CST$)
4200  PRINT
4220  IF I > PTR THEN 4280: REM  SEARCH COMPLETED
4240  ITEM$ = " ": SERIAL$ = " ": CST$ = " ": DESC$ = " "
4260  GOTO 4120: REM  CONTINUE SEARCH
4280  PRINT
4300  PRINT "TOTAL VALUE FOR "; SRCH$; "  =  "; TLROOM
4320  PRINT : GOSUB 9000: REM  HOUSEKEEPING
4340  GOTO 40: REM  MENU
4970 :
4980 :
4990 :
5000  REM **--SORT ALPHABETICALLY--**
5020  HOME : VTAB 5
5040  HTAB TB
5060  INVERSE : PRINT "WORKING--PLEASE DON'T TOUCH!!"
      : NORMAL
5080  Q = 1: REM  VALID RECORD COUNTER
5100  FOR I = 1 TO PTR
5120  PRINT D$; "READ INVENTORY, R"; I; ",B"; 0
5140  INPUT C$
5160  IF C$ = "D" THEN 5220
5180  C$(Q) = C$
5200  Q = Q + 1
5220  NEXT I
5240  N = Q - 1
5260  PRINT : PRINT : HTAB TB
```

```
5280 INVERSE : PRINT "STILL WORKING--PLEASE WAIT!"
     : NORMAL
5300 GOSUB 20000: REM SORT ROUTINE
5320 REM DISPLAY RESULTS
5340 HOME : VTAB 5
5360 SPEED= 150
5380 FOR I = 1 TO Q - 1
5400 PRINT I;" ";C$(I)
5420 NEXT I
5440 PRINT
5460 GOSUB 9000: REM HOUSEKEEPING
5480 GOTO 40: REM MENU
5970 :
5980 :
5990 :
6000 REM **--SORT BY SERIAL #--**
6020 HOME : VTAB 5
6040 HTAB TB
6060 INVERSE : PRINT "WORKING--PLEASE DON'T TOUCH!!"
     : NORMAL
6080 Q = 1: REM VALID RECORD COUNTER
6100 FOR I = 1 TO PTR
6120 PRINT D$;"READ INVENTORY,R";I;",B";25
6140 INPUT C$
6160 IF C$ = "D" THEN 6280
6180 C$(Q) = C$
6200 PRINT D$;"READ INVENTORY,R";I;",B";0
6220 INPUT ITEM$
6240 C$(Q) = C$(Q) + "*" + ITEM$
6260 Q = Q + 1
6280 NEXT I
6300 N = Q - 1
6320 PRINT : PRINT : HTAB TB
6340 INVERSE : PRINT "STILL WORKING--PLEASE WAIT!"
     : NORMAL
6360 GOSUB 20000: REM SORT ROUTINE
6380 REM DISPLAY RESULTS
6400 HOME : VTAB 5
6420 J = 1
6440 FOR I = 1 TO Q - 1
6460 LN = LEN (C$(I))
6480 PRINT I;" ";
6500 IF MID$ (C$(I),J,1) = "*" THEN PRINT LEFT$ (C$(I),J - 1
     );: HTAB 20: PRINT MID$ (C$(I),J + 1,LN): GOTO 6540
```

```
6520 J = J + 1: GOTO 6500
6540 J = 1
6560 NEXT I
6580 PRINT
6600 GOSUB 9000: REM HOUSEKEEPING
6620 GOTO 40: REM MENU
6970 :
7000 REM **--RETURN TO HOME MENU--**
7020 PRINT D$; "CLOSE INVENTORY"
7040 PRINT D$; "RUN HOME MENU"
7970 :
7980 :
7990 :
9000 REM **--HOUSEKEEPING--**
9020 ITEM$ = ""
9040 SERIAL$ = ""
9060 CST$ = ""
9080 ROOM$ = ""
9100 DESC$ = ""
9120 PRINT D$; "PR#0"
9140 SPEED = 255
9160 PRINT D$: REM CANCEL INPUT FROM DISK
9180 INPUT "HIT RETURN TO CONTINUE "; L$
9900 RETURN
9970 :
9980 :
9990 :
10000 REM **-- SEARCH SUBROUTINE--**
10020 PRINT D$; "READ INVENTORY, R"; I; ", B"; BYTE
10040 INPUT FIND$
10060 IF FIND$ = "D" THEN 10100
10080 IF SRCH$ = FIND$ THEN 10200
10100 I = I + 1
10120 IF I < PTR OR I = PTR THEN 10000
10140 PRINT : HTAB TB
10160 PRINT "SEARCH COMPLETED! ": FOR K = 1 TO 1000:
      NEXT K
10180 RETURN
10200 PRINT D$; "READ INVENTORY, R"; I; ", B"; 0
10220 INPUT ITEM$
10240 PRINT D$; "READ INVENTORY, R"; I; ", B"; 25
10260 INPUT SERIAL$
10280 PRINT D$; "READ INVENTORY, R"; I; ", B"; 40
10300 INPUT CST$
```

```
10320  PRINT D$; "READ INVENTORY, R"; I; ", B"; 50
10340  INPUT ROOM$
10360  PRINT D$; "READ INVENTORY, R"; I; ", B"; 70
10380  INPUT DESC$
10400  I = I + 1
10420  PRINT D$: REM CANCEL INPUT FROM DISK
10440  RETURN
19970 :
19980 :
19990 :
20000  REM **--SORT SUBROUTINE--**
20020  M = N
20040  M = INT (M / 2)
20060  IF M = 0 THEN 20300
20080  J = 1: K = N - M
20100  I = J
20120  L = I + M
20140  IF C$(I) < C$(L) THEN 20240
20160  T$ = C$(I) : C$(I) = C$(L) : C$(L) = T$
20180  I = I - M
20200  IF I < 1 THEN 20240
20220  GOTO 20120
20240  J = J + 1
20260  IF J > K THEN 20040
20280  GOTO 20100
20300  RETURN
```

## CORRECT HOME INVENTORY

```
10  REM **--CORRECT HOME INVENTORY--**
11 :
12 :
20  D$ = CHR$ (4): REM  CONTROL D
40  TB = 8: REM  HTAB VALUE
60  PRINT D$;"OPEN INVENTORY, L100"
70  PRINT D$;"READ INVENTORY,R0"
80  INPUT PTR
90  PRINT D$: REM  CANCEL INPUT FROM DISK
95 :
96 :
100  REM **--MENU ROUTINE--**
120  HOME : VTAB 5
140  HTAB 12
160  PRINT "CORRECT/DELETE MENU"
180  PRINT : PRINT
200  HTAB TB
220  PRINT "C...CORRECT INVENTORY RECORD"
240  PRINT : HTAB TB
260  PRINT "D...DELETE INVENTORY RECORD"
280  PRINT : HTAB TB
300  PRINT "R...RETURN TO HOME MENU"
320  PRINT : HTAB TB
340  INPUT "WHICH LETTER PLEASE? ";LT$
360  IF LT$ = "C" THEN 1000
380  IF LT$ = "D" THEN 2000
400  IF LT$ = "R" THEN 3000
420  PRINT : HTAB TB
440  PRINT "INCORRECT CHOICE": GOTO 320
970 :
980 :
990 :
1000  REM **--CORRECT RECORD--**
1005  HOME
1010  POKE 32,7: POKE 34,7: REM  SET WINDOW
1020  HOME
1040  FLAG$ = "NO": REM  INFO HAS YET TO BE CHANGED
1050  PRINT "TYPE A '0' TO RETURN TO MENU": PRINT
1060  INPUT "CORRECT WHICH RECORD? ";REC
1070  IF REC = 0 THEN TEXT : GOTO 100: REM  MENU
1075  IF REC > PTR THEN PRINT "INCORRECT CHOICE"
       : GOTO 1060
```

```
1080 GOSUB 6000: REM  READ FILE
1120 REM  **--DISPLAY FOR CORRECTION--**
1140 HOME
1160 PRINT "1.  "; ITEM$
1180 PRINT "2.  "; SERIAL$
1200 PRINT "3.  "; CST$
1220 PRINT "4.  "; ROOM$
1240 PRINT "5.  "; DESC$
1260 PRINT : PRINT
1280 INPUT "IS THIS CORRECT ('Y' OR 'N') "; YES$
1290 IF YES$ = "Y" AND FLAG$ = "NO" THEN 1000
1300 IF YES$ = "Y" AND FLAG$ = "YES" THEN 7000: REM FILE
     ROUTINE
1320 INPUT "WHICH NUMBER IS WRONG "; NB
1330 IF NB < 1 OR NB > 5 THEN PRINT "INCORRECT CHOICE"
     : GOTO 1320
1340 IF NB = 1 THEN SP = 25
1360 IF NB = 2 THEN SP = 15
1380 IF NB = 3 THEN SP = 10
1400 IF NB = 4 THEN SP = 20
1420 IF NB = 5 THEN SP = 30
1440 PRINT
1460 PRINT "TYPE IN CORRECT INFO"
1480 INPUT CT$(NB)
1500 IF LEN (CT$(NB)) > SP THEN PRINT "TOO LONG--TRY
     AGAIN PLEASE": GOTO 1460
1520 IF NB = 1 THEN ITEM$ = CT$(NB)
1540 IF NB = 2 THEN SERIAL$ = CT$(NB)
1560 IF NB = 3 THEN CST$ = CT$(NB)
1580 IF NB = 4 THEN ROOM$ = CT$(NB)
1600 IF NB = 5 THEN DESC$ = CT$(NB)
1610 FLAG$ = "YES": REM  INFO HAS BEEN CHANGED
1620 GOTO 1120: REM  CHECK AGAIN
1997 :
1998 :
1999 :
2000 REM  **--DELETE RECORD--**
2020 HOME
2040 POKE 32,7: POKE 34,7: REM  SET WINDOW
2060 HOME
2100 PRINT "TYPE A '0' TO RETURN TO MENU": PRINT
2120 INPUT "DELETE WHICH RECORD "; REC
2140 IF REC = 0 THEN TEXT : GOTO 100: REM  MENU
```

```
2150 IF REC > PTR THEN PRINT "INCORRECT CHOICE":
     GOTO 2120
2160 GOSUB 6000: REM  READ RECORD
2180 HOME
2200 PRINT ITEM$
2220 PRINT SERIAL$
2240 PRINT CST$
2260 PRINT ROOM$
2280 PRINT DESC$
2300 PRINT : PRINT
2320 INPUT "DELETE THIS RECORD? ";YES$
2340 IF YES$ = "Y" THEN 2380
2360 TEXT : GOTO 2000
2380 PRINT "ARE YOU SURE? ": PRINT
2390 INPUT "TYPE 'YES' TO DELETE RECORD ";YES$
2400 IF YES$ = "YES" THEN 2440
2420 TEXT : GOTO 2000
2440 ITEM$ = "D"
2460 SERIAL$ = "D"
2480 CST$ = "D"
2500 ROOM$ = "D"
2520 DESC$ = "D"
2540 GOTO 7000: REM  FILE ROUTINE
2970 :
2980 :
2990 :
3000 REM **--RETURN TO HOME MENU--**
3020 TEXT : PRINT D$;"CLOSE INVENTORY"
3040 PRINT D$;"RUN HOME MENU"
3970 :
3980 :
3990 :
6000 REM **--READ FILE ROUTINE--**
6020 PRINT D$;"READ INVENTORY,R";REC;",B";0
6040 INPUT ITEM$
6060 PRINT D$;"READ INVENTORY,R";REC;",B";25
6080 INPUT SERIAL$
6100 PRINT D$;"READ INVENTORY,R";REC;",B";40
6120 INPUT CST$
6140 PRINT D$;"READ INVENTORY,R";REC;",B";50
6160 INPUT ROOM$
6180 PRINT D$;"READ INVENTORY,R";REC;",B";70
6200 INPUT DESC$
6220 PRINT D$ : REM  CANCEL INPUT FROM DISK
```

```
6240  RETURN
6970  :
6980  :
6990  :
7000  REM **--FILE ROUTINE--**
7020  TEXT
7040  PRINT D$;"WRITE INVENTORY,R";REC;",B";0
7060  PRINT ITEM$
7080  PRINT D$;"WRITE INVENTORY,R";REC;",B";25
7100  PRINT SERIAL$
7120  PRINT D$;"WRITE INVENTORY,R";REC;",B";40
7140  PRINT CST$
7160  PRINT D$;"WRITE INVENTORY,R";REC;",B";50
7180  PRINT ROOM$
7200  PRINT D$;"WRITE INVENTORY,R";REC;",B";70
7220  PRINT DESC$
7240  PRINT D$;"CLOSE INVENTORY"
7260  PRINT D$;"OPEN INVENTORY,L100"
7280  PRINT D$
7300  GOTO 100: REM MENU
```

# MENU

```
10  REM  **--BACK ORDER SYSTEM--**
11 :
12 :
20  D$ = CHR$ (4): REM  CONTROL D
40  TB = 8: REM  HTAB VALUE
60 :
80 :
100  REM  **--MENU ROUTINE--**
120  HOME : VTAB 5
140  HTAB TB
160  PRINT "BACK ORDER SYSTEM"
180  PRINT : PRINT : PRINT
200  HTAB TB
220  PRINT "1. WRITE RECORD"
240  PRINT : HTAB TB
260  PRINT "2. READ RECORD"
280  PRINT : HTAB TB
300  PRINT "3. SEARCH RECORDS"
320  PRINT : HTAB TB
340  PRINT "4. CORRECT RECORD"
360  PRINT : HTAB TB
380  PRINT "5. END"
400  PRINT : HTAB TB
420  INPUT "WHICH NUMBER ";NUMBER
440  IF NUMBER < 1 OR NUMBER > 5 THEN 400
460  IF NUMBER = 1 THEN 1000
480  IF NUMBER = 2 THEN 2000
500  IF NUMBER = 3 THEN 3000
520  IF NUMBER = 4 THEN 4000
540  IF NUMBER = 5 THEN END
560 :
580 :
1000  REM  **--WRITE RECORD--**
1020  PRINT D$;"RUN CREATE BACK ORDER"
1998 :
1999 :
2000  REM  **--READ RECORD--**
2020  PRINT D$;"RUN DISPLAY BACK ORDER"
2998 :
2999 :
3000  REM  **--SEARCH RECORDS--**
```

```
3020 PRINT D$; "RUN SEARCH BACK ORDER"
3998 :
3999 :
4000 REM  **--CORRECT RECORDS--**
4020 PRINT D$; "RUN CORRECT BACK ORDER"
```

## CREATE BACK ORDER

```
10  REM  **--CREATE BACK ORDER--**
11 :
12 :
20  D$ = CHR$ (4): REM  CONTROL D
40  TB = 8: REM HTAB VALUE
60 :
80 :
100 REM  **--INPUT ROUTINE--**
120 HOME : VTAB 5
140 HTAB TB
160 PRINT "CREATE BACK ORDER"
180 PRINT : PRINT
185 VTAB 20: HTAB 8
190 PRINT "CR = LAST RECORD": PRINT : HTAB 8:
    PRINT "TYPE '-' FOR NO VALUE"
200 POKE 34,7: POKE 32,7: POKE 35,19
220 :
240 :
260 HOME
280 PRINT "ITEM NAME PLEASE. "
300 PRINT : PRINT
320 SP = 25
340 GOSUB 5000: REM  INPUT SUBROUTINE
360 INPUT ""; ITEM$
380 IF LEN (ITEM$) > SP THEN PRINT " ": GOTO 260 :
    REM 5 CTRL G'S
385 IF ITEM$ = "" THEN ITEM$ = A$
390 A$ = ITEM$
400 :
420 :
440 HOME
460 PRINT "ITEM DESCRIPTION PLEASE. "
480 PRINT : PRINT
500 SP = 30
```

# CREATE BACK ORDER

```
520 GOSUB 5000: REM  INPUT SUBROUTINE
540  INPUT "";DESC$
560  IF LEN (DESC$) > SP THEN PRINT "": GOTO 440 :
     REM 5 CTRL G'S
575  IF DESC$ = "" THEN DESC$ = B$
576 B$ = DESC$
580 :
600 :
620  HOME
640  PRINT "INDIVIDUAL'S NAME PLEASE."
660  PRINT : PRINT
680 SP = 20
700  GOSUB 5000: REM  INPUT SUBROUTINE
720  INPUT "";NAME$
740  IF LEN (NAME$) > SP THEN PRINT "": GOTO 620 :
     REM 5  CTRL G'S
755  IF NAME$ = "" THEN NAME$ = C$
756  C$ = NAME$
760 :
780 :
800  HOME
820  PRINT "PHONE #"
840  PRINT : PRINT
860 SP = 20
880  GOSUB 5000: REM  INPUT SUBROUTINE
900  INPUT "";PHNE$
920  IF LEN (PHNE$) > SP THEN PRINT "": GOTO 800 :
     REM 5 CTRL G'S
935  IF PHNE$ = "" THEN PHNE$ = D1$
936 D1$ = PHNE$
940 :
960 :
980  HOME
1000  PRINT "DATE REQUEST WAS MADE"
1020  PRINT : PRINT
1040 SP = 10
1060  GOSUB 5000: REM  INPUT SUBROUTINE
1080  INPUT "";DTE$
1100  IF LEN (DTE$) > SP THEN PRINT " ": GOTO 980:
      REM 5 CTRL G'S
1105  IF DTE$ = "" THEN DTE$ = E$
1110 E$ = DTE$
1120 :
1140 :
```

```
1160 HOME
1180 PRINT "ORDERED YET ('Y' OR 'N')"
1200 PRINT : PRINT
1220 SP = 1
1240 GOSUB 5000: REM  INPUT SUBROUTINE
1260 INPUT ""; OD$
1280 IF LEN (OD$) > SP THEN PRINT "" : GOTO 1160:
     REM  5 CTRL G'S
1285 IF OD$ = "" THEN OD$ = F$
1290 F$ = OD$
1291 :
1292 :
1300 HOME
1310 PRINT "AMOUNT DEPOSITED"
1320 PRINT : PRINT
1330 SP = 10
1340 GOSUB 5000 REM  INPUT SUBROUTINE
1350 INPUT ""; AMT$
1360 IF LEN (AMT$) > SP THEN PRINT "": GOTO 1300:
     REM  5 CTRL G'S
1365 IF AMT$ = "" THEN AMT$ = G$
1370 G$ = AMT$
1391 :
1392 :
1400 REM **--DISPLAY FOR CORRECTION--**
1410 TEXT : POKE 34,7: POKE 32,7
1420 HOME
1440 PRINT "1. "; ITEM$
1460 PRINT "2. "; DESC$
1480 PRINT "3. "; NAME$
1500 PRINT "4. "; PHNE$
1520 PRINT "5. "; DTE$
1525 PRINT "6. "; OD$
1530 PRINT "7. "; AMT$
1540 PRINT : PRINT
1560 INPUT "IS THIS CORRECT ('Y' OR 'N') "; YES$
1580 IF YES$ = "Y" THEN 2000: REM  FILE ROUTINE
1600 INPUT "WHICH NUMBER IS WRONG "; NB
1610 IF NB < 1 OR NB > 7 THEN PRINT "INCORRECT
     CHOICE": GOTO 1600
1620 IF NB = 1 THEN SP = 25
1640 IF NB = 2 THEN SP = 30
1660 IF NB = 3 THEN SP = 20
1680 IF NB = 4 THEN SP = 20
```

# CREATE BACK ORDER

```
1700  IF NB = 5 THEN SP = 10
1705  IF NB = 6 THEN SP = 1
1710  IF NB = 7 THEN SP = 10
1720  PRINT
1740  PRINT "TYPE IN CORRECT INFO "
1760  INPUT CT$(NB)
1780  IF LEN (CT$(NB)) > SP THEN PRINT "TOO LONG- -TRY
      AGAIN PLEASE": GOTO 1740
1800  IF NB = 1 THEN ITEM$ = CT$(NB)
1820  IF NB = 2 THEN DESC$ = CT$(NB)
1840  IF NB = 3 THEN NAME$ = CT$(NB)
1860  IF NB = 4 THEN PHNE$ = CT$(NB)
1880  IF NB = 5 THEN DTE$ = CT$(NB)
1885  IF NB = 6 THEN OD$ = CT$(NB)
1890  IF NB = 7 THEN AMT$ = CT$(NB)
1900  GOTO 1400: REM CHECK AGAIN
1998  :
1999  :
2000  REM **- -FILE ROUTINE- -**
2020  TEXT
2040  ONERR GOTO 2580: REM FIRST USE ONLY
2060  PRINT D$; "OPEN BACKORDER, L120"
2080  PRINT D$; "READ BACKORDER, R0"
2100  INPUT PTR
2120  PTR = PTR + 1: POKE 216,0: REM RESET ERROR FLAG
2140  PRINT D$; "WRITE BACKORDER, R"; PTR; ",B"; 0
2160  PRINT ITEM$
2180  PRINT D$; "WRITE BACKORDER, R"; PTR; ",B"; 25
2200  PRINT DESC$
2220  PRINT D$; "WRITE BACKORDER, R"; PTR; ",B"; 55
2240  PRINT NAME$
2260  PRINT D$; "WRITE BACKORDER, R"; PTR; ",B"; 75
2280  PRINT PHNE$
2300  PRINT D$; "WRITE BACKORDER, R"; PTR; ",B"; 95
2320  PRINT DTE$
2325  PRINT D$; "WRITE BACKORDER, R"; PTR; ",B"; 105
2330  PRINT OD$
2335  PRINT D$; "WRITE BACKORDER, R"; PTR; ",B"; 110
2337  PRINT AMT$
2340  PRINT D$; "WRITE BACKORDER, R0"
2360  PRINT PTR
2380  PRINT D$; "CLOSE BACKORDER"
2400  TEXT : HOME
2420  VTAB 5
```

```
2440 PRINT "DO YOU WANT TO ADD MORE ITEMS?"
2460 PRINT
2480 INPUT "TYPE 'NO' TO STOP ";NO$
2500 IF NO$ = "NO" THEN PRINT D$;"RUN MENU"
2520 GOTO 100: REM  BEGIN AGAIN
2540 :
2560 :
2580 REM  **--FIRST USE ONLY--**
2600 POKE 216,0: REM  RESET ERROR FLAG
2620 PRINT D$;"WRITE BACKORDER,R0"
2640 PRINT "0"
2660 PRINT D$;"CLOSE BACKORDER"
2680 GOTO 2000: REM  BEGIN FILE ROUTINE AGAIN
2700 :
2720 :
5000 REM  **--SUBROUTINE--**
5040 HTAB 1
5060 FOR I = 1 TO SP
5080 PRINT CHR$ (95);: REM  UNDERLINE
5100 NEXT I
5120 HTAB 1
5160 RETURN
```

## DISPLAY BACK ORDER

```
10  REM **--DISPLAY BACK ORDER--**
11  :
12  :
20  D$ = CHR$ (4): REM  CONTROL D
40  PRINT D$; "OPEN BACKORDER, L120"
60  PRINT D$; "READ BACKORDER, R0"
80  INPUT PTR
90  HOME
100 FOR I = 1 TO PTR
120 PRINT D$; "READ BACKORDER, R"; I; ", B"; 0
140 INPUT ITEM$
160 PRINT D$; "READ BACKORDER, R"; I; ", B"; 25
180 INPUT DESC$
200 PRINT D$; "READ BACKORDER, R"; I; ", B"; 55
220 INPUT NAME$
240 PRINT D$; "READ BACKORDER, R"; I; ", B"; 75
260 INPUT PHNE$
280 PRINT D$; "READ BACKORDER, R"; I; ", B"; 95
300 INPUT DTE$
305 PRINT D$; "READ BACKORDER, R"; I; ", B"; 105
310 INPUT OD$
315 PRINT D$; "READ BACKORDER, R"; I; ", B"; 110
317 INPUT AMT$
320 PRINT D$
340 PRINT I; "   "; ITEM$;
360 HTAB 15: PRINT DESC$
380 PRINT NAME$;
400 HTAB 15: PRINT PHNE$
420 PRINT DTE$;: HTAB 12: PRINT OD$;: HTAB 15: PRINT AMT$
440 PRINT : PRINT
460 NEXT I
480 PRINT D$; "CLOSE INVENTORY"
510 PRINT
520 INPUT "HIT RETURN TO RETURN TO MENU "; L$
540 PRINT D$; "RUN MENU"
```

## SEARCH BACK ORDER

```
10  REM  ***--SEARCH/SORT RECORDS--***
11 :
12 :
13 :
20  D$ = CHR$ (4): REM  CONTROL D
22  PRINT D$;"OPEN BACKORDER, L120"
24  PRINT D$;"READ BACKORDER, R0"
26  INPUT PTR
28  PRINT D$: DIM C$(PTR)
30 :
32 :
40  REM  **--MENU ROUTINE--**
50  HOME : VTAB 3
60  TB = 8: HTAB 12
80  PRINT "SEARCH/SORT MENU"
100 PRINT : PRINT
120 HTAB TB
140 PRINT "1. SEARCH FOR ITEM"
160 PRINT : HTAB TB
180 PRINT "2. SEARCH FOR NAME"
200 PRINT : HTAB TB
220 PRINT "3. SEARCH FOR DATE"
240 PRINT : HTAB TB
260 PRINT "4. ITEMS NOT YET ORDERED"
280 PRINT : HTAB TB
300 PRINT "5. SORT ITEMS ALPHABETICALLY"
320 PRINT : HTAB TB
340 PRINT "6. SORT BY NAME"
360 PRINT : HTAB TB
380 PRINT "7. RETURN TO MAIN MENU"
400 PRINT : HTAB TB
420 INPUT "WHICH NUMBER ";NUMBER
440 IF NUMBER < 1 OR NUMBER > 7 THEN PRINT
    "INCORRECT NUMBER!": GOTO 400
510 IF NUMBER = 1 THEN 1000
520 IF NUMBER = 2 THEN 2000
530 IF NUMBER = 3 THEN 3000
540 IF NUMBER = 4 THEN 4000
550 IF NUMBER = 5 THEN 6000
560 IF NUMBER = 6 THEN 6000
570 IF NUMBER = 7 THEN 7000
```

# SEARCH BACK ORDER

```
970 :
980 :
990 :
1000 REM **--SEARCH FOR ITEM--**
1020 HOME : VTAB 5
1040 HTAB TB
1060 INPUT "WHICH ITEM? ";SRCH$
1080 I = 1:BYTE = 0
1090 GOSUB 8000: REM PRINT ROUTINE
1100 GOSUB 10000: REM SEARCH ROUTINE
1120 PRINT ITEM$;: HTAB 25: PRINT NAME$
1140 PRINT DTE$;: HTAB 10: PRINT PHNE$;: HTAB 30:
     PRINT OD$
1160 PRINT DESC$
1180 PRINT : ITEM$ = "":NAME$ = "":DTE$ = "" :PHNE$ =
     "":DESC$ = "":OD$ = "":AMT$ = ""
1200 IF I > PTR THEN 1260
1220 INPUT "SEARCH FOR MORE? ";YES$
1240 IF YES$ = "Y" THEN GOTO 1100
1260 PRINT
1280 GOSUB 9000: REM HOUSKEEPING
1300 GOTO 40: REM MENU
1970 :
1980 :
1990 :
2000 REM **--SEARCH FOR NAME--**
2020 HOME : VTAB 5
2040 HTAB TB
2060 INPUT "WHICH NAME? ";SRCH$
2070 GOSUB 8000: REM PRINT ROUTINE
2080 I = 1:BYTE = 55
2100 GOSUB 10000: REM SEARCH ROUTINE
2120 PRINT ITEM$;: HTAB 25: PRINT NAME$
2140 PRINT DTE$;: HTAB 10: PRINT PHNE$;: HTAB 30 :
     PRINT OD$
2160 PRINT DESC$
2180 PRINT : ITEM$ = "":NAME$ = "" :DTE$ = "":PHNE$ =
     "":DESC$ = "":OD$ = "":AMT$ = ""
2200 IF I = PTR OR I > PTR THEN 2260
2220 INPUT "SEARCH FOR MORE? ";YES$
2240 IF YES$ = "Y" THEN GOTO 1100
2260 PRINT
2280 GOSUB 9000: REM HOUSKEEPING
2300 GOTO 40: REM MENU
```

```
2970 :
2980 :
2990 :
3000 REM **--SEARCH FOR DATE--**
3020 HOME : VTAB 5: BYTE = 95 : FIND$ = ""
3040 HTAB 14
3060 PRINT "SEARCH FOR ITEMS..."
3080 PRINT : HTAB TB
3100 PRINT "A... AFTER A CERTAIN DATE"
3120 PRINT : HTAB TB
3140 PRINT "B... BEFORE A CERTAIN DATE"
3160 PRINT : HTAB TB
3180 INPUT "WHICH LETTER 'A' OR 'B' ";LT$
3190 IF LT$ = "A" THEN 3220
3200 IF LT$ = "B" THEN 3500
3210 PRINT "INCORRECT CHOICE ": GOTO 3160
3211 :
3212 :
3220 REM **--ITEMS AFTER DATE--**
3230 HTAB TB
3240 INPUT "AFTER WHICH DATE? ";DS$
3245 GOSUB 8000: REM PRINT ROUTINE
3250 HOME : VTAB 2: HTAB 14
3260 PRINT "ITEMS AFTER ";DS$
3270 FOR I = 1 TO PTR
3280 PRINT D$;"READ BACKORDER,R";I;",B";BYTE
3290 INPUT FIND$
3300 IF FIND$ = "D" THEN 3360
3310 IF DS$ > FIND$ THEN 3360
3320 PRINT D$;"READ BACKORDER,R";I;",B";0
3330 INPUT ITEM$
3340 PRINT D$;"READ BACKORDER,R";I;",B";55
3345 INPUT NAME$
3350 PRINT ITEM$;: HTAB 30: PRINT FIND$
3355 PRINT NAME$
3357 PRINT
3360 NEXT I
3370 PRINT
3380 REM
3390 PRINT : GOSUB 9000: REM HOUSEKEEPING
3400 GOTO 40: REM MENU
3496 :
3497 :
3500 REM **--ITEMS BEFORE DATE--**
```

```
3510 HTAB TB
3520 INPUT "BEFORE WHICH DATE ";DS$
3525 GOSUB 8000: REM PRINT ROUTINE
3530 HOME : VTAB 2: HTAB 14
3540 PRINT "ITEMS BEFORE ";DS$
3550 FOR I = 1 TO PTR
3560 PRINT D$; "READ BACKORDER,R"; I; ",B";BYTE
3570 INPUT FIND$
3580 IF FIND$ = "D" THEN 3640
3590 IF VAL (DS$) < VAL (FIND$) THEN 3640
3600 PRINT D$; "READ BACKORDER,R"; I; ",B"; 0
3610 INPUT ITEM$
3615 PRINT D$; "READ BACKORDER,R"; I; ",B"; 55
3620 INPUT NAME$
3630 PRINT ITEM$;: HTAB 30: PRINT FIND$
3635 PRINT NAME$
3637 PRINT
3640 NEXT I
3650 PRINT
3660 REM
3670 PRINT
3680 GOSUB 9000: REM HOUSEKEEPING
3690 GOTO 40: REM MENU
3970 :
3980 :
3990 :
4000 REM **--SEARCH FOR ROOM ITEMS--**
4020 HOME : VTAB 5
4040 HTAB TB
4060 SRCH$ = "N"
4070 GOSUB 8000: REM PRINT ROUTINE
4080 I = 1: BYTE = 105: HOME : VTAB 5
4100  HTAB 10: PRINT "ITEMS NOT ORDERED YET": PRINT :
     PRINT
4120 GOSUB 10000: REM SEARCH ROUTINE
4140 PRINT ITEM$;: HTAB 25: PRINT NAME$
4150 PRINT PHNE$;: HTAB 20: PRINT DTE$;: HTAB 30: PRINT
     AMT$
4160 PRINT DESC$
4180 TLROOM = TLROOM + VAL (CST$)
4200 PRINT
4220 IF I > PTR THEN 4280: REM SEARCH COMPLETED
4240 ITEM$ = "": DESC$ = "": NAME$ = "" : PHNE$ = "": DTE$
     = "": AMT$ = ""
```

```
4260 GOTO 4120: REM CONTINUE SEARCH
4280 PRINT
4300 REM
4320 PRINT : GOSUB 9000: REM HOUSEKEEPING
4340 GOTO 40: REM MENU
4970 :
4980 :
4990 :
5000 REM **--SORT ALPHABETICALLY--**
5020 HOME : VTAB 5
5030 IF NUMBER = 5 THEN BYTE = 0
5035 IF NUMBER = 6 THEN BYTE = 55
5040 HTAB TB
5060 INVERSE : PRINT "WORKING--PLEASE DON'T TOUCH!!":
     NORMAL
5080 Q = 1: REM VALID RECORD COUNTER
5100 FOR I = 1 TO PTR
5120 PRINT D$;"READ BACKORDER,R";I;",B";BYTE
5140 INPUT C$
5160 IF C$ = "D" THEN 5220
5180 C$(Q) = C$
5200 Q = Q + 1
5220 NEXT I
5240 N = Q - 1
5260 PRINT : PRINT : HTAB TB
5280 INVERSE : PRINT "STILL WORKING--PLEASE WAIT!" :
     NORMAL
5300 GOSUB 20000: REM SORT ROUTINE
5320 REM DISPLAY RESULTS
5340 HOME : VTAB 5
5360 SPEED= 150
5380 FOR I = 1 TO Q - 1
5400 PRINT I;" ";C$(I)
5420 NEXT I
5440 PRINT
5460 GOSUB 9000: REM HOUSEKEEPING
5480 GOTO 40: REM MENU
5970 :
5980 :
5990 :
6000 REM **--SORT BY SERIAL #--**
6010 GOSUB 8000: REM PRINT ROUTINE
6020 HOME : VTAB 5
6030 IF NUMBER = 5 THEN BYTE = 0
```

# SEARCH BACK ORDER

```
6035  IF NUMBER = 6 THEN BYTE = 55
6040  HTAB TB
6060  INVERSE : PRINT "WORKING- -PLEASE DON'T TOUCH!!":
      NORMAL
6080  Q = 1: REM  VALID RECORD COUNTER
6100  FOR I = 1 TO PTR
6120  PRINT D$; "READ BACKORDER, R"; I; ", B"; BYTE
6140  INPUT C$
6160  IF C$ = "D" THEN 6280
6180  C$(Q) = C$
6200  REM
6220  REM
6240  C$(Q) = C$(Q) + "*" + STR$(I)
6260  Q = Q + 1
6280  NEXT I
6300  N = Q - 1
6320  PRINT : PRINT : HTAB TB
6340  INVERSE : PRINT "STILL WORKING- -PLEASE WAIT!":
      NORMAL
6360  GOSUB 20000: REM  SORT ROUTINE
6380  REM DISPLAY RESULTS
6400  HOME : VTAB 5
6420  J = 1
6440  FOR I = 1 TO Q - 1
6460  LN = LEN (C$(I))
6480  PRINT I;"  ";
6500  IF MID$ (C$(I), J, 1) = "*" THEN PRINT LEFT$ (C$(I),
      J - 1): K$ = MID$ (C$(I), J + 1, LN): GOTO 6540
6520  J = J + 1: GOTO 6500
6540  K = VAL (K$)
6560  PRINT D$; "READ BACKORDER, R"; K; ", B"; 0
6570  INPUT ITEM$
6580  PRINT D$; "READ BACKORDER, R"; K; ", B"; 25
6590  INPUT DESC$
6600  PRINT D$; "READ BACKORDER, R"; K; ", B"; 55
6610  INPUT NAME$
6620  PRINT D$; "READ BACKORDER, R"; K; ", B"; 75
6630  INPUT PHNE$
6640  PRINT D$; "READ BACKORDER, R"; K; ", B"; 95
6650  INPUT DTE$
6660  PRINT D$; "READ BACKORDER, R"; K; ", B"; 105
6670  INPUT OD$
6680  PRINT D$; "READ BACKORDER, R"; K; ", B"; 110
6690  INPUT AMT$
```

```
6700 PRINT D$: REM  CANCEL INPUT FROM DISK
6710 PRINT ITEM$
6720 PRINT DESC$
6730 PRINT NAME$
6740 PRINT PHNE$
6750 PRINT DTE$
6760 PRINT OD$
6770 PRINT AMT$
6780 PRINT : PRINT
6790 J = 1
6800 NEXT I
6820 PRINT
6840 GOSUB 9000: REM  HOUSEKEEPING
6860 GOTO 40: REM  MENU
6970 :
6980 :
6990 :
7000 REM  **- -RETURN TO HOME MENU- -**
7020 PRINT D$; "CLOSE BACKORDER"
7040 PRINT D$; "RUN MENU"
7970 :
7980 :
7990 :
8000 REM  **- -PRINT ROUTINE- -**
8010 PRINT : HTAB TB
8020 INPUT "DO YOU WANT A PRINTOUT? "; YES$
8040 IF YES$ = "Y" THEN 8080
8050 PRINT
8060 RETURN
8080 PRINT D$; "PR#1"
8090 PRINT
8100 RETURN
8980 :
8990 :
8995 :
9000 REM  **- -HOUSEKEEPING- -**
9020 ITEM$ = ""
9040 NAME$ = ""
9050 PHNE$ = ""
9060 DTE$ = ""
9070 OD$ = ""
9080 AMT$ = ""
9100 DESC$ = ""
9120 PRINT D$; "PR#0"
```

# SEARCH BACK ORDER

```
9140  SPEED= 255
9160  PRINT D$: REM  CANCEL INPUT FROM DISK
9180  INPUT "HIT RETURN TO CONTINUE ";L$
9900  RETURN
9970 :
9980 :
9990 :
10000 REM **- - SEARCH SUBROUTINE- -**
10020 PRINT D$; "READ BACKORDER, R"; I; ",B"; BYTE
10040 INPUT FIND$
10060 IF FIND$ = "D" THEN 10100
10080 IF SRCH$ = FIND$ THEN 10200
10100 I = I + 1
10120 IF I < PTR OR I = PTR THEN 10000
10140 PRINT : HTAB TB
10160 PRINT "SEARCH COMPLETED!": FOR K = 1 TO 1000:
      NEXT K
10180 RETURN
10200 PRINT D$; "READ BACKORDER, R"; I; ",B"; 0
10220 INPUT ITEM$
10240 PRINT D$; "READ BACKORDER, R"; I; ",B"; 25
10260 INPUT DESC$
10280 PRINT D$; "READ BACKORDER, R"; I; ",B"; 55
10300 INPUT NAME$
10320 PRINT D$; "READ BACKORDER, R"; I; ",B"; 75
10340 INPUT PHNE$
10360 PRINT D$; "READ BACKORDER, R"; I; ",B"; 95
10380 INPUT DTE$
10385 PRINT D$; "READ BACKORDER, R"; I; ",B"; 105
10390 INPUT OD$
10395 PRINT D$; "READ BACKORDER, R"; I; ",B"; 110
10397 INPUT AMT$
10400 I = I + 1
10420 PRINT D$: REM  CANCEL INPUT FROM DISK
10440 RETURN
19970 :
19980 :
20000 REM **- -SORT SUBROUTINE- -**
20020 M = N
20040 M = INT (M / 2)
20060 IF M = 0 THEN 20300
20080 J = 1: K = N - M
20100 I = J
20120 L = I + M
```

```
20140   IF C$(I) < C$(L) THEN 20240
20160   T$ = C$(I):C$(I) = C$(L):C$(L) = T$
20180   I = I - M
20200   IF I < 1 THEN 20240
20220   GOTO 20120
20240 J = J + 1
20260   IF J > K THEN 20040
20280   GOTO 20100
20300   RETURN
```

## CORRECT BACK ORDER

```
10  REM  **CORRECT BACK ORDER--**
11 :
12 :
20  D$ = CHR$ (4): REM  CONTROL D
40  TB = 8: REM  HTAB VALUE
60  PRINT D$;"OPEN BACKORDER, L120"
70  PRINT D$;"READ BACKORDER, R0"
80  INPUT PTR
90  PRINT D$: REM  CANCEL INPUT FROM DISK
95 :
96 :
100  REM  **--MENU ROUTINE--**
120  HOME : VTAB 5
140  HTAB 12
160  PRINT "CORRECT/DELETE MENU"
180  PRINT : PRINT
200  HTAB TB
220  PRINT "C...CORRECT BACKORDER RECORD"
240  PRINT : HTAB TB
260  PRINT "D...DELETE BACKORDER RECORD"
280  PRINT : HTAB TB
300  PRINT "R...RETURN TO MENU"
320  PRINT : HTAB TB
340  INPUT "WHICH LETTER PLEASE? ";LT$
360  IF LT$ = "C" THEN 1000
380  IF LT$ = "D" THEN 2000
400  IF LT$ = "R" THEN 3000
420  PRINT : HTAB TB
440  PRINT "INCORRECT CHOICE": GOTO 320
970 :
```

## CORRECT BACK ORDER

```
980 :
990 :
1000 REM  **--CORRECT RECORD--**
1005 HOME
1010 POKE 32,7: POKE 34,7: REM  SET WINDOW
1020 HOME
1040 FLAG$ = "NO": REM  INFO HAS YET TO BE CHANGED
1050 PRINT "TYPE A '0' TO RETURN TO MENU" : PRINT
1060 INPUT "CORRECT WHICH RECORD? ";REC
1070 IF REC = 0 THEN TEXT : GOTO 100: REM  MENU
1075 IF REC > PTR THEN PRINT "INCORRECT CHOICE": GOTO
     1060
1080 GOSUB 6000: REM  READ FILE
1120 REM  **--DISPLAY FOR CORRECTION--**
1140 HOME
1160 PRINT "1.  ";ITEM$
1180 PRINT "2.  ";DESC$
1200 PRINT "3.  ";NAME$
1220 PRINT "4.  ";PHNE$
1240 PRINT "5.  ";DTE$
1245 PRINT "6.  ";OD$
1250 PRINT "7.  ";AMT$
1260 PRINT : PRINT
1280 INPUT "IS THIS CORRECT ('Y' OR 'N') "; YES$
1290 IF YES$ = "Y" AND FLAG$ = "NO" THEN 1000
1300 IF YES$ = "Y" AND FLAG$ = "YES" THEN 7000: REM
     FILE ROUTINE
1320 INPUT "WHICH NUMBER IS WRONG ";NB
1330 IF NB < 1 OR NB > 7 THEN PRINT "INCORRECT CHOICE":
     GOTO 1320
1340 IF NB = 1 THEN SP = 25
1360 IF NB = 2 THEN SP = 30
1380 IF NB = 3 THEN SP = 20
1400 IF NB = 4 THEN SP = 20
1420 IF NB = 5 THEN SP = 10
1425 IF NB = 6 THEN SP = 1
1430 IF NB = 7 THEN SP = 10
1440 PRINT
1460 PRINT "TYPE IN CORRECT INFO "
1480 INPUT CT$(NB)
1500 IF LEN (CT$(NB)) > SP THEN PRINT "TOO LONG--TRY
     AGAIN PLEASE": GOTO 1460
1520 IF NB = 1 THEN ITEM$ = CT$(NB)
1540 IF NB = 2 THEN DESC$ = CT$(NB)
```

```
1560 IF NB = 3 THEN NAME$ = CT$(NB)
1580 IF NB = 4 THEN PHNE$ = CT$(NB)
1600 IF NB = 5 THEN DTE$ = CT$(NB)
1603 IF NB = 6 THEN OD$ = CT$(NB)
1606 IF NB = 7 THEN AMT$ = CT$(NB)
1610 FLAG$ = "YES": REM  INFO HAS BEEN CHANGED
1620 GOTO 1120: REM  CHECK AGAIN
1997 :
1998 :
1999 :
2000 REM **--DELETE RECORD--**
2020 HOME
2040 POKE 32,7: POKE 34,7: REM  SET WINDOW
2060 HOME
2100 PRINT "TYPE A '0' TO RETURN TO MENU": PRINT
2120 INPUT "DELETE WHICH RECORD ";REC
2140 IF REC = 0 THEN TEXT : GOTO 100: REM MENU
2150 IF REC > PTR THEN PRINT "INCORRECT CHOICE": GOTO
     2120
2160 GOSUB 6000: REM  READ RECORD
2180 HOME
2200 PRINT ITEM$
2220 PRINT DESC$
2240 PRINT NAME$
2260 PRINT PHNE$
2280 PRINT DTE$
2285 PRINT OD$
2290 PRINT AMT$
2300 PRINT : PRINT
2320 INPUT "DELETE THIS RECORD? ";YES$
2340 IF YES$ = "Y" THEN 2380
2360 TEXT : GOTO 2000
2380 PRINT "ARE YOU SURE?": PRINT
2390 INPUT "TYPE 'YES' TO DELETE RECORD ";YES$
2400 IF YES$ = "YES" THEN 2440
2420 TEXT : GOTO 2000
2440 ITEM$ = "D"
2460 NAME$ = "D"
2480 DTE$ = "D"
2500 PHNE$ = "D"
2520 DESC$ = "D"
2525 OD$ = "D"
2530 AMT$ = "D"
2540 GOTO 7000: REM  FILE ROUTINE
```

# CORRECT BACK ORDER

```
2970 :
2980 :
2990 :
3000 REM **--RETURN TO HOME MENU--**
3020 TEXT : PRINT D$; "CLOSE BACKORDER"
3040 PRINT D$; "RUN MENU"
3970 :
3980 :
3990 :
6000 REM **--READ FILE ROUTINE--**
6020 PRINT D$; "READ BACKORDER, R"; REC; ", B"; 0
6040 INPUT ITEM$
6060 PRINT D$; "READ BACKORDER, R"; REC; ", B"; 25
6080 INPUT DESC$
6100 PRINT D$; "READ BACKORDER, R"; REC; ", B"; 55
6120 INPUT NAME$
6140 PRINT D$; "READ BACKORDER, R"; REC; ", B"; 75
6160 INPUT PHNE$
6180 PRINT D$; "READ BACKORDER, R"; REC; ", B"; 95
6200 INPUT DTE$
6205 PRINT D$; "READ BACKORDER, R"; REC; ", B"; 105
6210 INPUT OD$
6215 PRINT D$; "READ BACKORDER, R"; REC; ", B"; 110
6217 INPUT AMT$
6220 PRINT D$: REM  CANCEL INPUT FROM DISK
6240 RETURN
6970 :
6980 :
6990 :
7000 REM **--WRITE FILE ROUTINE--**
7020 TEXT
7040 PRINT D$; "WRITE BACKORDER, R"; REC; ", B"; 0
7060 PRINT ITEM$
7080 PRINT D$; "WRITE BACKORDER, R"; REC; ", B"; 25
7100 PRINT DESC$
7120 PRINT D$; "WRITE BACKORDER, R"; REC; ", B"; 55
7140 PRINT NAME$
7160 PRINT D$; "WRITE BACKORDER, R"; REC; ", B"; 75
7180 PRINT PHNE$
7200 PRINT D$; "WRITE BACKORDER, R"; REC; ", B"; 95
7220 PRINT DTE$
7225 PRINT D$; "WRITE BACKORDER, R"; REC; ", B"; 105
7230 PRINT OD$
7235 PRINT D$; "WRITE BACKORDER, R"; REC; ", B"; 110
```

```
7237 PRINT AMT$
7240 PRINT D$; "CLOSE BACKORDER"
7260 PRINT D$; "OPEN BACKORDER, L120"
7280 PRINT D$
7300 GOTO 100: REM  MENU
```

# APPENDIX I.

## STOCKMARKET SYSTEM PROGRAMS

**STOCK MENU**

```
10  REM  ***--STOCK MENU--**
11 :
12 :
20  D$ = CHR$ (4) : REM  CONTROL D
40  TB = 8
100 HOME : VTAB 3
120 HTAB 15
140 PRINT "STOCK MENU"
160 PRINT : PRINT
180 HTAB TB
200 PRINT "1. ADD STOCK INFO. "
220 PRINT
240 HTAB TB
260 PRINT "2. DISPLAY STOCK INFO. "
280 PRINT
300 HTAB TB
320 PRINT "3. DISPLAY HI/LOW VALUES"
340 PRINT
360 HTAB TB
380 PRINT "4. CREATE/CORRECT HI/LOW"
400 PRINT
420 HTAB TB
440 PRINT "5. CORRECT DATA"
460 PRINT
480 HTAB TB
500 PRINT "6. CATALOG"
520 PRINT
540 HTAB TB
560 PRINT "7. END"
```

```
580  PRINT
600  HTAB TB
620  INPUT "WHICH NUMBER ";NB
640   IF NB  <  1 OR NB  >  7 THEN PRINT : HTAB TB: PRINT
      "INCORRECT CHOICE": GOTO 600
660  IF NB = 1 THEN PRINT D$;"RUN ADD STOCKS"
680  IF NB = 2 THEN PRINT D$;"RUN DISPLAY STOCKS"
700  IF NB = 3 THEN PRINT D$;"RUN DISPLAY HI/LOW"
720  IF NB = 4 THEN PRINT D$;"RUN CREATE HI/LOW"
740  IF NB = 5 THEN PRINT D$;"RUN CORRECT STOCKS"
760  IF NB = 6 THEN PRINT D$;"CATALOG"
780  IF NB = 7 THEN END
800  PRINT
820  INPUT "HIT RETURN TO CONTINUE ";L$
840  GOTO 100
```

## ADD STOCKS

```
10  REM ***--ADD STOCK INFO--***
11 :
12 :
13  REM **--VARIABLES LIST--**
14  REM  STK$  = STOCK SYMBOL
15  REM  HI$   = CURRENT HI PRICE
16  REM  LOW$  = CURRENT LOW PRICE
17  REM  PE    = P/E RATIO
18  REM  VOL   = SALES VOLUME
19  REM  H     = DAILY HIGH PRICE
20  REM  L     = DAILY LOW PRICE
21  REM  C     = DAILY CLOSING PRICE
22  REM  CR    = CORRECTED FIGURE
27 :
28 :
29  REM **--INITIALIZATION--**
30  D$ = CHR$ (4): REM  CONTROL D
40  PRINT D$;"OPEN STOCKS,L260"
60  PRINT D$;"READ STOCKS,R0"
80  INPUT PTR
100  PRINT D$;"CLOSE STOCKS"
120  PTR = PTR + 1
140  PRINT D$;"OPEN STOCKS HI/LOW"
160  PRINT D$;"READ STOCKS HI/LOW"
180  FOR I = 0 TO 9
```

## ADD STOCKS

```
200  INPUT STK$(I)
220  INPUT HI$(I)
240  INPUT LOW$(I)
260  NEXT I
280  PRINT D$;"CLOSE STOCKS HI/LOW"
297 :
298 :
299  REM **--KEYBOARD INPUT--**
300  I = 0
320  HOME : VTAB 5
330  INPUT "TODAY'S DATE ";DT$
335  HOME : VTAB 5: PRINT DT$: PRINT : PRINT
340  PRINT STK$(I)
350  PRINT : PRINT
360  INPUT "TODAY'S P/E RATIO ";PE
370  PRINT
380  INPUT "TODAY'S VOLUME ";VOL
390  GOSUB 7000
400  INPUT "TODAY'S HIGH ";H
410  GOSUB 7000
420  INPUT "TODAY'S LOW ";L
430  GOSUB 7000
440  INPUT "TODAY'S CLOSE ";C
457 :
458 :
459  REM **--CORRECTION ROUTINE--**
460  HOME : VTAB 3
470  HTAB 10
480  PRINT STK$(I)
490  PRINT : PRINT
500  PRINT "1. TODAY'S P/E RATIO-- ";PE
510  PRINT
520  PRINT "2. TODAY'S VOLUME----- ";VOL
530  PRINT
540  PRINT "3. TODAY'S HIGH------- ";H
550  PRINT
560  PRINT "4. TODAY'S LOW-------- ";L
570  PRINT
580  PRINT "5. TODAY'S CLOSE------ ";C
600  PRINT
620  INPUT "ARE THESE FIGURES CORRECT? ";YES$
640  IF YES$ = "N" THEN 680
660  GOTO 900
680  PRINT
```

```
700  INPUT "WHICH NUMBER IS WRONG ";NB
720  IF NB < 1 OR NB > 5 THEN PRINT "INCORRECT CHOICE":
     GOTO 680
740  INPUT "THE CORRECT FIGURE = ";CR
760  IF NB = 1 THEN PE = CR
780  IF NB = 2 THEN VOL = CR
800  IF NB = 3 THEN H = CR
820  IF NB = 4 THEN L = CR
840  IF NB = 5 THEN C = CR
860  GOTO 460
897 :
898 :
899  REM **--EXCHANGE HI/LOW--**
900  IF H > VAL (HI$(I)) THEN HI$(I) = STR$ (H)
920  IF L < VAL (LOW$(I)) THEN LOW$(I) = STR$ (L)
940  IF VAL (LOW$(I)) = 0 THEN LOW$(I) = STR$ (L)
957 :
958 :
959  REM **--FILE UPDATE--**
960  PRINT D$;"OPEN STOCKS,L260"
970  PRINT D$;"WRITE STOCKS,R";PTR;",B";0
975  PRINT DT$
980  PRINT D$;"WRITE STOCKS,R";PTR;",B";(I * 25) + 10 + 0
1000 PRINT PE
1020 PRINT D$;"WRITE STOCKS,R";PTR;",B";(I * 25) + 10 + 4
1040 PRINT VOL
1060 PRINT D$;"WRITE STOCKS,R";PTR;",B";(I * 25) + 10 + 10
1080 PRINT H
1100 PRINT D$;"WRITE STOCKS,R";PTR;",B";(I * 25) + 10 + 15
1120 PRINT L
1140 PRINT D$;"WRITE STOCKS,R";PTR;",B";(I * 25) + 10 + 20
1160 PRINT C
1180 PRINT D$;"CLOSE STOCKS"
1200 I = I + 1
1220 IF I < 10 THEN 335
1237 :
1238 :
1239 REM **--NEW HI/LOW FILE--**
1240 PRINT D$;"OPEN STOCKS HI/LOW"
1260 PRINT D$;"WRITE STOCKS HI/LOW"
1280 FOR K = 0 TO 9
1300 PRINT STK$(K)
1320 PRINT HI$(K)
```

## ADD STOCKS

```
1340  PRINT LOW$ (K)
1360  NEXT K
1380  PRINT D$; "CLOSE STOCKS HI/LOW"
1397 :
1398 :
1399  REM **--FILE POINTER--**
1400  PRINT D$; "OPEN STOCKS, L260"
1420  PRINT D$; "WRITE STOCKS, R0"
1440  PRINT PTR
1460  PRINT D$; "CLOSE STOCKS"
1998 :
1999 :
5000  REM **--RETURN TO STOCK MENU
5020  PRINT D$; "RUN STOCK MENU"
6997 :
6998 :
6999  REM **--REMINDER SUBROUTINE--**
7000  HOME : VTAB 3
7020  PRINT "***--REMEMBER--***"
7040  PRINT "YOU MUST ADD THE FRACTION"
7060  PRINT "AS THE LAST DIGIT"
7080  PRINT "1/8------= 1"
7100  PRINT "1/4------= 2"
7120  PRINT "3/8------= 3"
7140  PRINT "1/2------= 4"
7160  PRINT "5/8------= 5"
7180  PRINT "3/4------= 6"
7200  PRINT "7/8------= 7"
7220  PRINT "EVEN-----= 0"
7222  PRINT : PRINT
7225  PRINT "****---IMPORTANT---****"
7230  PRINT
7235  PRINT "IF THE NUMBER HAS NO"
7240  PRINT
7245  PRINT "FRACTION, PLEASE ENTER"
7250  PRINT
7255  PRINT "A '0' AFTER THE NUMBER."
7260  PRINT
7280  RETURN
```

## DISPLAY STOCKS

```
10  REM  ***--DISPLAY STOCK HISTORY--***
11 :
12 :
13  REM  **--VARIABLES LIST--**
14  REM  STK$   = STOCK SYMBOL
15  REM  HI$    = CURRENT HI PRICE
16  REM  LOW$   = CURRENT LOW PRICE
17  REM  PE     = P/E RATIO
18  REM  VOL    = SALES VOLUME
19  REM  H      = DAILY HIGH PRICE
20  REM  L      = DAILY LOW PRICE
21  REM  C      = DAILY CLOSING PRICE
22  REM  DT$    = DATE
23  REM  F$     = FRACTION
24  REM  AV     = AVERAGE VOLUME
25  REM  AP     = AVERAGE PRICE
26  REM  CV     = CLOSING PRICE W/O CONV.
27  REM  C1     = 1ST CLOSE PRICE
28  REM  C2     = LAST CLOSE PRICE
29  REM  CD     = DIFF. BETWEEN C2
30  REM  M      = COMMON VAR. CONV
31  REM  L1     = COMMON VAR. CONV
32  REM  W      = TEMP. STOCK #
46 :
47 :
48 :
49  REM **--INITIALIZATION--**
50  D$ = CHR$ (4): REM CONTROL D
55  PRINT D$;"OPEN STOCKS,L260"
60  PRINT D$;"READ STOCKS,R0"
80  INPUT PTR
100 PRINT D$;"CLOSE STOCKS"
116 :
117 :
118 :
119 REM **--SET UP--**
120 HOME : VTAB 5
140 PRINT D$;"OPEN STOCKS HI/LOW"
160 PRINT D$;"READ STOCKS HI/LOW"
180 FOR I = 0 TO 9
200 INPUT STK$(I)
220 INPUT HI$(I)
```

## DISPLAY STOCKS

```
230 M = VAL (HI$(I)): GOSUB 8000: HI$(I) = STR$ (M) + " "
    + F$
240 INPUT LOW$(I)
245 M = VAL (LOW$(I)): GOSUB 8000: LOW$(I) = STR$ (M) +
    " " + F$
250 PRINT I + 1;".";: HTAB 5: PRINT STK$(I)
260 NEXT I
270 STK$(10) = "STOCK MENU"
275 PRINT "11.";: HTAB 5: PRINT STK$(10)
280 PRINT D$;"CLOSE STOCKS HI/LOW"
285 PRINT
290 INPUT "WHICH STOCK ";W
291 IF W < 1 OR W > 11 THEN PRINT "INCORRECT CHOICE" :
    GOTO 290
292 IF W= 11 THEN PRINT D$;"RUN STOCK MENU"
293 I = W - 1
294 :
295 REM ***--TITLES--***
296 HOME : VTAB 5: HTAB 18: PRINT STK$(I): PRINT : PRINT
297 PRINT "DATE";: HTAB 10: PRINT "VOL";: HTAB 15: PRINT
    "HI";: HTAB 23: PRINT "LOW";: HTAB 31: PRINT "CLOSE"
298 :
299 :
300 REM **--DISK INPUT ROUTINE--**
320 PRINT D$;"OPEN STOCKS,L260"
330 FOR K = 1 TO PTR
335 PRINT D$;"READ STOCKS,R";K;",B";0
337 INPUT DT$
340 PRINT D$;"READ STOCKS,R";K;",B";(I * 25) + 10
360 INPUT PE
380 PRINT D$;"READ STOCKS,R";K;",B";(I * 25) + 10 + 4
400 INPUT VOL
420 PRINT D$;"READ STOCKS,R";K;",B";(I * 25) + 10 + 10
440 INPUT H
460 PRINT D$;"READ STOCKS,R";K;",B";(I * 25) + 10 + 15
480 INPUT L
500 PRINT D$;"READ STOCKS,R";K;",B";(I * 25) + 10 + 20
520 INPUT C
536 :
537 :
538 :
539 REM **--DISPLAY ROUTINE--**
540 PRINT DT$;
600 HTAB 10
```

```
620 PRINT VOL;
740 HTAB 15
750 M = H: GOSUB 8000: H = M
760 PRINT H; " "; F$;
780 HTAB 23
790 M = L: GOSUB 8000: L = M
800 PRINT L; " "; F$;
820 HTAB 31
825 IF K = 1 THEN C1 = C: V1 = VOL
827 IF K = PTR THEN C2 = C
828 CV = C: L1 = CV: GOSUB 9000: CV = L1
830 M = C: GOSUB 8000: C = M
840 PRINT C; " "; F$
850 AV = AV + VOL
860 AP = AP + CV
880 NEXT K
884 PRINT D$
885 PRINT : INPUT "HIT RETURN TO CONTINUE "; L$
886 :
887 :
888 :
889 REM **--DISPLAY SECOND PAGE--**
890 HOME: VTAB 5
895 HTAB: PRINT STK$(I)
899 PRINT : PRINT
900 PRINT "CURRENT P/E RATIO = "; PE
920 PRINT
940 PRINT "CURRENT HIGH = "; HI$(I)
945 AV = AV / (K - 1)
950 AP = AP / (K - 1)
955 PRINT
960 PRINT "CURRENT LOW = "; LOW$(I)
962 PRINT
965 PRINT "AVERAGE VOL. = "; AV
967 PRINT
970 PRINT "AVERAGE PRICE = "; AP
972 PRINT
975 L1 = C2: GOSUB 9000: C2 = L1
976 L1 = C1: GOSUB 9000: C1 = L1
977 CD = C2 - C1
980 PRINT "PRICE DIFF. FROM 1ST REC. = "; CD
985 PRINT
990 PRINT "LAST PRICE = "; C; " "; F$
996 :
```

```
997 :
998 :
999  REM  **--ANOTHER STOCK--**
1000  PRINT D$
1010  PRINT
1020  INPUT "HIT RETURN TO CONTINUE ";L$
1040  HOME : VTAB 5
1050  AV = 0:AP = 0
1060  FOR I = 1 TO 11
1080  PRINT I;".";: HTAB 5: PRINT STK$(I - 1)
1100  NEXT I
1120  PRINT
1140  GOTO 290
1997 :
1998 :
1999 :
5000  REM  **--RETURN TO STOCK MENU"
5020  PRINT D$;"RUN STOCK MENU"
6996 :
6997 :
6998 :
8000  REM  **--CONVERT TO FRACTION--**
8005  F = M - INT (M / 10) * 10
8010  M = INT (M / 10)
8020  IF F = 0 THEN F$ = ""
8040  IF F = 1 THEN F$ = "1/8"
8060  IF F = 2 THEN F$ = "1/4"
8080  IF F = 3 THEN F$ = "3/8"
8100  IF F = 4 THEN F$ = "1/2"
8120  IF F = 5 THEN F$ = "5/8"
8140  IF F = 6 THEN F$ = "3/4"
8160  IF F = 7 THEN F$ = "7/8"
8200  RETURN
8997 :
8998 :
8999 :
9000  REM  **--CONVERT TO DECIMAL--**
9010  L1 = L1 / 10:S1 = INT (L1):D1 = L1 - S1
9020  D1 = (D1 * 10) / 8:L1 = S1 + D1:L1 = INT (L1 * 1000
      + .5) / 1000
9040  RETURN
```

## CREATE HI/LOW

```
10  REM  ***--STOCKS HI/LOW--***
11 :
12 :
13  REM  **--VARIABLES LIST--**
14  REM  STK$ = STOCK SYMBOL
15  REM  HI$  = CURRENT HIGH PRICE
16  REM  LOW$ = CURRENT LOW PRICE
18 :
19 :
20 D$ = CHR$ (4): REM  CONTROL D
27 :
28 :
30  REM  **--KEYBOARD INPUT--**
40  FOR I = 0 TO 9
60  HOME : VTAB 5
80  INPUT "STOCK SYMBOL ";STK$(I)
85  PRINT
90  PRINT "IF YOU ARE NOT SURE OF THE"
92  PRINT
94  PRINT "HI OR LOW, ENTER A '0'."
95  PRINT
100  INPUT "HI VALUE ";HI$(I)
110  PRINT
120  INPUT "LOW VALUE ";LOW$(I)
128 :
129 :
130  REM  **--CORRECTION ROUTINE--**
140  HOME
160  VTAB 5
180  PRINT "1. ";STK$(I)
200  PRINT "2. ";HI$(I)
220  PRINT "3. ";LOW$(I)
240  PRINT
260  INPUT "IS THIS CORRECT? ";YES$
280  IF YES$ = "N" THEN 320
300  GOTO 500
320  PRINT
340  INPUT "WHICH NUMBER IS WRONG? ";NB
360  IF NB < 1 OR NB > 3 THEN PRINT "INCORRECT CHOICE": GOTO 320
```

## CREATE HI/LOW

```
380  IF NB = 1 THEN INPUT "CORRECT STOCK NAME PLEASE "
     ; STK$(I)
400  IF NB = 2 THEN INPUT "CORRECT HI VALUE PLEASE "; HI$(I)
420  IF NB = 3 THEN INPUT "CORRECT LOW VALUE PLEASE "
     ; LOW$(I)
440  GOTO 140
500  NEXT I
1198 :
1199 :
1200 REM **--CREATE HI/LOW FILE--**
1240 PRINT D$; "OPEN STOCKS HI/LOW"
1260 PRINT D$; "WRITE STOCKS HI/LOW"
1280 FOR K = 0 TO 9
1300 PRINT STK$(K)
1320 PRINT HI$(K)
1340 PRINT LOW$(K)
1360 NEXT K
1380 PRINT D$; "CLOSE STOCKS HI/LOW"
1398 :
1399 :
1400 REM **--RETURN TO STOCK MENU--**
1420 PRINT D$; "RUN STOCK MENU"
```

## DISPLAY HI/LOW

```
10  REM **--READ HI/LOW FILE--**
11 :
12 :
20  D$ = CHR$ (4) : REM CONTROL D
100 REM **--DISK INPUT--**
120 PRINT D$; "OPEN STOCKS HI/LOW"
140 PRINT D$; "READ STOCKS HI/LOW"
160 FOR I = 1 TO 10
180 INPUT STK$(I)
200 INPUT HI$(I)
220 INPUT LOW$(I)
240 NEXT I
260 PRINT D$; "CLOSE STOCKS HI/LOW"
297 :
298 :
300 REM **--DISPLAY ROUTINE--**
305 HOME : VTAB 5
310 PRINT "STOCK SYMBOL";: HTAB 18: PRINT "HI";: HTAB 28: PRINT "LOW"
315 PRINT
320 FOR I = 1 TO 10
330 PRINT I;".";
335 HTAB 5
340 PRINT STK$(I);
360 HTAB 18
370 M = VAL (HI$(I)): GOSUB 8000: HI$(I) = STR$ (M)
380 PRINT HI$(I);" ";F$;
390 M = VAL (LOW$(I)): GOSUB 8000: LOW$(I) = STR$ (M)
400 HTAB 28
420 PRINT LOW$(I);" ";F$
460 NEXT I
480 PRINT : INPUT "HIT RETURN TO CONTINUE ";L$
497 :
498 :
500 REM **--RETURN TO STOCK MENU--**
520 PRINT D$; "RUN STOCK MENU"
8000 REM **--CONVERT TO FRACTION--**
8005 F = M - INT (M / 10) * 10
8010 M = INT (M / 10)
```

```
8020  IF F = 0 THEN F$ = ""
8040  IF F = 1 THEN F$ = "1/8"
8060  IF F = 2 THEN F$ = "1/4"
8080  IF F = 3 THEN F$ = "3/8"
8100  IF F = 4 THEN F$ = "1/2"
8120  IF F = 5 THEN F$ = "5/8"
8140  IF F = 6 THEN F$ = "3/4"
8160  IF F = 7 THEN F$ = "7/8"
8200  RETURN
```

## STOCK CORRECTION

```
10  REM ***--DISPLAY STOCK HISTORY--***
11 :
12 :
13  REM **--VARIABLES LIST--**
14  REM  STK$ = STOCK SYMBOL
15  REM  HI$  = CURRENT HI PRICE
16  REM  LOW$ = CURRENT LOW PRICE
17  REM  PE   = P/E RATIO
18  REM  VOL  = SALES VOLUME
19  REM  H    = DAILY HIGH PRICE
20  REM  L    = DAILY LOW PRICE
21  REM  C    = DAILY CLOSING PRICE
22  REM  DT$  = DATE
46 :
47 :
48 :
49  REM **--INITIALIZATION--**
50  D$ = CHR$ (4): REM CONTROL D
55  PRINT D$;"OPEN STOCKS,L260"
60  PRINT D$;"READ STOCKS,R0"
80  INPUT PTR
100  PRINT D$;"CLOSE STOCKS"
116 :
117 :
118 :
119  REM **--SET UP--**
120  HOME : VTAB 5
140  PRINT D$;"OPEN STOCKS HI/LOW"
160  PRINT D$;"READ STOCKS HI/LOW"
180  FOR I = 0 TO 9
```

```
200  INPUT STK$(I)
220  INPUT HI$(I)
240  INPUT LOW$(I)
250  PRINT I + 1;". ";: HTAB 5: PRINT STK$(I)
260  NEXT I
270  STK$(10) = "STOCK MENU"
275  PRINT "11. ";: HTAB 5: PRINT STK$(10)
280  PRINT D$;"CLOSE STOCKS HI/LOW"
285  PRINT
290  INPUT "WHICH STOCK ";W: PRINT : PRINT "WHICH RECORD?
     1 TO " ;PTR;" ";: INPUT K: IF K > PTR THEN 290
291  IF W < 1 OR W > 11 THEN PRINT "INCORRECT CHOICE":
     GOTO 290
292  IF W = 11 THEN 5000
293  I = W - 1
294 :
295  REM ***--TITLES--***
296  HOME : VTAB 5: HTAB 18: PRINT STK$(I): PRINT : PRINT
298 :
299 :
300  REM **--DISK INPUT ROUTINE--**
320  PRINT D$;"OPEN STOCKS,L260"
335  PRINT D$;"READ STOCKS,R";K;",B";0
337  INPUT DT$
340  PRINT D$;"READ STOCKS,R";K;",B";(I * 25) + 10
360  INPUT PE
380  PRINT D$;"READ STOCKS,R";K;",B";(I * 25) + 10 + 4
400  INPUT VOL
420  PRINT D$;"READ STOCKS,R";K;",B";(I * 25) + 10 + 10
440  INPUT H
460  PRINT D$;"READ STOCKS,R";K;",B";(I * 25) + 10 + 15
480  INPUT L
500  PRINT D$;"READ STOCKS,R";K;",B";(I * 25) + 10 + 20
520  INPUT C
536 :
537 :
538 :
539  REM **--CORRECTION ROUTINE--**
540  PRINT "1. DATE   = ";DT$
545  PRINT
550  PRINT "2. P/E    = ";PE
555  PRINT
560  PRINT "3. VOL.   = ";VOL
565  PRINT
```

```
570 PRINT "4. HIGH   = ";H
575 PRINT
580 PRINT "5. LOW    = ";L
585 PRINT
590 PRINT "6. CLOSE  = ";C
592 PRINT
594 PRINT "0. ALL CORRECT"
595 PRINT : PRINT D$
600 INPUT "WHICH NUMBER IS WRONG ";NB
610 IF NB > 6 THEN PRINT "INCORRECT CHOICE": GOTO 600
615 IF NB = 0 THEN 710
620 INPUT "CORRECT INFORMATION = ";CR$
630 IF NB = 1 THEN DT$ = CR$
640 IF NB = 2 THEN PE  = VAL (CR$)
650 IF NB = 3 THEN VOL = VAL (CR$)
660 IF NB = 4 THEN H   = VAL (CR$)
670 IF NB = 5 THEN L   = VAL (CR$)
680 IF NB = 6 THEN C   = VAL (CR$)
690 HOME : VTAB 5: HTAB 18: PRINT STK$(I)
695 PRINT : PRINT
700 GOTO 540
707 :
708 :
709 REM **--WRITE CORRECTED FILE--**
710 PRINT D$;"WRITE STOCKS,R";K;",B";0
720 PRINT DT$
730 PRINT D$;"WRITE STOCKS,R";K;",B";(I * 25) + 10
740 PRINT PE
750 PRINT D$;"WRITE STOCKS,R";K;",B";(I * 25) + 10 + 4
760 PRINT VOL
770 PRINT D$;"WRITE STOCKS,R";K;",B";(I * 25) + 10 + 10
780 PRINT H
790 PRINT D$;"WRITE STOCKS,R";K;",B";(I * 25) + 10 + 15
800 PRINT L
810 PRINT D$;"WRITE STOCKS,R";K;",B";(I * 25) + 10 + 20
820 PRINT C
996 :
997 :
998 :
999 REM **--ANOTHER STOCK--**
1000 PRINT D$
1010 PRINT
1020 INPUT "HIT RETURN TO CONTINUE ";L$
```

```
1040 HOME : VTAB 5
1060 FOR I = 1 TO 11
1080 PRINT I;".";: HTAB 5: PRINT STK$(I - 1)
1100 NEXT I
1120 PRINT
1140 GOTO 290
1997 :
1998 :
1999 :
5000 REM **--RETURN TO STOCK MENU
5010 PRINT D$;"CLOSE STOCKS"
5020 PRINT D$;"RUN STOCK MENU"
```

# APPENDIX J.

## MISCELLANEOUS PROGRAMS

### HELLO

```
20  PRINT "HELLO"
40  PRINT "MY NAME IS APPLE II"
60  PRINT "I AM A SMART COMPUTER"
```

### APPLE

```
20   D$ = CHR$ (4) : REM  CONTROL D
40   PRINT D$; "OPEN ADDRESS FILE"
60   PRINT D$; "WRITE ADDRESS FILE"
80   PRINT "APPLE II IS A BRIGHT COMPUTER"
100  PRINT D$; "CLOSE ADDRESS FILE"
120  HOME
140  PRINT D$; "OPEN ADDRESS FILE"
160  PRINT D$; "READ ADDRESS FILE"
180  INPUT NAME$
200  PRINT D$; "CLOSE ADDRESS FILE"
220  VTAB 10
240  PRINT NAME$
```

# APPENDIX K.

## MISCELLANEOUS INFORMATION

1. POKE 216,0----RESET THE ERROR FLAG
2. POKE 32,n-----SET LEFT WINDOW MARGIN
3. POKE 33,n-----SET RIGHT WINDOW MARGIN
4. POKE 34,n-----SET TOP WINDOW MARGIN
5. POKE 35,n-----SET BOTTOM WINDOW MARGIN
6. CALL -151-----ENTER MONITOR
7. CALL 768------BEGIN ROUTINE AT $300

   48K SYSTEM
8. PEEK (43634)--LOW ORDER BYTE OF BLOAD FILE STARTING ADDRESS
9. PEEK (43635)--HIGH ORDER BYTE OF BLOAD FILE STARTING ADDRESS
10. PEEK (43616)--LOW ORDER BYTE OF BLOAD FILE LENGTH
11. PEEK (43617)--HIGH ORDER BYTE OF BLOAD FILE LENGTH
12. CHR$(4)-------CONTROL D
13. CHR$(34)-----QUOTATION MARK
14. CHR$(95)-----UNDERLINE

# Index

A (type of file)   1–12
APPEND   29, 37, 43, 44, 116, 117
Apple   1, 11, 18, 21, 96, 97, 100, 176, 199, 259, 263, 274
Applesoft   1, 3, 6, 8, 10, 15, 16, 18, 27, 96, 100–102, 199, 259, 274
Arrays   31, 37, 102, 153, 172, 176, 196–198, 200, 274
ASC   274
ASCII   18, 21
Assembly   263

B (type of file)   1, 2, 3, 5
BASIC   1, 4, 6, 8, 18, 19, 21, 45, 58, 67, 68, 96, 97, 100, 102, 259, 260, 262, 267, 271
Binary   1, 3, 5, 27, 96, 100, 101, 258–267
BLOAD, BSAVE, BRUN   101, 103, 258, 259, 262–265, 267
Booting the system   2, 8, 28, 263
BOT   149, 153
Byte   87–89, 168–170, 172–173, 176, 178, 192, 195, 198, 240, 263, 265, 271

CALL   101, 261, 262, 264, 267, 410
CATALOG   2, 3, 5, 7, 9, 10, 20, 21, 34, 35, 41, 49, 56, 70, 86, 99, 261, 270, 272, 273

CHAIN   89, 96, 100–103
CHR$(34)   97, 99, 151, 410
CHR$(4)   18, 410
CHR$(95)   119, 191, 410
CLOSE   19, 22, 26, 33, 54
Cobol   6, 7
Code   48, 60, 61, 63, 69, 70, 102, 152, 191, 194, 238, 241, 263, 265
Colon   21, 30, 147, 150
Comma   21, 30, 172
Computer Instruction   *See* Program
Concatenate   198
Constant   173
Control D   16–18, 21, 23, 26, 27, 30, 40, 41, 43, 58, 172, 194, 202
Counter   30–32, 34, 43, 47, 48, 85, 102, 194, 196
CTRL   17, 18
Cursor   9, 19, 22, 28, 30, 32, 99, 192, 261, 263, 271

Data Base   190, 200, 201, 238, 241
Data Section   145, 146, 148–152, 154–156
Decimal   14, 18, 239, 264, 271
DEF FN   119
Default   23
Deferred mode   17, 26
DEL   43
DELETE   11, 14, 24, 29, 33, 42, 90, 97

411

## INDEX

Delimit   30, 168, 170, 239
DIF   144–155, 201
DIM (DIMension)   30, 37, 40, 41, 48, 49, 60, 69, 70, 102
Disk   1–3, 8, 12, 21, 200
Disk Drive   2, 9, 12, 45, 86, 87, 173, 191, 196, 240, 271, 274
Diskette   1, 2, 10, 20, 22, 34, 49, 70, 84, 87, 96, 99, 100–102, 169, 172, 176, 200, 239–241, 261, 262, 264, 270–273
Dollar sign ($)   18, 26, 33, 258
DOS (Disk Operating System)   8, 9, 11, 12, 14, 16, 18, 19, 23, 27, 29, 44, 49, 85, 96, 115, 144, 191, 201, 260, 261

Edit on screen (ESC I, J, K, M)   192
END   43
EOD   146, 149, 153, 154, 156
Error   23, 29, 85, 102, 116–118, 148, 171, 191, 192, 196, 240
EXEC   35, 37, 89, 96–100, 103, 151

FID   101
Field   169
File name   117, 173, 174
File pointer   87, 88, 168, 171, 172, 175, 191–193, 240, 241
FLASH   196
Floating point   7
FOR–NEXT   32, 37, 118
Fortran   6, 7

GOSUB   61, 73, 155, 175, 191, 194, 198
GOTO   44, 50, 63, 64, 116–118, 121

Header Section   145–147, 150–152, 155, 156
**Hexadecimal**   18, 258, 260–263, 265, 267, 270, 273

HOME   22, 26, 30, 32, 43
HTAB   58, 73

I (type of file)   1–12
IF–THEN   30, 32, 33, 37, 46, 47, 58, 63, 64, 66, 198–199
INIT   9, 23, 261
INPUT   22, 23, 26, 30, 40, 41, 43, 171, 193, 194, 198
INT   119, 121
Integer   1, 3, 6, 15, 16, 18, 27, 96, 100–102, 259
Interface card   1, 19, 46, 47
Intermediate mode   17, 26
INVERSE   196

Key   16, 21, 23, 27, 43, 58

LEFT$   67, 68, 70, 73, 197, 198
LEN   70, 73, 119, 198
Line numbers   8, 48, 101, 175
LIST   8–11, 20, 22, 42, 60, 261
LOAD   7, 8, 11, 12, 16, 17, 59, 60, 86, 100
LOCK   11, 14
Loops   44, 47, 48, 85, 87, 118, 153, 174, 196, 197

Memory   7–12, 18, 22, 23, 30–32, 41, 48, 56, 58, 70, 200, 258–259, 261–263, 271
Micros   18, 155
MID$   67, 68, 70, 73, 197, 198
Mnemonics   263
MON   35, 97, 98
MONITOR   261, 262, 265

NEW   8, 11, 14, 20, 261
NOMON   37, 99
NORMAL   196
Numeric variables   35, 58, 69, 172

# INDEX

ONERR 116–118, 121, 171, 191–192, 240
OPEN 19, 21, 22, 26, 29, 33, 43, 85, 86, 97, 116, 117, 168–170

Parameters (R, B, or L) 83, 87–90, 167–170, 172, 173, 176, 178
PEEK 264, 267, 410
Peripheral 47, 264
POKE–216,0 116, 118, 171, 172, 191, 410
POKE–32,n **191, 410**
POKE–33,n **410**
POKE–34,n **191, 410**
POKE–35,n **410**
POSITION 88, 90, 167, 178
PR# 1, 19, 46, 47
PRINT 8, 21–23, 30, 172, 173, 193
Printer 46, 47, 61, 62, 65, 70, 241
Program 1, 3, 6–9, 12, 15, 17, 27, 45, 56, 58, 96, 100–102, 114, 259
Programming 3, 4, 45, 59, 60, 114, 169, 199, 200, 238, 259–260, 264

Quotation mark 21, 30, 86, 97, 99, 103, 151, 172, 173, 191

Random access 16, 18, 23, 26, 27, 45, 119, 155, 167–169, 172, 173–174, 176, 190, 192, 200, 240
READ 19, 22, 26, 41, 43, 193, 194, 271
RECALL 274
Record 34, 87, 88, 89, 167–175, 193–199, 238, 240
REM 19, 21, 26, 116
RENAME 11, 14, 24, 49, 85, 86, 90
Reset 263
RETURN 34, 61, 73, 194
RETURN key 2, 8, 9, 19, 21, 29, 34, 35, 46, 261, 263
RIGHT$ 67, 68, 70, 73

RND 119
Routine 30, 32, 64–67, 170, 176, 195, 197, 241
RUN 3, 7, 8, 11, 12, 14, 16, 17, 20–22, 34, 41, 43, 56, 58, 99, 100, 101
RWTS 261, 265, 270–273

SAVE 7, 8, 11, 12, 14, 16, 17, 20, 22, 49, 58, 59, 70, 86, 145
Screen 21, 23, 191, 194, 241
Search 61, 63–65, 175, 193, 194, 195
Sectors 2, 49, 50, 70, 270, 273
Semicolon 31, 172
Separator 31, 33, 47, 64, 145, 169
Sequential access 16, 18, 23, 26, 27, 45, 70, 87, 102, 114, 119, 155, 167–170, 176, 178, 200, 240, 274
SGN 119
Slot 19, 46, 47
Sorting 61, 66, 67, 69, 193, 195, 196–198
Sorting—Bubble sort 67, 73
Sorting—Quicksort 67, 68, 69, 73
Sorting—Shell–Metzner 67, 73, 193, 195
SPEED 97–99, 103, 197
STORE 274
STR$ 70, 73, 119, 121
String variables 18, 21, 23, 26, 32, 48, 49, 63, 66, 67, 69, 70, 102, 117, 168, 172–174, 176, 196, 199, 200
Strings 29, 83, 168, 198
Subroutine 70, 174, 193–197, 199, 264
System 59, 86, 96, 197, 200, 201, 238

T (type of file) 1–5, 15–23
Tab 22, 23, 30, 32, 43, 58

Tape  8, 12, 14, 274–277
TEXT  202
Text files  1, 15–24, 26, 27, 33, 259
Track  270, 273
Tuples  145–149
Type of file  1–4, 27, 258, 272–273

UNLOCK  11, 14

VAL  70, 73, 119, 121

Variables  18, 26, 31, 43, 44, 60, 101, 102, 117, 152, 173, 175, 176, 238, 239
Vectors  145–148
VERIFY  11, 14
VTAB  22, 23, 30, 32, 43

WRITE  19, 21, 22, 26, 33, 44, 116, 172, 193, 194, 271